Challenges in Physics Education

Series Editor

Marisa Michelini, Dipartimento di Fisica, University of Udine, Udine, Italy

This book series covers the many facets of physics teaching and learning at all educational levels and in all learning environments. The respective volumes address a wide range of topics, including (but not limited to) innovative approaches and pedagogical strategies for physics education; the development of effective methods to integrate multimedia into physics education or teaching/learning; innovative lab experiments; and the use of web-based interactive activities. Both research and experienced practice will feature prominently throughout.

The series is published in cooperation with GIREP, the International Research Group on Physics Teaching, and will include selected papers from internationally renowned experts, as well as monographs. Book proposals from other sources are entirely welcome.

Challenges in Physics Education addresses professionals, teachers, researchers, instructors and curriculum developers alike, with the aim of improving physics teaching and learning, and thereby the overall standing of physics in society.

Book proposals for this series my be submitted to the Publishing Editor: Marina Forlizzi; email: Marina.Forlizzi@springer.com

Marilena Streit-Bianchi · Marisa Michelini ·
Walter Bonivento · Matteo Tuveri
Editors

New Challenges and Opportunities in Physics Education

Springer

Editors
Marilena Streit-Bianchi
ARSCIENCIA
Vienna, Austria

Walter Bonivento
INFN Sezione di Cagliari
Monserrato (Cagliari), Italy

Marisa Michelini
Physics Education Research Unit
University of Udine
Udine, Italy

Matteo Tuveri
Physics Department
University of Cagliari
Cagliari, Italy

INFN Sezione di Cagliari
Complesso Universitario Monserrato
Monserrato, Italy

ISSN 2662-8422 ISSN 2662-8430 (electronic)
Challenges in Physics Education
ISBN 978-3-031-37386-2 ISBN 978-3-031-37387-9 (eBook)
https://doi.org/10.1007/978-3-031-37387-9

This Springer imprint is published by the registered company Springer Nature Switzerland AG
The registered company address is: Gewerbestrasse 11, 6330 Cham, Switzerland

Paper in this product is recyclable.

Preface from the Editors

The motivation for this book stems from the need and opportunity to contribute to the link between informal and formal physics education that arises from the digital challenges that teachers and students have faced, are facing, and will face. The important role played by digital information and internationally disseminated connections, as well as the outcomes and lessons learned from distance education due to the recent pandemic, are a source of important considerations in the field of education and in the search for digitally available programs that can be used in educational contexts. It is an important task for institutional academics in physics education, program developers, and teachers to prepare today's young people to become active players in the high-tech society in which they will live and work.

In recent years, Big Science organizations in the field of high-energy physics and astrophysics have implemented outreach programs to explain to the public the research they are conducting, along with appropriately organized student visits and teacher's programs.

In 2022, the Cagliari Division of the Istituto Nazionale di Fisica Nucleare (INFN) in Italy developed a project called Gravitas, a dialog with renowned philosophers and scientists of different schools of thought and methods, coming from European universities and research institutions. The aim of the project was to interest students in the topics of STEM and to inform them about the unresolved questions of physics and philosophy of science. The main topic was gravity, a subject that we thought we had understood since Newton and that represent to this day a missing piece in the puzzle of unifying the forces governing our universe. The dialogs, in Italian, are available on YouTube. In the fall of 2022, a two-day festival called GravitasFest was also organized, with the participation of students from secondary schools and open to the public, to familiarize people with the many topics addressed by contemporary physics and philosophy of science, as well as with the link between creativity, art, and science, which illustrates the importance of interdisciplinarity.

The universe as it presents itself to us today is full of mysteries, full of things we struggle to understand, such as dark matter and dark energy, inflation, black holes, and their origins. All these topics, which are rarely explained in secondary school curricula, raise big questions, not only to scientists, but also to those who want to think about the meaning of basic science work and improve education STEM.

The science questions addressed in this book have clear implications that go beyond physics. The authors of the various chapters have competence in physics education, research, science communication, history and philosophy of physics, experts in business and innovation, and artists.

The book is divided into two parts:

1. Communicating Contemporary Physics
2. Digital Challenges for Physics Learning

The first part explains where we stand today with our knowledge of general relativity, quantum physics, particle physics, and cosmology. It is easy to be fascinated by the Higgs discovery, the black hole image, the Hubble or James Webb telescope images, but understanding the physics underlying the quest and the implications for knowledge, education, and society is something else. The book also looks at the role of informal education in relation to language, science communication, and the history of physics. Interest in art and theater developments from an educational perspective is illustrated with examples of projects; more could be done in both areas to promote science culture and education. The challenges and opportunities of digital visual programs or ongoing developments are also presented. It is hoped that the extensive information can serve as a stimulus for classroom discussions in which even students who are not specifically interested in science topics can participate.

The second part of the book illustrates the different strategies developed and/or implemented in various countries from Italy to Finland, Poland, the USA, Vietnam, Australia, and North Macedonia. A chapter is reserved for teaching and learning physics with digital technologies and the implications and changes needed in classroom education. The challenges represented by distance learning and the many initiatives and programs available for both students and teachers to promote physics education are presented by a very active school teacher, together with the illustration of the IPPOG research database and initiatives aiming at understanding scientific reasoning, methods, and updated knowledge. It has been also considered interesting to illustrate to school teachers how interdisciplinary challenge-based innovation courses are developed for students in physics at the university and secondary school levels to stimulate innovation. Furthermore, conceptual, creative, and critical thinking is today an important asset in science as well as in art, philosophy, and business. How to develop and stimulate their start in the classroom. The Science Gateway project where physics education programs at the edge between informal to formal physics education are being developed in the CERN research environment is presented. The last very small contribution has been focused to the challenges and opportunities in physics education from artificial intelligence and the most recent chatbot development sources of expectations and fears.

To summarize, the book is based on research contributions relevant to update teachers' awareness of some of the main results in recent physics and astronomy research, presents strategies, tools, and digital programs available to facilitate teaching/learning school activities, discusses the problems a teacher might encounter, and tries to provide clues on how to overcome it. It also shows the digital experience as a learning facilitator and how it is possible to integrate digital tools into physics education.

Scientific methods and commonsense challenge opinions, thoughts, and reasoning. The book with its education goals aims at secondary school and college teachers as well as physics education program developers, education state administrators, and interested students struggling with the understanding of contemporary, modern, and classical physics. It shows how interdisciplinarity and transdisciplinarity, with the help also of informal digital education and the setup of specific programs involving teachers and students, can be a new way to go in science education. Understanding where we come from, what the Universe we live in is made of, and what laws are governing it is a fundamental part of mankind.

We hope the book will reach the objective we had in mind as editors:

- be a source of inspiration in the daily work of many teachers
- make their educational task easier
- contribute to improving the understanding and interest of students toward scientific subjects.

Vienna, Austria — Walter Bonivento
Udine, Italy — Marisa Michelini
Cagliari, Italy — Marilena Streit-Bianchi
Cagliari, Italy — Matteo Tuveri

Contents

Communicating Contemporary Physics

From Atoms to the Higgs Boson 3
Pierluigi Bortignon, Steven Goldfarb, Michael Gregory,
Suchita Kulkarni, and Konstantinos Nikolopoulos

About Teaching Quantum Mechanics in High Schools 25
Paola Verrucchi

History of Physics for Education: The Scientific Contributions of Enrico Fermi 37
Salvatore Esposito

Three Tales of Gravity 51
Ugo Moschella

Gravity Between Physics and Philosophy 65
Mariano Cadoni and Mauro Dorato

Einstein's Theory at the Extremes: Gravitational Waves and Black Holes 81
Mariafelicia De Laurentis, Paolo Pani, and Michele Punturo

The Dark Universe 93
Riccardo Murgia, Walter M. Bonivento, and Cristiano Galbiati

Gravitational Waves: An Historical Perspective 107
Adele La Rana

Notes on Revolutions in Physics in the Twentieth Century 123
Francesco Vissani

Physics and Cultural Industry 135
Maria Chiara Di Guardo, Emiliano Ilardi, and Laura Poletti

Teaching, Communication, and Dissemination for Society 145
Matteo Tuveri, Elisabetta Gola, and Matteo Serra

Use of Art in Scientific Education 159
Margarita Cimadevila and Wolfgang Trettnak

Theatre: The Other Side of Physics 175
Marco Giliberti

Digital Challenges for Physics Learning

Research-Based Contribution on ICT as Learning Challenges in Physics Education 193
Marisa Michelini and Alberto Stefanel

Project-Based Learning in Secondary Science: Digital Experiences in Finnish Classroom 219
Anna Lager, Jari Lavonen, and Kalle Juuti

The New Methodologies in e-Learning and the Italian Experience in the Physics Teaching Field and STEM 237
Pierpaolo Limone and Giusi Antonia Toto

Applications of Technology to Promote Active Learning with Examples from Acceleration and Gravity 247
David R. Sokoloff

Future Competencies in Physics Education and Learning with Multimedia in Poland 259
Tomasz Greczyło

Digitalization in Early School Education in North Macedonia 275
Marina Vasileva Connell and Darko Taleski

Digital Transformation for Vietnam Education: From Policy to School Practices 293
Duy Hai Tuong, Trinh-Ba Tran, and Duc Dat Nguyen

Teaching and Learning Physics with Digital Technologies—What Digitalization-Related Competencies Are Needed? 313
Lars-Jochen Thoms, Sebastian Becker, and Erik Kremser

Distance Learning in Physics: Potential and Challenges 327
Michael Gregory and Steven Goldfarb

Education and Interdisciplinarity: New Keys for Future Generations and the Challenge-Based Learning 343
Christine Thong, Aaron Down, and Anita Kocsis

The Quantum Bit Woman: Promoting Cultural Heritage with Quantum Games 361
Maria Luisa Chiofalo, Jorge Yago Malo, and Laura Gentini

Conceptual, Creative and Critical Thinking for Science, Philosophy, Business and Art 381
Anita Kocsis, Claudia Sciarma, Konstantinos Nikolopoulos, Christine Thong, and Grace McCarthy

CERN Science Gateway: Example of Informal Contemporary Physics Education in an Authentic Research Environment 409
Daria Dvorzhitskaia, Patricia Verheyden, Julia Woithe, and Annabella Zamora

Artificial Intelligence: New Challenges and Opportunities in Physics Education 427
Rüdiger Wink and Walter M. Bonivento

Epilogue from the Editors 435

About the Editors

Marilena Streit-Bianchi received a doctorate in Biological Sciences from the University of Rome and joined CERN, the European Organization for Nuclear Research in Geneva, Switzerland, in 1969. She has been a pioneer in the study of high-energy particles produced by accelerators for cancer treatment. She has held managerial positions on safety training and technology transfer, has been a senior honorary staff member at CERN, and is actively engaged in multi-disciplinarity. She has been a book editor and curator of exhibitions in Europe and Mozambique promoting art and science in particular. She is the vice president of the international association ARSCIENCIA and a member of the Italian Physics Society (SIF). She has been the editor of the booklet *Cern: Science Bridging Cultures*, available in Zenodo Open Access, and co-editor of the book *Mare Plasticum-The Plastic Sea—Combatting Plastic Pollution Through Science and Art* Springer 2020, and the book *Advances in Cosmology: Science-Art-Philosophy* Springer Nature 2022.

Marisa Michelini is Senior Professor of Physics Education in the Department of Mathematics, Computer Science and Physics (DMIF) of the University of Udine, Italy. She was from 1994 to 2022 Rector Delegate for different areas and for GEO University Consortium that she headed since 2014. She founded in 1992 and headed the Physics Education Research Unit (URDF). She has headed different EU projects and the Italian IDIFO project series of PLS on Innovation in Physics Education involving 20 Italian universities from 2006. At the international level, since 2012 she is the President of International Research Group on Physics Education (GIREP) and, since 2016, a board member of PED Section of European Physical Society (EPS), after being (2014–2022) a board member of Multimedia Physics Teaching Learning (MPTL). She was the head of the Physics Department and director of the University School of Specialization for Secondary School Teachers of the Udine University. Her research activity was on electrical transport properties of metal and superconductors from 1985 to 1994 and during her entire career on physics education research, including innovative physics education paths on modern physics. She received the

Italian Physical Society Award for the Exhibit Games Experiments Ideas in 1989 and the IUPAP-ICPE International Award for the research in physics education in 2018.

Walter Bonivento is an experimental physicist in astroparticle and particle physics. He is Director of Research at INFN of Cagliari, Italy, and has been working at CERN for 20 years (DELPHI, LHCb, with many leadership roles) and more recently on the Darkside experiment at LNGS. He is a member of the executive board of this experiment and the Scientific Director of the Aria project in Sardinia for the distillation of argon for dark matter searches. He is also a lecturer at the Physics Department of the University of Cagliari and INFN Manager of the dissemination project on Dark Matter DARK. Additionally, he is the project leader and coordinator, together with Matteo Tuveri and Viviana Fanti, of the Gravitas project, and conversations between physics and philosophy held online from December 2021 to April 2022 and of the Festival GravitasFest.

Matteo Tuveri has a Master's degree in theoretical physics and a Ph.D. with research subjects including black holes, holography, dark matter, and emergent gravity. He is a researcher at the University of Cagliari (Unica), Italy, in the field of physics education and scientific dissemination. He is a professor of the History of Physics of the Twentieth Century, Physics laboratory in Primary Education Science, and New Methods in Teaching Physics at the same university. In his research, he develops communication strategies and new methods for teaching and learning physics by combining art, technology, and science to promote interdisciplinary approaches. He works closely with institutions, festivals, and schools in Italy. Since February 2019, he has been Vice President of IdeAS (Incontri di Divulgazione e Astrofisica in Sardegna). He is responsible for outreach activities at the Cagliari Division of The National Institute of Nuclear Physics (INFN). He is also the project leader, together with Walter Bonivento and Viviana Fanti, and coordinator of the Gravitas INFN project, and the organizer of associated public events. He is a member of SIF and GIREP.

Communicating Contemporary Physics

Communicating
Contemporary
Physics
Springer

From Atoms to the Higgs Boson

Pierluigi Bortignon, Steven Goldfarb, Michael Gregory, Suchita Kulkarni, and Konstantinos Nikolopoulos

Abstract Secondary science education is vital, not only for students considering careers in science and technology-based fields, but for everyone. Its methodology teaches us how to extract information from the world around us and to find pertinent solutions to complex problems. Unfortunately, current curricula often end at the physics of the beginning of the twentieth century, only touching on relativity and quantum mechanics. Yet, we have come far since then and current advances, including the discovery of the Higgs boson and gravitational waves, present exciting concepts that can ignite the interest of students, while still teaching them the fundamentals of physics. This chapter attempts to address this century-long gap in science education. It begins with a recap of the history of the atom which, for centuries, was considered the fundamental building block of matter. We then describe how exploring the details of the atom led us to realise that our understanding of physics at those dimensions required a rethink. From there, we describe our journey down the subatomic rabbit hole, discovering, organising, and re-organising the microscopic world of elementary particles until arriving at an understanding that is central to today's research, the

P. Bortignon
Department of Physics, University of Cagliari, Complesso Universitario di Monserrato, S.P. Monserrato-Sestu Km 0, 700-09042 Monserrato, CA, Italy
e-mail: pierluigi.bortignon@dsf.unica.it

S. Goldfarb (✉)
School of Physics, University of Melbourne, Swanston Street and Tin Alley, Parkville, VIC 3010, Australia
e-mail: steven.goldfarb@cern.ch

M. Gregory
My Favourite Experiments, Paris, France
e-mail: myfavouriteexperiments@gmail.com

S. Kulkarni
Institute of Physics, NAWI Graz, University of Graz, Universitaetsplatz 5, A-8010 Graz, Austria
e-mail: suchita.kulkarni@uni-graz.at

K. Nikolopoulos
School of Physics and Astronomy, University of Birmingham, Edgbaston, Birmingham B15 2TT, UK
e-mail: k.nikolopoulos@bham.ac.uk

M. Streit-Bianchi et al. (eds.), *New Challenges and Opportunities in Physics Education*, Challenges in Physics Education, https://doi.org/10.1007/978-3-031-37387-9_1

Standard Model. From there, we present the shortcomings of that model and pose questions we are striving to answer. We hope this chapter will spark the interest of science teachers, so that they too can pass the excitement of this research on to their students, the discoverers and inventors of tomorrow.

Keywords Teaching · Physics education · Scientific literacy · Quantum mechanics · Relativity · Atom · Higgs boson · Particle physics

1 Introduction

The discovery of the Higgs boson by the ATLAS (2012) and CMS (2012) experiments at CERN in 2012 depicted in Fig. 1 was a giant leap forward for particle physics and human understanding of our universe. Such discoveries help to increase global awareness of the importance of fundamental scientific research and provide educators with an opportunity to spark the interest of their students in science by providing them with concrete examples of its methods and applications.

Unfortunately, these lessons are often minimal and short-lived, as many of the secondary classrooms around the world have not yet advanced beyond early twentieth century physics. Of course, the level of advancement depends on the background of the teacher, the curriculum and priorities set by the educational system, and the availability of time, resources, and material. In many cases, teachers who could take their students further are constrained by time and/or obligation to prepare students with specific backgrounds for standardised exams. In other cases, teachers perfectly capable of teaching more advanced concepts might be intimidated by perceived complexity of the topics or not have the time or authority to extend beyond the established programme. For whatever reason, it is rare to find secondary classrooms that provide their students with an overview of particle physics, the Standard Model and current questions under exploration in the field.

The authors of this chapter comprise experimental particle physicists, theorists, and educators active in public engagement and informal science education. In the past few decades, we have developed and participated in a variety of methods and programmes designed to bring current physics topics into the classroom. Our goal has been to effectively bridge the historical gap between early twentieth century physics, when modern concepts such as quantum mechanics and special relativity were introduced to describe the atom, and today's headline results in particle physics and cosmology.

Our objectives are straightforward: Bringing recent topics into the classroom motivates both students and teachers by sharing the excitement of exploration and by demonstrating the relevance of their studies. In many cases, standard physics topics relate directly to current research: charges moving through magnetic fields and measurement of particle momentum, invariant mass and the discovery of the Higgs boson, conservation of momentum, and the search for Dark Matter. Furthermore, by introducing the data-driven methods employed by scientific collaborations,

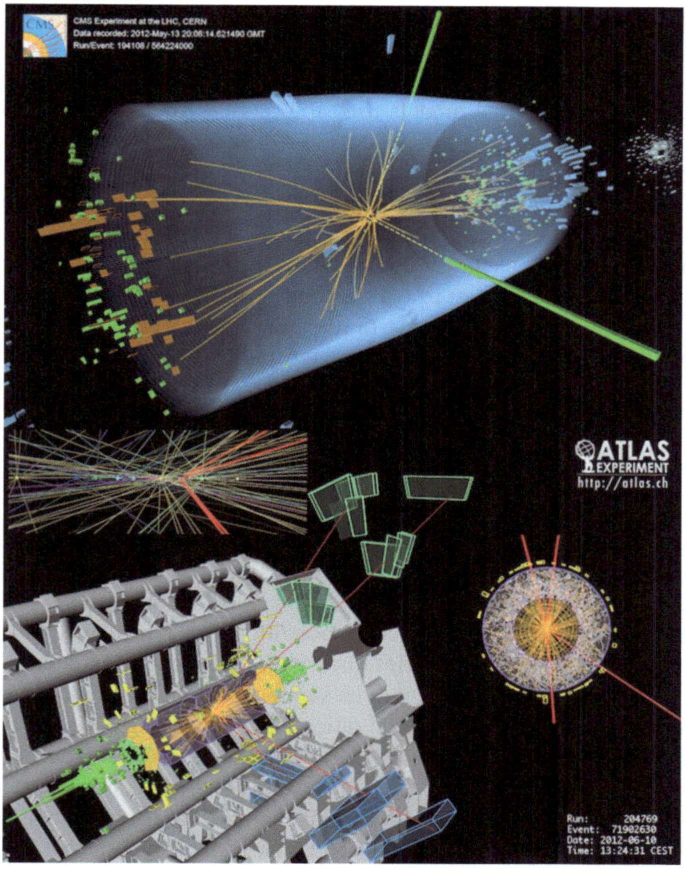

Fig. 1 Candidate Higgs boson events reconstructed in the CMS (top) and ATLAS (bottom) detectors at CERN. *Image* CERN, CC-BY-SA-4.0

students gain an appreciation for our reliance on measurement to attain information and to improve our understanding of our universe. That is, they learn how truth is extracted, not according to opinion, but through objective, iterative, and peer-reviewed procedures. This is an especially valuable lesson to have in today's social media rich environment.

In this chapter, we provide a brief overview of the advancements that have come through our attempts to understand and describe the basic building blocks of our universe. We explore the evolution of atomic physics from the early postulates of Leucippus and Democritus (development of atomic theory) to the quantum mechanical description proposed by Niels Bohr (1913) and the experiments that paved further development. We examine the exploration of ever smaller spatial dimensions by the usage of higher energy devices, allowing us to delve deeper into the atom and its nucleus on our quest to find its most basic constituents.

These discussions reveal how scientific methodology forges a continual dialogue between theorists and experimentalists, inevitably pushing the boundaries of research. In many cases, theories developed to explain phenomena measured in the data force a reshaping of the models used to describe nature. Particles thought to be elementary are found to be composite and forces thought to be independent are found to be interlinked.

In one well-known case, the description of a mechanism postulated to explain the existence of mass in elementary particles sends researchers on a decades-long quest, the search for the Higgs boson. We describe the Higgs' importance to the theory and what it teaches us about the rest of our universe. We also briefly examine the major questions that remain, topics enticing to researchers, but even more importantly to the students that represent humanity's hope for finding the answers.

2 A Brief History of Atoms

The concept that matter is made up of tiny indivisible particles appeared in several ancient cultures, based on philosophical reasoning rather than empirical evidence. The word atom itself is derived from the Greek word ἄτομος which means "the one that cannot be split," while the corresponding term in ancient Indian sciences was *parmanu*. In the following, a brief history of the concept of atoms and how we reached modern day atomic theory will be summarised.

The efforts towards a systematic and rational understanding of nature began at or before the sixth century BCE in ancient Greece. The philosopher Thales of Miletus (seventh and sixth centuries BCE), often referred to as "the father of science," refused to accept supernatural explanations for natural phenomena affirming that they should have a natural cause. This shift in approach heralded the beginning of the scientific methodology. Heraclitus (540–480) at around 500 BCE proposed that the basic law governing the Universe was the principle of change, where nothing remains the same indefinitely. In the following centuries, the classical period in Greece and the Hellenistic times, natural philosophy developed into a flourishing field of study.

Leucippus and Democritus (c. 460–c. 370 BCE) are typically credited with the introduction of an atomic theory of the universe. Their theory held that everything is composed of "atoms," which are physically indivisible, with the space between atoms being empty. These atoms were considered indestructible and in eternal motion. The world would be composed by atoms of different shapes and sizes. Although this atomic theory bears some resemblance to the nineteenth century understanding of atomic structure, the supporting reasoning was very different and without empirical basis. For example, the atom of Leucippus and Democritus is an inert solid interacting and connecting with other atoms mechanically.

Aristotle (384–322 BCE) was the first to refer to this field of study as "Physics." He promoted the idea that observation of physical phenomena could lead to the discovery of the natural laws governing them. He attempted to explain ideas such as motion and gravity with a system known as Aristotelian physics, where all matter was made up of aether, or some combination of four elements: earth, water, air, and fire. Aristotelian physics, through translations from Greek to Arabic and then to Latin, became the established scientific paradigm in Europe until the scientific revolution of the sixteenth and seventeenth century CE.

In parallel to these developments, atomistic views of matter were developed also in ancient Indian sciences. For example, Maharishi Kanada (https://en.wikipedia.org/wiki/Ka%E1%B9%87%C4%81da_(philosopher) is credited as the first to systematically develop a theory of atoms at around 200 BCE.[1] Pakudha Kaccayana (https://en.wikipedia.org/wiki/Pakudha_Kaccayana), a sixth century BCE Indian philosopher also put forward ideas about the atomic constitution of the world. The particle of matter that could not be divided further was termed *parmanu*, and it was considered indestructible and eternal.

It is worth noting in passing, that already in these times it was common knowledge that the Earth is spherical and Eratosthenes (276–194194 BCE) had accurately estimated its circumference. Furthermore, in contrast to Aristotle's geocentric views, Aristarchus of Samos (c. 310–c. 230 BCE) (310) presented a heliocentric model of the Solar System. This was supported by his follower Seleucus of Seleucia (c. 190–c. 150 BCE), who is said by Plutarch to be the first to prove the heliocentric system through reasoning; alas, the specific arguments he used are lost.

In the seventh to fifteenth centuries CE, scientific progress continued in the Islamic world, including significant achievements in mechanics and optics. Many classic works in various languages, including Indian and Greek, were translated into Arabic. These included the works of Aristotle. During this period, Ibn al-Haytham (965–1040) and Ibn Al-Bîrûnî (973–c. 1052) were early proponents of the scientific method. Ibn al-Haytham is considered the "father of the modern scientific method" due to his emphasis on experimental data and reproducibility of results. Al-Bîrûnî introduced early scientific methods for several different fields of inquiry during the 1020s and 1030s. His methodology resembled the modern scientific method, particularly in his emphasis on repeated experimentation.

During the medieval times, ancient works became known to European scholars through translations from Arabic to Latin. They sought to reconcile the philosophy of the classical philosophers with Christian theology and proclaimed Aristotle the greatest thinker of the ancient world. Aristotelian physics became the foundation for the physical explanations of the European Churches. However, in the sixteenth and seventeenth centuries, a large advancement of scientific progress known as the Scientific revolution took place in Europe, with natural philosophers turning to mathematics for the description of Nature.

Towards the end of the eighteenth century, the empirical evidence for the atomic hypothesis started to accumulate. Antoine Lavoisier (1743–1794) formulated the law of conservation of mass in 1784, stating that the mass of the products of a reaction is the same as the mass of the reactants. Subsequently, Joseph Louis Proust (1754–1826) proposed the law of definite proportions: the masses of elements in a compound always occur in the same proportion. Although Lavoisier and Proust did not explicitly

[1] Some scholars place him as early as the sixth century BCE.

refer to atoms, their work formed the basis for the work of John Dalton (1766–1844), an English chemist, physicist, and meteorologist. Dalton compiled experimental data that he and others collected and observed a pattern now known as the "law of multiple proportions." This law states that if two elements form more than one compound, then the ratios of the masses of the second element which combine with a fixed mass of the first element will always be ratios of small whole numbers. This suggested that each element combines with other elements by a basic unit of weight, called "atom" by Dalton. He proposed that each chemical element consists of a single type of atom that could not be destroyed by chemical means. The publication of his findings in 1805 marks the beginning of the modern atomic theory. The novelty of Dalton's atomic theory is that it provided a method of calculating relative atomic weights for the chemical elements, which gives the means for the assignment of molecular formulas for chemical substances.

In 1811, Amedeo Avogadro (1776–1856) improved on Dalton's theory by proposing that equal volumes of gases at equal temperature and pressure contain the same number of particles. Avogadro's law enabled the accurate estimation of atomic masses of elements and made the distinction between atoms and molecules clear. Another significant contribution to atomic theory was made in 1827 by botanist Robert Brown (1773–1858), when he noticed the random movement of dust particles floating in water. The origin of this Brownian motion, as it came to be known, remained unexplained until 1905, when Albert Einstein (1879–1955) postulated that it was due to the movement of water molecules. This was confirmed experimentally soon after by Jean Perrin (1870–1942) offering further support to the atomic theory.

In 1897, atoms lost their place as the fundamental undivided building blocks of matter, with J. J. Thomson (1856–1940) discovering the electron. Because the electron carried a negative charge, while the atom as a whole is electrically neutral, Thomson proposed the *plum pudding* model of the atom, in which electrons were embedded in a mass of positive charge to yield an electrically neutral atom. However, Ernest Rutherford (1871–1937), one of Thomson's students, disproved the *plum pudding* model in 1909 through the experiments performed by Geiger and Marsden. He described a *planetary* model, in which electrons orbited a small, positive-charged nucleus, which was further refined by Niels Bohr (1885–1962) in 1913 with a model which states that electrons only orbit the nucleus at specific distances. As a result, electrons would not spiral into the nucleus as expected by classical physics, but could make quantum leaps between energy levels. This model explained the spectral lines of hydrogen but did not extend to the behaviour of atoms with multiple electrons. Several discoveries expanded the understanding of atoms.

In parallel, Louis de Broglie (1892–1947) in 1924 proposed a wave-like behaviour of moving particles. Erwin Schrödinger (1887–1961) in 1926 developed the equation that bears his name, which describes the evolution of matter waves in space and time. The uncertainty principle, stating that it is not possible to simultaneously determine with infinite precision both the position and momentum of a particle, was formulated by Werner Heisenberg in 1927 (1901–1976). The electron can potentially be found anywhere in the atom but there are areas, e.g., around the classical Bohr radius for hydrogen, that can be found with the greatest probability. Rather than the circular orbits of the Bohr-Rutherford model, modern atomic theory describes orbitals that may be spherical, dumbbell-shaped, etc. These developments are described further in the following section, as these were key to the early understanding of elementary particle physics. Now, we know that the fundamental constituents of matter, up to the energy scales we have probed, are the quarks and the leptons. Nevertheless, the atom remains the smallest unit of matter that cannot be divided using chemical means.

3 From Atoms to Elementary Particles and the Higgs Boson

Simultaneous to these advancements in atomic theory, major leaps were made at the end of the nineteenth century in the understanding of electromagnetism and light. In fact, this major revolution in the field, led by an increasingly more rigorous mathematical interpretation of the measurements, had convinced many scientists that there was almost nothing left to be discovered.

Physicists at the end of the 1800s believed that atomism was a correct interpretation of matter, that it must be composed of small unbreakable "bricks." As described above, this theory—developed in ancient Greece—had an overall acceptance in science. It was only in the last half of the nineteenth century, as chemistry was transitioning to a more quantitative approach, that scientists realised there is a fundamental difference between composed and mixed material. The rules extracted from the composition of substances soon revealed a simple and regular way to classify material, the periodic table. The elements were perfect candidates for the different kinds of atoms. In the physics community, this brought about a revamp of atomistic theories, like the microscopic interpretation of thermodynamics, which was yet to be demonstrated. Thanks to the combined work of physicists and chemists, atomic theory was evolving rapidly. A crucial step of this journey was measurement of the Avogadro number, and the discovery of radioactivity. The Avogadro number, the number of elements in a mole of substance, gives proof that any substance is composed of a definite number of small objects. At the same time, Thomson discovered the electron, the first elementary particle.

In the following decades, effects observed in the interaction of light with matter, led to the idea that energy could be exchanged by light only in discrete amounts, called quanta. Strong evidence of the quantum nature of light came from the interpretation of the photoelectric effect. Two plates of conducting metal are placed face to face in a vacuum glass pipe, as shown in Fig. 2. They are connected to a potential difference, acquiring electrical charge, one positive one negative. An external source of light is directed to the negatively charged plate, the cathode. When light with frequency above a certain threshold, which depends on the material of the cathode, hits the plate, electrons are emitted by the cathode and are accelerated towards the other metal plate, the anode. An electric current appears in the circuit. The intensity of the current is proportional to the intensity of the light, but the energy of the electrons is not. Albert Einstein provided an explanation of this effect showing that the electron energy is proportional to the frequency of the external light source, interpreting light

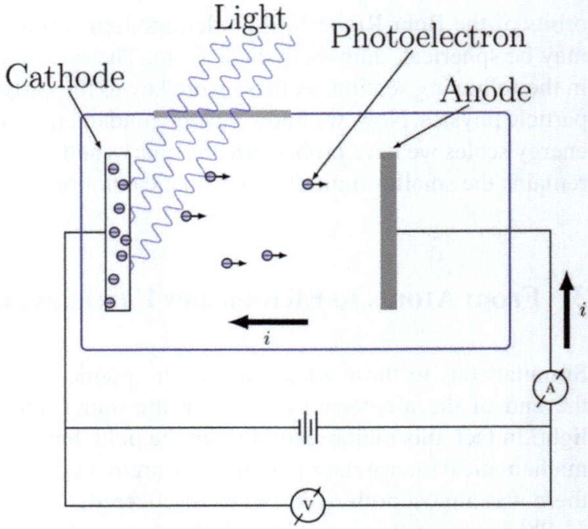

Fig. 2 Experiment for the study of the photoelectric effect. *Image* K. Nikolopoulos

as particles absorbed by the electrons at the surface of the metal plate. Einstein's explanation helped to solidify the theory of the "quanta of light" in the community, but opened another hard question: is light a particle or a wave? In the past experiments, e.g. Young's double slit experiment demonstrated that light was a wave. It created interference patterns when shone through small openings. New evidence showed that it behaves like small packages of energy. Which interpretation is correct?

A further step down the rabbit hole happened during the development of the atomic model. Radioactivity showed that a large amount of energy was naturally coming out from (radioactive) elements: A transformation should occur and elements could mutate, but the mechanism was still obscure.

In 1897 J. J. Thomson discovered the electron, which has negative electrical charge. Since matter is not electrically charged, there should be a positively charged counterpart in the atom. Thomson proposed the *plum pudding* model of the atom, where negative electrons are like pieces of raisin in a plum pudding. The plum pudding represents here the positively charged part of the atom. Hans Geiger (1882–1945) and Ernest Marsden (1889–1970), under the suggestion of Rutherford, demonstrated in 1909 that instead the atom is mostly empty. They shot particles coming from radioactive elements at a gold foil to see how they would be deflected. Most of the particles were passing through undeflected, but a few were scattered through very large angles, and some even scattered in the backward direction. The *plum pudding* model predicted that only weak and uniform electrical forces would be exerted on the incoming particles. The very energetic alpha particles should have passed through without large scattering. After the experiment of Geiger and Marsden, it was clear that the *plum pudding* model was wrong. Soon after, in 1911, Rutherford proposed the *planetary model*, where the atom consists of a small positively charged nucleus at its centre, where most of the mass of the atom resides, with electrons orbiting around it, like planets in the solar system.

Most of the experimental information about the structure of the atoms was coming from spectroscopy. Many elements were studied and their spectra recorded with very good precision. But, the *planetary model* had problems to explain the very precise and regular structure of the spectra. Moreover, an electron orbiting a nucleus should lose most of its energy almost instantly and collapse into the nucleus, therefore emitting light for a very short amount of time producing a continuous spectrum different from what was observed.

At the time, Niels Bohr was a young theoretical physicist working at the laboratory of Rutherford at Manchester. He was exposed to various discussions about the model and the lab results. He joined the challenge. He created a model where there is a fundamental energy state, which represented the minimum energy the electron and the nucleus can have, and excited states, which can only have certain quantised energies. Photon emission would only happen when an electron would transition from one excited energy state to another one, therefore at very specific frequencies.

A step further in the understanding of the structure of the atom came from Louis de Broglie. De Broglie suggested that electrons in the atom could behave like waves, similarly to the analogy presented by Einstein that the light could be considered as a particle. The energy states of the atoms were stationary waves of the electron. Using this assumption, the model was coherent with the Bohr model, and started to become even more predictive. A wave equation was found for simple atoms with not many electrons, like hydrogen. Considering that electrons were behaving like waves inside the atoms was a breakthrough, and, at the same time, a very confusing idea. Matter seems to behave like particles and waves at the same time.

Soon after, Erwin Schrödinger proposed a mathematically more complete model to explain the work of Bohr and de Broglie. He found a differential equation that had as solution, called the "wave function," the waves of de Broglie. Finding the analytical form of the solutions of this wave equation allowed Schrödinger to present the mathematical form of the stationary waves of the electron in the atom and the derivation of the excited energy states. This new mathematical model allowed also to reconcile Maxwell's theory with atomic theory and gave a clear picture of what was happening at the atomic scale. Figure 3 illustrates the evolution of the atomic model from Thomson to Schrödinger.

On one hand, it was acceptable to consider the electron as a stationary wave in the confined space of an atom. On the other hand, extending it to a free moving object was conceptually more difficult. How could a wave equation explain the behaviour of free electrons which have a defined position and speed like particles? This question kept most physicists in the second half of the twentieth century very busy.

Max Born (1882–1970), a theoretical physicist at Gottingen Institute of Physics, was looking to find a connection between the particle and wave interpretations of electrons. He proposed that the connection could be statistical. The square of the wave function could be proportional to the probability to find the particle in a determined volume. Determinism was completely abandoned. From now on physics could not

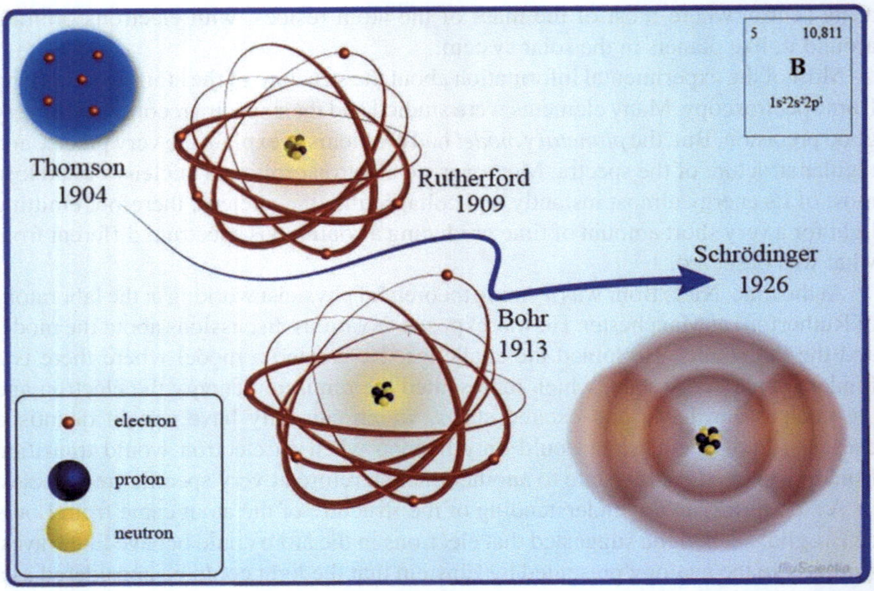

Fig. 3 Evolution of atomic models. *Photo Credit* IlluScientia/Wikimedia commons

determine where particles are, but just calculate a probability for them to be found in a region of space. And this is not because of some unknown parameters, but a fundamental feature of Nature.

Matter is an expression of waves, or better fields, that fill the space. Physicists found an end to the "seeds inside seeds" chain of the Greek philosopher Anaxagoras, that there are fundamental, unbreakable building blocks called quanta. At the same time, these basic constituents of matter are not localised, but permeate the entire space, through their associated fields.

In the 1930s, physicists had a coherent theory that could predict the observations at atomic scale, but they also knew they were still missing a theory that could explain these effects. In a way, they built the kinematics of the atom, but they were missing a theory for its dynamics. The step forward would have been to include the newly accepted photons, the quantum of light, to the electromagnetic field theory of Maxwell.

In 1947 Richard Feynman (1918–1994) and Julian Schwinger (1918–1994), and independently Shinichiro Tomonaga (1906–1979), completed the first quantum field theory, quantum electrodynamics, also known as QED. The theory explains the interaction between charged particles as the exchange of photons. It is able to explain at the same time the quantum effects of the atom's spectra and the electromagnetic behaviours of light and charged particles of the Maxwell theory. In QED, the force that two charged particles experience is brought from one to another by one photon, so the photon is the carrier of the electromagnetic interaction. In nature, there are many different elementary interactions, and each has their own carrier. The task of physics in the second half of the twentieth century was to bring this theory to all fundamental interactions.

An interesting aspect of the quantum fields proposed by Paul Dirac (1902–1984) already in 1928 is the prediction of an antiparticle. Each particle should have an associated antiparticle with the same mass but opposite charge, so the electron, he predicted, should have a positively charged antiparticle, a *positron*. Only a few years later Carl Anderson (1905–1991) discovered the positron and it appears the prediction is valid for all known particles.

The distinction between particles and antiparticles leads also to a different particle classification. Particles can be either *fermions* or *bosons*. An electron is a fermion, while a photon is a boson. In quantum field theory, fermions can be created only if their antiparticle is also created at the same time and place. And they can be destroyed only if they meet their antiparticle. Bosons can instead be created and destroyed individually, if there is enough energy. The fermions constitute the fundamental element of ordinary matter, while bosons, on the other hand, are carriers of interactions.

Particle colliders are the tool used by physicists to investigate the details of elementary particles. Particle colliders accelerate charged particles and make them collide. Part of the energy they carry is converted to new exotic and unstable particles that fly away from the interaction point. Particle detectors surround the collision point and measure the properties of these newly produced particles. Thanks to these sophisticated tools physicists in the last 100 years discovered that protons and neutrons, constituents of the atoms, are not elementary particles but are in fact composed of lighter particles that they named *quarks*. Quarks carry a fractional electrical charge and are always bound together in heavier structures, like neutrons and protons, through a force which is of a different kind compared to what was known before. They named this new force the *strong force*. *Quantum chromodynamics (QCD)* is the quantum field theory associated with it. This force is much stronger compared to the electromagnetic and gravitational ones, and does not allow quarks to be alone or free. It is carried by an elementary boson named *gluon*, which is electrically neutral, but has a different property, called a *colour charge*. While an electrical charge is the ability of a particle to emit and absorb photons, a *colour charge* is the ability of a particle to emit and absorb gluons.

Another fundamental interaction discovered in the second half of the twentieth century was the *weak interaction*. This interaction is responsible for radioactivity. In nature, only a few elementary particles are stable, with an infinitely long lifetime. Most ordinary matter, protons and electrons are stable. Radioactive elements and most exotic particles produced at particle accelerators have a limited lifetime. Shortly after being produced, they start a decay chain until a stable element or particle is reached. The weak interaction has more than one carrier and, contrary to photons and gluons, its carriers have a mass. The mass makes the interaction very short ranged. This interaction and its bosons, the W and Z, were predicted in the 1960s but only in the 1980s were produced and observed for the first time at CERN (The W and Z bosons 1983), a particle physics laboratory in Geneva, Switzerland.

Fig. 4 Portrait of Emmy Noether circa 1900. *Image* Public domain

Explaining how the W and the Z boson obtained their mass, in contrast to the massless photons and gluons, was a significant puzzle. A new non-directional or "scalar" field, had to be introduced in the theory to explain this effect.[2] It was proposed in 1964 by Brout and Englert (1964), and Peter Higgs (1964). This field also provides a natural explanation to the origin of the masses of fermions. The Higgs field interacts with elementary particles; the stronger the interaction, the larger the mass of the particle. This field has its own boson, the Higgs boson, which itself has mass, indicating that the Higgs field interacts with itself. The Higgs boson was included in the Standard Model of particle physics in the 1960s, but observed experimentally only in 2012 by the ATLAS and CMS experiments at CERN.

It is important to note that these highly successful and now well-supported quantum field theories are all deeply rooted in mathematics first developed by Emmy Noether (1882–1935) (Fig. 4) in the 1930s. The theorem she has proven and bears her name, that every symmetry in nature corresponds to a conservation law, is a cornerstone to the models used by particle physicists today. That she is rarely listed in current textbooks is a bad oversight, both concerning her influence on mathematics and her value as role model for young women studying mathematics and physics.

4 What the Higgs Boson Teaches Us and Questions that Remain

After a long and winding journey through the discoveries made by experimentalists and efforts by theorists, scientists working at the CERN Large Hadron Collider (LHC) observed the Higgs boson, about a decade ago. This discovery was unique: not only did we find the last missing piece of the Standard Model, but we also found a very special kind of particle.

[2] Historically referred to as the "Higgs field".

The Higgs boson is unlike any other known elementary particle, as it is associated to the field that gives mass to other particles. Since its discovery, we have been focusing on the measurement of its properties. Much progress has been achieved, but this is a long-term research programme that will require patience and effort. Soon after the Big Bang, a phenomenon called spontaneous symmetry breaking took place (and we are still trying to understand why), which resulted in the Higgs field appearing everywhere in the universe. It interacts with the other elementary particles, providing them with mass. Why does the Higgs behave this way and more importantly, is this the only particle doing that? These questions will shape the careers of future generations of physicists.

Along with the Higgs boson, particle physicists have discovered several interesting things. For example, only in the past few decades have particle physicists obtained clear evidence that neutrinos have mass (1998). Our best particle physics theories of the day, namely, the Standard Model (https://home.cern/science/physics/standard-model) (Fig. 5), cannot explain this evidence. Even more surprisingly, it cannot explain why the Universe exists in its current form, i.e. it cannot explain why there is so much more matter than antimatter in the Universe (https://home.cern/science/physics/matter-antimatter-asymmetry-problem). Nor does it explain the evidence for the so-called Dark Matter (https://home.cern/science/physics/dark-matter). Dark matter is an invisible mass in the universe that helps keep galaxies together, but we do not see it because it does not interact with light. If dark matter is made of particles, we do not know their mass nor how they interact. However, scientists are working on experiments to directly detect weak interactions of dark matter with sensitive targets buried underground on earth or placed on the International Space Station. They also seek indirect evidence for the production of dark matter in the experiments of the LHC or more energetic colliders in the future. Mysteries like this are what keep the field exciting.

When looking at the list of open questions, one would naturally ask, then what have you learned? The honest answer is, we have learned a lot and yet very little about the Universe. We can very successfully describe the world around us starting with atoms to the elementary particles. Even better, we know how elementary particles make atoms. These relationships have taken decades to be understood. They also required the tireless work of hundreds of physicists from all over the world. The resulting theory, which is known as the Standard Model of particle physics, is very good indeed. It is so good that despite the efforts of thousands of scientists, at CERN and elsewhere, so far it has not been proven wrong, albeit it is understood to be incomplete.

We have also understood how to perform incredibly complex experiments and measure properties of elementary particles very precisely. Accelerators, such as the LHC, achieve very high energies by colliding highly energetic particles like protons. Accelerating many protons to very high energies, with velocities approaching that of light, at the same time is a difficult task. They either get accelerated to high energies but do not stay together or stay together but do not accelerate. In order to maintain both, high magnetic fields are necessary and for that powerful magnets are developed. This is a vivid demonstration of the challenges involved in performing experiments.

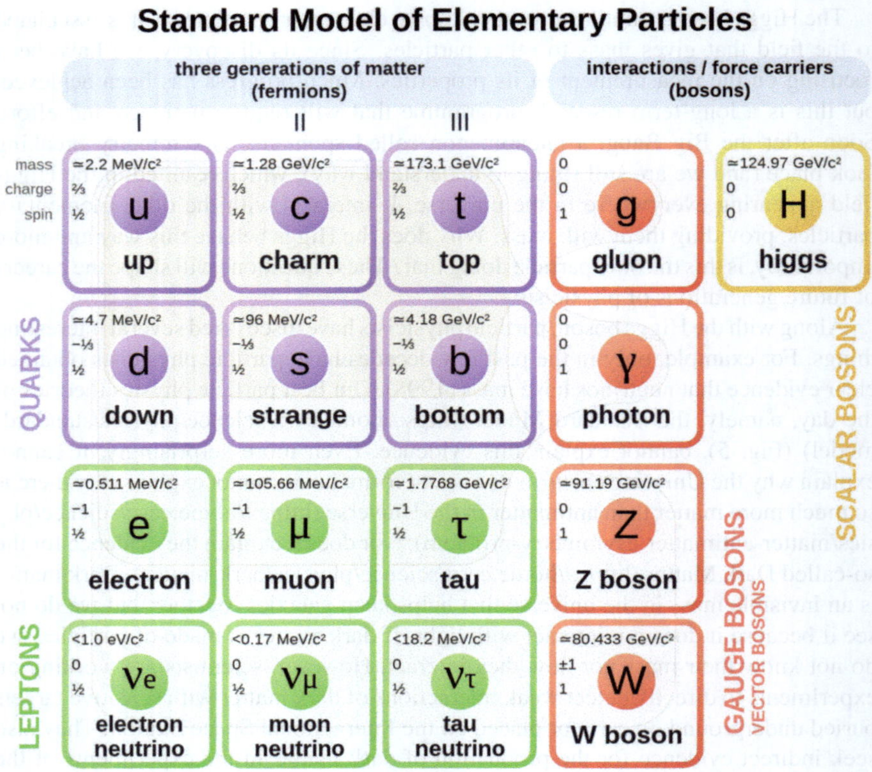

Fig. 5 Standard Model of particle physics. *Image* Wikimedia commons

Today, the physicists have pushed the limit of existing technologies to build the LHC, and the experiments of tomorrow will take this further. It also means that particle physics has been responsible for important new technological developments.

Using these new technologies, particle physicists have helped develop new industrial applications, such as ultrafast electronics and transparent touch screens (Lowe 1974). The World Wide Web was first developed by Tim Berners-Lee in 1989 at CERN (Berners-Lee 2022) as a means to share science documentation, then offered freely to the public. In addition, the vast majority of accelerators around the world are located at medical facilities, with precision detectors used for diagnostics and particle beams used for treatment, such as particle beam therapy (Matsumoto et al. 2021).

Finally, we have improved the methods by which we learn and explore. This is no small task. Understanding experimental findings and developing mathematical language to describe it was not easy. Theorists who write mathematical models and experimentalists who search for these models need to talk to each other. We understand how to transform the abstract mathematical world into real life collisions of particles. We will use this methodology and the knowledge it yields going further.

5 A Proton and an Electron Walk into A Bar; A Dialogue on Endless Interactions

The following conversation might or might not have been heard in the H bar, a popular rest stop somewhere out there in intergalactic space.

Proton: Well, that was quite a trip.
Electron: Seriously? It took forever and I swear it is getting colder and darker out there.
Bartender: Accelerated expansion, I tell you. The whole universe is getting bigger, faster, and faster. What'll you have?
Electron: I'll take a glass of photons.
Proton: Not too quickly, now. Last time, you left me.
Electron: That was 2 billion years ago. Could you get over it, by now? Ionisation happens!
Proton: Sorry. It is not easy being alone. I did a stint in a helium nucleus and I can tell you, it is not my thing.
Electron: O.K. O.K. What're you having?
Proton: Just a couple gluons, please. I don't want any strangeness messing up my inherent charm.
Electron: Charm. Yea right. You're all up and down, at least when you're with me.
Bartender: Here you go. No charge for the gluons. I gave you my favourite colour combinations.
Proton: Thanks. So, what'd you think of earth?
Electron: Nice place. The life forms are a bit weird, though.
Proton: Those were scientists. They like studying us.
Electron: Is that why they had you spinning in circles?
Proton: 11,000 times a second. You know how many times I had to show my passport?
Electron: Seems excessive.
Proton: They like to take a lot of data to make sure what they say is correct.
Electron: That's not like the other humans, you know the ones on social media.
Proton: Indeed. The scientists are much more careful about what they say. Even when they are sure, they ask their colleagues to check their results. Several times.
Electron: Obsessed with accuracy?
Proton: For the humans, these days, they seem like the only ones who you can trust.
Electron: So, the scientists are the ones that measure.

Proton: Well, those are the experimentalists. There are also theorists who look at the results and try to draw conclusions about the universe.
Electron: Do they find the answers?
Proton: Mainly more questions. Then the experimentalists have to take more data to seek answers.
Electron: Sounds like another loop.
Proton: More like a helix. It does have a positive direction. It is how these humans learn.
Electron: So, what happened to you in that spinning machine?
Proton: The LHC? I was put into a group of about 100 billion other protons, called a bunch.
Electron: That's a big bunch.
Proton: They used magnets to keep us close to each other and going in the right direction. In my case, it was counter clockwise. Electric fields accelerated us up to high energy, then they made our bunch go through other bunches moving clockwise at four different places on the ring.
Electron: Right through each other? Didn't you all collide?
Proton: Only about 60 of us collided each time we passed. In the end, I only glanced off another proton and got pushed through the end of the detector. That's how I could make it back to you.
Electron: That's nice, but what about the others?
Proton: Some did have some strong interactions, others weak, others electromagnetic. Some got transformed into new particles.
Electron: How's that?
Proton: Sometimes quarks or gluons run right into each other, then they interact and might form a new, more massive particle.
Electron: How can they become more massive?
Proton: Their energy gets converted into matter. You know, kind of like the opposite of stars.
Electron: Alpha Centauri was hot, baby.
Proton: Sure. I think you're just star struck.
Electron: Very funny. So, were any interesting new particles formed?
Proton: The Higgs boson is probably the most well-known.
Electron: You mean you could have become a Higgs boson?
Proton: Only for an instant. After that, I would have transformed into less massive particles, like bottom quarks or Z bosons. Eventually I could have returned to being a proton.
Electron: What about Dark Matter? Could you have become a Dark Matter particle?
Proton: Well, maybe, but we don't know what a Dark Matter particle is, or even if Dark Matter is composed of elementary particles.
Electron: It does take up a lot of our Universe.
Proton: It is about 85% of all matter, but we still don't know what it is.
Electron: Well, I find it attractive.

Proton: Only through gravitation, though. Not the kind of electromagnetic attraction we have.
Electron: Awww!
Proton: Don't get too excited, now. You'll start shooting out photons, again.
Bartender: Hey! You guys keep your binding to yourselves. This is a public bar.

6 Meanwhile, Back on Earth, A Scientist and a Physicist Meet

A chance encounter happens between a science teacher named Joe and a particle physicist named Mo, both hiking in the Jura Mountains overlooking Geneva. Joe, out of breath from the climb up Crête de la Neige walks up to Mo, who is already admiring the view.

Joe: Beautiful view!
Mo: Sure is!
Joe looks around, trying to spot different landmarks, noticing the water jet in the lake, the Geneva cathedral, etc.…
Joe: Excuse me—are you from around here?
Mo: I've lived in the area for a while, sure. What's up?
Joe: I'm a science teacher, and I'm excited to see the LHC at CERN. I heard it is pretty big, so I should be able to see it from up here, right?
Mo: (laughs) Well… it is underground, so there's not much you can see from above, but you can see some of the experimental halls above the 27 km diameter ring. (Points out a couple of sites).
Joe: That's pretty cool! How do you know so much about it?
Mo: I actually work down there at CERN—I'm a particle physicist… you said you're a science teacher, right? I'm curious—what do you teach your students about our work?
Joe: (thinks for a moment) …. Hmmm… I guess it doesn't really come up much. We just kind of know it is a big place with a lot of scientists, and figure it must be a cool place to visit someday.
Mo: Really??? Particle physics never comes up?
Joe: Well, there's an option for students to learn about it in my grade 12 physics course, but I've never really studied it, so most years I pick an easier option, like astrophysics or medical applications. The kids like it, it is easier to teach, and I even know some cool experiments to include.
Mo: That's a shame, because there's so much cool science going on in particle physics and we're so eager to share what we're up to!
Joe: I wouldn't even know where to start, and I'm kind of intimidated by the subject.

Mo: Even if you're a beginner, a great starting point is to visit some of the impressive research facilities, either right here at CERN or at many institutes around the world. Most offer free visits, both online and in person (https://visit.cern).
Joe: That sounds like fun, I'll check it out when I get off of this mountain. But a visit or two won't get me up to speed on all of particle physics, will it?
Mo: (laughs again) No, of course not. But it can be a great eye-opener for you or for your students. If you want to go further, there are a number of teacher programmes, where you can spend a week or more learning alongside other teachers. Many are even free of charge or subsidised! (https://teacher-programmes.web.cern.ch, https://perimeterinstitute.ca/einsteinplus).
Joe: Thanks, that's a great idea! But I had a friend who applied and wasn't accepted—maybe it is too exclusive for me.
Mo: That's a good point… offering free professional development to thousands of teachers a year can sometimes spread us a bit thin, but there are also a number of high-quality resources available online for free.
Joe: There are so many resources online for any science topic, but it can take ages sorting out which ones might actually work in the classroom – do you have any ideas where to start?
Mo: Sure! The International Particle Physics Outreach Group has recently redesigned their website, including major updates to their curated resource database, which is on track to becoming the number one place where educators and researchers engaged in outreach go to find resources (https://ippog.org/ippog-resource-database).
Joe: This sounds great, do you have any favourites?
Mo: Sure, one of my current favourites is from the Perimeter Institute of Theoretical Physics, in Waterloo, Canada. They're always making something new to make fundamental physics more accessible. Just last year they came out with a new particle physics escape game: Igniting the Orbitron! (2022).
Joe: My students love escape games, but it is so time consuming to plan a good one.
Mo: This one from Perimeter is all ready to go, so all you need to do is download the files or have your students play through online. Plus there are teacher guides with tips on how to run it well, and all of the files are fully modifiable so if you like you can adapt the resource to fit your own needs.
Joe: Sounds great! But all this is still limited to the short time I might spend on a particle physics option for my oldest physics students. Otherwise it just doesn't fit into the curriculum, and I don't have time to add much extra material.
Mo: Maybe, but I think it depends how you look at it… even very basic scientific principles are used on a daily basis in our experiments. For example, what are some of the concepts you're currently teaching?
Joe: Well, often I start off the year with kinematics, Newton's laws of motion, momentum, collisions, …
Mo: COLLISIONS! That's what I do all day every day!
Joe: What do you mean?

Mo: I work at the ATLAS Experiment, and it is one of the detectors where the LHC collides particles. There, as at all detectors, conservation of momentum and energy are two of our most powerful tools to determine which particles come from a collision. Between that and calculating the force on a charged particle in a magnetic field, we can identify almost any particle in the Standard Model.
Joe: So... are you saying that with all of your fancy equipment, you're still only using high school physics?
Mo: Not exactly... there is a lot more to it than that, but yes, much of our work can be understood by using fairly simple principles put into the right context.
Joe: Wow! This opens up a whole new world for me. So, I can use particle physics examples to teach general physics topics?
Mo: Even the search for Dark Matter at the LHC comes down to conservation of momentum.
Joe: Excellent!
Mo: And it can be easier than you think once you know where to start. Many of us are enthusiastic and want to help out, and it is just a matter of connecting the outreach efforts of enthusiastic scientists with teachers who are eager to bring the exciting world of modern science into their classrooms!
Joe: Thank you! I'm sure my students will be as excited as I am to explore the amazing world of particle physics!

7 Conclusions

I seem to have been only like a boy playing on the seashore, and diverting myself in now and then finding a smoother pebble or a prettier shell than ordinary, whilst the great ocean of truth lay all undiscovered before me.

Isaac Newton

Bringing the modern science curriculum up to date should not be a burden for a secondary classroom science teacher. Rather, it is only a matter of adding the next sequel to the already fascinating story of humanity's quest to understand the world we live in. It appears, at least for now, to be a never-ending journey. All of our exploration, in terms of elementary particle physics, has brought us an understanding of about 5% of our universe. And, even then, we are missing major components, such as a microscopic understanding of gravity or an explanation of the imbalance between matter and antimatter.

Yet, to put this in perspective, we should not forget that we inhabit but one planet, orbiting a star that is but one of one hundred billion stars in a galaxy that is but one of one hundred billion galaxies in our universe. And we aren't even sure if our universe is the only one. So, understanding 5% is pretty impressive.

The authors of this chapter hope its readers will appreciate not only the spectacular nature of our quest, but its ability to spark the interest of young students. These students need not leave school with a detailed comprehension of quantum field theory, be able to calculate interaction cross sections, or even understand what a penguin is doing in a Feynman diagram. However, the stories behind our exploration of the most fundamental components of matter can give them an appreciation of the intricate interplay between experimentation and theory. It can teach them how precise measurement and careful interpretation drive our understanding of nature and convince them of the importance of international collaboration to solve some of the most complex problems known to humanity. And these are lessons they can take with them regardless of the paths they choose.

References

Amalie Emmy Noether (1882–1935) German mathematician. https://www.britannica.com/biography/Emmy-Noether. Accessed 18 Mar. 2023
Aristarchus of Samos (c. 310 BCE–c. 230 BCE), Greek astronomer. https://www.britannica.com/biography/Aristarchus-of-Samos. Accessed 16 Mar. 2023
Aristotle, Greek philosopher (384–322 BCE). https://www.britannica.com/biography/Aristotle. Accessed 16 Mar. 2023
ATLAS Collaboration, Observation of a new particle in the search for the Standard Model Higgs boson with the ATLAS detector at the LHC. Phys. Lett. B716 (2012a) 1
A. Avogadro (1776–1856), Italian mathematical physicist. https://www.britannica.com/biography/Amedeo-Avogadro. Accessed 17 Mar. 2023
M. Born (1882–1970), German physicist. https://www.britannica.com/biography/Max-Born. Accessed 17 Mar. 2023
R. Brown (1773–1858), Scottish botanist. https://www.britannica.com/biography/Robert-Brown-Scottish-botanist. Accessed 17 Mar. 2023
Carl David Anderson (1905–1991), American physicist. https://www.britannica.com/biography/Carl-David-Anderson. Accessed 17 Mar. 2023
CERN Teacher Programs. Geneva, Switzerland. https://teacher-programmes.web.cern.ch. Accessed 13 Mar. 2023
CERN Visits. Geneva, Switzerland. https://visit.cern. Accessed 13 Mar. 2023
C.M.S. Collaboration, Observation of a new boson at a mass of 125 GeV with the CMS experiment at the LHC. Phys. Lett. B **716**, 30 (2012)
J. Dalton (1766–1844), English meteorologist and chemist. https://www.britannica.com/biography/John-Dalton. Accessed 17 Mar. 2023
Dark matter, CERN physics explainer. https://home.cern/science/physics/dark-matter. Accessed 17 Mar. 2023
Development of atomic theory. https://www.britannica.com/science/atom/Development-of-atomic-theory. Accessed 13 Mar. 2023
Einstein Plus Teacher Program. Perimeter Institute, Waterloo, Canada. https://perimeterinstitute.ca/einsteinplus Accessed 13 Mar. 2023
A. Einstein (1879–1955), German-born physicist. https://www.britannica.com/biography/Albert-Einstein. Accessed 17 Mar. 2023
F. Englert, R. Brout, Broken symmetry and the mass of gauge vector mesons. Phys. Rev. Lett. **13**(9), 321–323 (1964)
Eratosthenes (276 BCE – 194 BCE), Greek scientist. https://www.britannica.com/biography/Eratosthenes. Accessed 16 Mar. 2023
Ernest, Baron Rutherford of Nelson (1871–1937), New Zealand-born British physicist. https://www.britannica.com/biography/Ernest-Rutherford. Accessed 17 Mar. 2023

Heraclitus, Greek philosopher (540–480 BCE). https://www.britannica.com/biography/Heraclitus. Accessed 16 Mar. 2023
P. Higgs, Broken symmetries and the masses of gauge bosons. Phys. Rev. Lett. **13**(16), 508–509 (1964)
Ibn al-Bīrūnī (973–1052), Muslim astronomer, mathematician, ethnographist, anthropologist, historian, and geographer. https://www.britannica.com/biography/al-Biruni. Accessed 17 Mar. 2023
Ibn al-Haytham (c. 965–c. 1040), Egyptian mathematician and astronomer. https://www.britannica.com/biography/Ibn-al-Haytham. Accessed 17 Mar. 2023
Igniting the Orbitron, Perimeter Institute, Waterloo, Canada. (2022). https://resources.perimeterinstitute.ca/products/igniting-the-orbitron-breakout-activity. Accessed 13 Mar. 2023
Johannes Wilhelm Geiger (1882–1945), German physicist. https://www.britannica.com/biography/Hans-Geiger. Accessed 17 Mar. 2023
Julian Seymour Schwinger (1918–1994), American physicist. https://www.britannica.com/biography/Julian-Schwinger. Accessed 17. Mar. 2023
Pakudha Kaccayana (c. 6th century BCE), Indian teacher and philosopher. https://en.wikipedia.org/wiki/Pakudha_Kaccayana. Accessed 16 Mar. 2023
Maharishi Kanada, Indian natural scientist and philosopher (between 6th and 2nd century BCE). https://en.wikipedia.org/wiki/Ka%E1%B9%87%C4%81da_(philosopher). Accessed 16 Mar. 2023
A. Lavoisier (1743–1794), French chemist. https://www.britannica.com/biography/Antoine-Lavoisier. Accessed 17 Mar. 2023
Louis-Victor-Pierre-Raymond, 7e duc de Broglie (1892 – 1987), French physicist. https://www.britannica.com/biography/Louis-de-Broglie. Accessed 17–03–2023.
J.F. Lowe, Computer creates custom control panel. *Design News*, (18 November 1974), pp. 54–55
Y. Matsumoto, N. Fukumitsu, H. Ishikawa, K. Nakai, H. Sakurai, A critical review of radiation therapy: from particle beam therapy (proton, carbon, and BNCT) to beyond. J. Personalized Med. **11**(8), 825 (2021)
Neutrino mass discovered, *Physics World Magazine*, 1 July 1998. https://physicsworld.com/a/neutrino-mass-discovered. Accessed 17 Mar. 2023
Niels Henrik David Bohr (1885–1962), Danish physicist. https://www.britannica.com/biography/Niels-Bohr. Accessed 17 Mar. 2023
On the Constitution of Atoms and Molecules, N. Bohr, The London, Edinburgh, and Dublin Philosophical Magazine and Journal of Science, sixth series, July 1913.
Paul Adrien Maurice Dirac (1902–1984), English theoretical physicist. https://www.britannica.com/biography/Paul-Dirac. Accessed 17 Mar. 2023
J.-B. Perrin (1870–1942), French physicist. https://www.britannica.com/biography/Jean-Perrin. Accessed 17 Mar. 2023
J.-L. Proust (1754–1826), French chemist. https://www.britannica.com/biography/Joseph-Louis-Proust. Accessed 17 Mar. 2023
Richard Phillips Feynman (1918–1988), American theoretical physicist. https://www.britannica.com/biography/Richard-Feynman. Accessed 17 Mar. 2023
Seleucus of Seleucia (c. 190 BCE–c. 150 BCE), Hellenistic astronomer and philosopher. https://en.wikipedia.org/wiki/Seleucus_of_Seleucia. Accessed 17 Mar. 2023
Sir Joseph John Thomson (1856 1940), English physicist. https://www.britannica.com/biography/J-J-Thomson. Accessed 17 Mar. 2023
Sir Ernest Marsden (1889–1970), English-New Zealand physicist. https://en.wikipedia.org/wiki/Ernest_Marsden. Accessed 17 Mar. 2023
The Standard Model, *CERN physics explainer*. https://home.cern/science/physics/standard-model. Accessed 17 Mar. 2023
The W and Z bosons, discovered at CERN in 1983. https://en.wikipedia.org/wiki/W_and_Z_bosons. Accessed 17 Mar. 2023
The IPPOG Resource Database. https://ippog.org/ippog-resource-database. Accessed 26 Feb. 2023

The matter-antimatter asymmetry problem, CERN physics explainer. https://home.cern/science/physics/matter-antimatter-asymmetry-problem. Accessed 17. Mar. 2023

B.-L., Tim, Information management: a proposal. w3.org. *The World Wide Web Consortium*. Retrieved 12 Feb. 2022

Tomonaga Shin'ichirō (1906–1979), Japanese physicist. https://www.britannica.com/biography/Tomonaga-Shinichiro. Accessed 17 Mar. 2023

Werner Karl Heisenberg (1901–1976), German physicist and philosopher. https://www.britannica.com/biography/Werner-Heisenberg. Accessed 17 Mar. 2023

About Teaching Quantum Mechanics in High Schools

Paola Verrucchi

Abstract Teaching quantum mechanics in classrooms other than those of (some) university departments, let alone speaking about it to the public, has been a taboo for almost one century. However, in the last few decades, it has become clear that it is time to release this taboo, both for an urgent need for professional figures that can understand the principles of quantum information and computation without being physicists, and for the intellectual honesty that obliges us to tell anyone, and young students in particular, the truth about how quantum mechanics shapes our world. In this chapter, we will first discuss the reasons that have relegated quantum mechanics into scientific academic education for such a long time and present our opinion about why these reasons are not valid anymore. We will then present a proposal to introduce the two main postulates of quantum mechanics, namely the state- and measurement-postulates, using a simple formalism that yet allows us to discuss some of the most revolutionary aspects of the theory. The proposal is then revisited to give some guidelines to design formally correct and yet educationally effective approaches to quantum mechanics. Finally, we will comment upon possible strategies to let the principles of quantum mechanics enter at least some high-school programs.

Keywords Classical physics · Quantum mechanics · Qubit · Vectors · Quantum physics for high schools

1 Quantum Versus Classical

Dealing with the introduction of quantum mechanics in high-school education is a process that extends well beyond its own declared objective, in so far as it requires understanding the way classical and quantum physics have talked to each other

P. Verrucchi (✉)
ISC-CNR, UOS Dipartimento di Fisica e Astronomia, Università degli Studi di Firenze, 50019 Sesto Fiorentino, FI, Italy
e-mail: verrucchi@fi.infn.it

INFN Sezione di Firenze, Polo Scientifico e Tecnologico, Dipartimento di Fisica e Astronomia, INFN, 50019 Sesto Fiorentino, FI, Italy

M. Streit-Bianchi et al. (eds.), *New Challenges and Opportunities in Physics Education*, Challenges in Physics Education, https://doi.org/10.1007/978-3-031-37387-9_2

for almost one century. The dialog has been adversarial from the very beginning when messages about the inadequacy of classical physics (CP) in describing certain observed phenomena started to gather up in physics laboratories. These messages said that a fundamentally different theory was necessary, despite CP being a successful theory upon which no one dared cast doubts. The first reaction developed around the attempt of setting quantum mechanics (QM) into CP, with the former seen as a special case of the latter, whose formalism so nicely corresponds to our way of perceiving the world around us. CP is about what we can observe, see, and touch… and this is something quite powerful in the relationship between human beings and science in general, and physics. On the contrary, QM arose to explain very unusual phenomena, utterly out of our everyday life and yet so fundamental that one cannot explain why our universe exists as it is, without understanding them.

Confronted with the necessity of describing quantum phenomena with the formal tools of CP, scientists came up with a very complicated mathematical arsenal, from which paradoxes emerged, as well as true oddities such as the wave-particle dualism, or the infamous "spooky action at a distance", as Einstein dubbed what is now recognized as one of the key ingredients of our entire universe, namely the *entanglement* (that will be introduced later).

Differential equations, probability distributions, and path integrals, just to name a few, are not, and will never be, part of the mathematical background of a scientific high-school student, let alone of any student. Therefore, if these tools are truly necessary for dealing with QM, then QM is doomed to remain a mysterious and unintelligible corner of science for most of the population, despite its huge relevance in understanding the world around us.

The last century tries to exit this blind alley have been proposed along two main lines: bringing as much as the above-mentioned formalism at least in scientific high schools or simplifying the original structure of the theory to make it more like CP. None of the two strategies seemed to have worked, and while the former failed without doing much damage, the latter is responsible for misconceptions and a conceptual derailing [see Ref. Krijtenburg-Lewerissa (2020) for a recent analysis and a thorough bibliography] that can be summarized into the newly created verb "to quantize". The verb, still very much used by physicists, refers to a formal procedure that transforms a known problem of CP into some "quantum analog", as if a classical-to-quantum flow is legitimate. There are special cases in which such analog at least corresponds to a problem that makes sense in QM, but that is not necessarily the case and, even so, it may have no physical correspondence with the classical problem from which one has started. Ultimately, QM cannot be traced back to CP. Rather the opposite is true: CP emerges from QM, as a special case of it. As obvious as it is, this statement only recently entered the phrasebook of physicists (not all of them, in fact), the reason being that it claims some sort of quantum supremacy over CP, which is very difficult to accept, given the above-mentioned correspondence between the latter and our everyday experience. QM shapes our world, from the elementary units of quantum information, the tiny qubit, to the most massive objects of our universe, the extraordinary black holes.

This evidence makes the discussion on the possibility of bringing QM into high-school classrooms essential for designing strategies for teaching modern physics. The discussion is very well represented by the Quantum Technology Education (QTEdu) Project of the European Community, whose purpose is *"to assist the European Quantum Flagship with the creation of the learning ecosystem necessary to inform and educate society about quantum technologies"* (from the homepage of the QTEdu official site). The project is based on the activity of five working groups, *School education, and Public outreach, Educational initiatives in higher education, Lifelong learning and workforce training, Educational research in Quantum Technology, Equity, and Inclusion for QT educational initiatives,* which are making valuable resources available online (there included the *European Competence Framework for Quantum Technologies*) such as programs, courses, training, and evaluation tools, for primary and secondary schools (follow the link "Resources for everyone" from the main menu of the official site).

It is not by chance that the qubit is mentioned above, as quantum information has a fundamental role in the story we are telling. At the end of the last century, QM started being viewed not only as a useful, though disturbing, physical theory, but also as the carrier of an original logical system, based on a small number of postulates, fully consistent, and extraordinarily powerful (Nielsen and Chuang 2011). In this context, the connection with CP was not necessary, and QM could speak its native language: its laws were finally expressed in the simplest possible way, and the cumbersome translation into classical-like expressions became needless. The ensuing process is quite impressive, in so far as it simplifies the mathematical formalism without depriving QM of any of its elements of novelty, as we try to explain in the next section, briefly introducing the two postulates of QM that are today recognized as the cornerstone of the theory, namely the state- and the measurement-postulates.

Before moving on, though, let us consider the following. The formulation of most scientific theories begins with some statement about the "state" of the system under scrutiny. The state is a generic concept: we say that a garden is in a poor state or a friend is in a very good state. However, the concept becomes scientifically useful when embodied in a formal tool, capable of conveying information about the properties of the system to which it refers. In CP such a tool is the "point", whose coordinates represent position and momentum, as in Fig. 1a–c, or pressure and volume, electric and magnetic field, or other possible sets[1] of physical quantities. Dealing with points implies working in the geometrical space where the points are defined, for instance, a two-dimensional plane or a sphere. This is neither the three-dimensional space where we move nor the $3 + 1$-dimensional space time used in relativity. Rather, it is a space, technically dubbed *phase-space*, whose shape is determined by some essential physical constraints and whose dimension is the minimum number of coordinates that are necessary to identify a point on it. Once the formal tool that describes the state is identified, the main goal of CP is that of providing equations, as simple and general as possible, via which one can predict how the point describing the initial state of the system will move, in the above-mentioned space, as time goes by; in

[1] By *possible* we here refer to sets of canonically conjugated variables, in the Hamiltonian formalism.

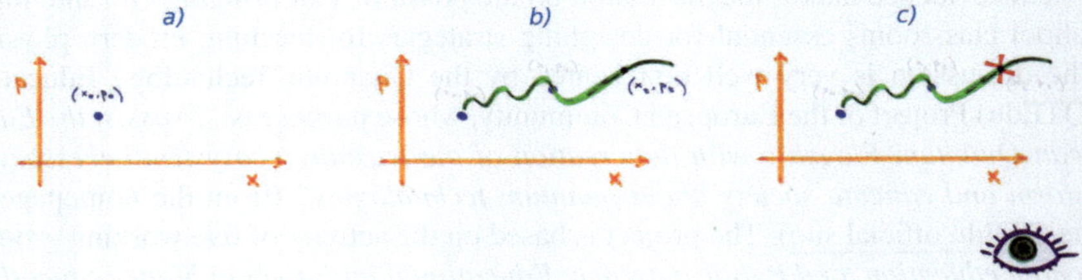

Fig. 1 The state of a system in classical physics is represented by a point: here is the case when the coordinates are the position x and momentum p of a particle moving along a one-dimensional path. **a** the state at some initial time t_0; **b** the trajectory that describes the dynamics of the state at any later time t; **c** the measurement process as a harmless snapshot

other terms, CP aims at drawing the so-called *trajectory* of the initial point, in so far doing allowing us to predict the dynamics of the system at whatever later time, shown as a line in Fig. 1b. Notice that once the trajectory is drawn, the theory has accomplished its task, as the possible observation of the system is just a practical implementation of some experimental procedure that does not alter the state of the system, as represented in Fig. 1c.

2 Simply Quantum

The above familiar picture is revolutionized in QM. The state-postulate of QM tells us that the state of a system Ψ is formally represented by a vector, as shown in Fig. 2a, whose length is set equal to 1, aka *unit vector*, which is consistently dubbed *state-vector*. The symbol $|\cdot\rangle$ was first introduced by Dirac to indicate state-vectors, and by him dubbed *ket*, from the noun *bra-ket*; the graphical sign $|\Psi\rangle$ is hence read "*ket* Ψ". Replacing points with vectors may seem a harmless process: vectors, like points, are simple mathematical objects, oriented segments introduced well before QM, largely used in CP for describing forces, velocities, angular momenta, or other properties, and consequently familiar to many of us. What makes the step revolutionary, and to some extent baffling, is the fact that a vector is now used to describe the state of a system and not one of its properties. Why baffling? Because vectors are algebraic objects, meaning that they can be added together. Some people may remember the parallelogram law of vector addition, shown in Fig. 2b; others may anyway recognize the horizontal and vertical components of an oblique vector, instinctively understanding that these components sum together to set an inclined direction; similarly, a blue paint can be added to a yellow one, and a green color is obtained. With these examples in

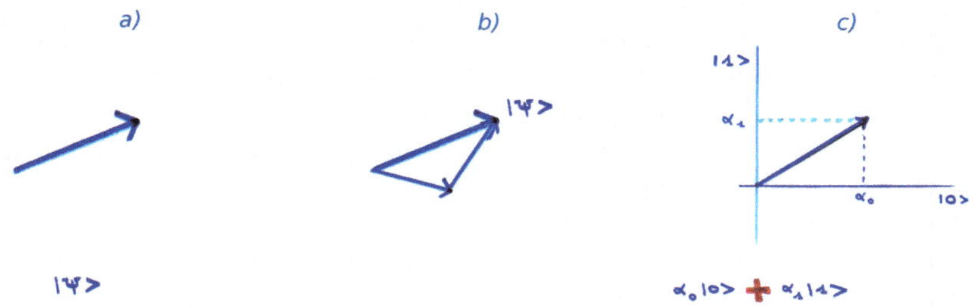

Fig. 2 The state of a system in quantum mechanics is represented by a unit vector: here the case of a qubit (see text). **a** The state; **b** the state as a sum of any two other vectors; **c** the state as a weighted sum of two other states

mind, particularly the last one, we see why the state-postulate of QM, here reasonably simplified into "the state of a system is described by a unit vector"[2] delivers the revolutionary *superposition principle*, according to which different states of a system can coexist, which is the same as saying that their respective state-vectors can be summed up. It is thus possible for a system to be at the same time horizontal AND vertical, yellow AND blue, 0 AND 1, or even alive AND dead, as the infamous Schrödinger cat. In fact, one of the most fascinating aspects of QM is the way, in which the mathematical language of the theory, which is called linear algebra, corresponds to a description of reality: the sign " + " that enters the algebraic sum between vectors shown in Fig. 2b precisely corresponds to the logical conjunction AND in the description of a world where systems can be simultaneously in different states. It is not a matter of being either horizontal OR vertical, perhaps yellow OR maybe blue, 0 OR more likely 1, as Einstein kept on thinking, never recognizing state-superposition as a real phenomenon (he rather thought it was an improper way to treat a lack of information of the same sort of that dealt with by any statistical analysis (Einstein et al. 1935, 1971)): QM informs us that a system can be in two or more different states at the same time, possibly weighted differently. This is nicely expressed by the most general form of the simplest possible state-vector

$$\Psi\rangle = \alpha_0|0\rangle + \alpha_1|1\rangle, \tag{1}$$

which is that of the simplest possible quantum system, today called *qubit*. A qubit is a system that can be in any superposition of two different states, as in Fig. 2c, here indicated by $|0\rangle$ and $|1\rangle$ with reference to the usual notation adopted in quantum information to describe qubits (usually called Alice, Bob, Charlie...). In Eq. (1), α_0 and α_1 are numbers, such that $|\alpha_0|^2 + |\alpha_1|^2 = 1$, to ensure $|\Psi\rangle$ is a unit vector and hence a proper state-vector. Once physical states are given their formal representation

[2] For those who know some basic linear algebra, the postulate says "to every system is associated a Hilbert space: Any normalized vector of this space describes a possible state of the system, and vice versa".

as unit vectors, QM deals with the dynamical evolution of systems along the same path of CP, i.e., providing equations to describe how the state-vector changes with time. Notice that, as its length is set equal to 1 by the postulate, the only variation possible for the state-vector resides in its *direction*, where the italic fonts serve to remind us that we are dealing with vectors in an abstract space, technically named *Hilbert space*, whose dimension can go from 2 to infinity, a space where the notion of direction still exists, but must be treated with a bit of care. It is not fair to say that QM deals with time evolution without introducing elements of novelty, in fact, the so-called *problem of time* in QM is at the heart of any attempt to reconcile QM with General Relativity (Anderson 2017). However, QM describes the dynamics of systems in quite a standard way, without tearing with respect to CP, and we will not dwell on this aspect, here, but rather move toward the observation stage, where things change drastically again.

The superposition principle cannot be satisfactorily processed without considering what happens when making a measurement of a system. Suppose we ask the oblique state-vector in Fig. 3a "are you horizontal or vertical"? Following strict logic, QM teaches us that all the different answers provided by the definition of our question are possible, and the predictive capacity of the theory consists in furnishing the probability with which each possible answer will be given. Moreover, when the specific answer is given, i.e., some result is produced by the experimental apparatus, a true change in the state-vector occurs: answering "I am horizontal (vertical)", the oblique state-vector will lay down (stand up), as seen in Fig. 3b, c. In fact, the quantum measurement process is the subject of the measurement postulate, which provides us with a formula, known as Born's rule, for getting the probability that each possible result (the answer) is obtained, given the specific measurement done (the question). The formula is simple: referring to Fig. 3a, the probability of getting the answer "I am horizontal" equals $|\alpha_0|^2$, whenever "I am vertical" consistently occurs with probability $|\alpha_1|^2$. Despite having been the subject of heated discussions and disparate interpretations, the formal content of this postulate is quite simple and embodies the predictive power of QM. Without entering the realm of QM interpretations, we notice that the idea that a measurement may alter the state of the experimentally tested system is quite reasonable: the information transfer implied by the question–answer scheme needs some sort of interaction, that may well cause a change in the state of the parties involved.

Getting back to the above-mentioned predictive power of QM, one may wonder whether it is too poor for giving the theory the status of a proper scientific tool. To this respect, besides considering the exceptionally precise predictions that QM provides for the most diverse experiments, it is to be noticed that there always exists a question, sometimes called "the right question", for which the answer is perfectly determined, in so far as the probability to get it equals 1. Referring to Fig. 4, for instance, if one asks the state-vector: "do you point toward northeast or northwest", the answer will be with absolute certainty "toward northeast", and the system will remain in its state after the measurement has occurred. Obviously, it is not always possible to guess the right question, unless one accurately designs and controls the dynamical evolution of the system, but the fact that such a specific question exists

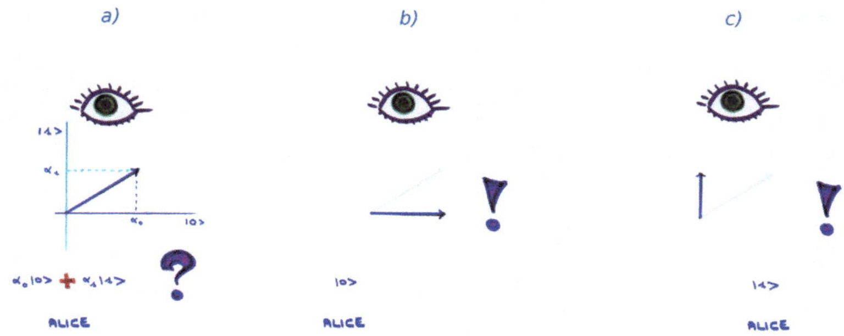

Fig. 3 The measurement process in QM: here is the case of a qubit (see text). **a** The measurement seen as a question; **b** the effect associated with the answer 0; **c** the effect associated with the answer 1

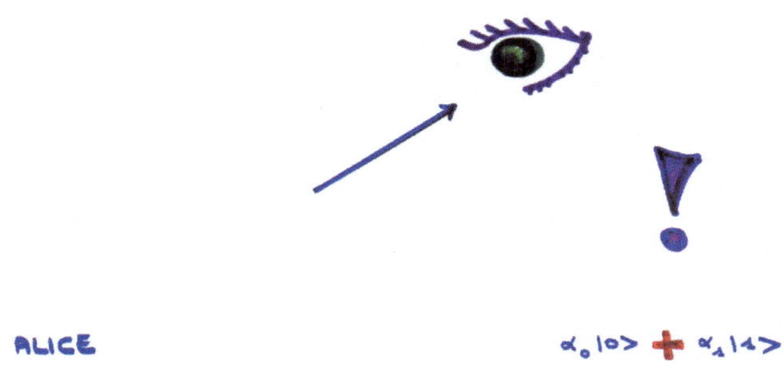

Fig. 4 The "right question" in quantum mechanics (see text)

allows, for instance, to obtain exact results from quantum devices such as quantum computers or simulators.

So far, we have only seen two postulates, and yet even textbooks that introduce QM via postulates[3] often refer to 4 or even more postulates [see for instance Ref. Kaye et al. (2006)]. In fact, most of them can be derived from either the state-postulate or the measurement one, which means that the others are not ultimately necessary to define QM,[4] which is why we will not consider them here.

Let us rather concentrate upon one of the most surprising consequences of the above-considered postulates, namely what happens when two systems, say the famous Alice and Bob that inhabit any quantum information textbook, are put

[3] QM can be alternatively presented by means of some principles, a choice that has been often adopted in the last century, but has been mainly dismissed in the last decades.

[4] Recent works on the problem of time (Foti et al. 2021) and on the way systems are jointly described (Carcassi et al. 2021) indicate that indeed QM can be formulated using only the state- and the measurement-postulates.

together. Suppose the two friends A and B are qubits, meaning that they can separately be in any superposition of the two basis states introduced in Eq. (1), namely $|0\rangle$ and $|1\rangle$; quite obviously, the system "Alice with Bob" will possibly be in any superposition of the four basis states $\{|00\rangle, |1\rangle, |10\rangle, |11\rangle\}$, where we may think that the digit on the left refers to Alice, and that on the right is for Bob, so that the state $|01\rangle$ sees Alice in $|0\rangle$ and Bob in $|1\rangle$. Among the infinite superpositions possible, some are characterized by the fact that despite representing a well-defined state of the AB-pair, they do not, and cannot, convey information about the individual state of each qubit separately. Consider for instance the following superposition.

$$|AB\rangle_{\uparrow\uparrow} = \frac{1}{\sqrt{2}}\left(|00\rangle + |11\rangle\right) , \tag{2}$$

or also

$$|AB\rangle_{\uparrow\downarrow} = \frac{1}{\sqrt{2}}\left(|01\rangle + |10\rangle\right) : \tag{3}$$

We cannot see what the state of Alice is, nor that of Bob, and yet we have a perfectly defined state for the two of them: the two friends now form a pair, where the state of Alice establishes the state of Bob and vice-versa.

The state of the two friends, now become a couple, is said to be *entangled* or, quite equivalently, we say that Alice and Bob are entangled, or quantumly correlated, where the adverb stands to recall that the bond between them holds up against any spatial separation and the passing of time unless some external disturbance occurs. In fact, in the above reasoning, there is no reference to the fact that Alice and Bob be close or distant, that they communicate or not, that they recently met or do not see each other for ages. And yet, by asking a question, Alice not only will change her state according to the answer received but will also affect the state of Bob.

In the case of the state reported in Eq. (2), for instance, when the result of a measurement done upon Alice tells us that she is in the state $|0\rangle$, we also know, without any further measurement or communication procedure, that Bob is also in the state $|0\rangle$. This fact made Einstein crazy, as he thought it meant that Bob was instantaneously informed about the result of the measurement on Alice, no matter how distant they were, via some sort of "spooky action at a distance" (Kaye et al. 2006) already mentioned. In fact, he never came to terms with the fact that Alice and Bob had already shared the whole information content of their pair-state, when prepared in such an entangled state. Whereupon they can be separated at will: if no event occurs that changes their pair-state in Eq. (2) they will stay quantumly correlated forever, and the above reasoning on the measurement will stay valid.

The state- and measurement-postulates introduced in this section can be, and have been, the subject of many, extremely relevant and interesting discussions and considerations, and further research is still needed to fully understand their physical meaning and their logical and philosophical content. Besides these possible developments, there is already quite a lot to digest in the simple narrative presented here, as briefly discussed in the next section.

3 Layered Strategies

In this last section, we want to consider possible strategies to speak about QM in high schools or, in general, to the public. No doubt the subject is difficult and may be tricky to deal with it, but this is no excuse for renouncing or, which is worse, for speaking about it in a simple but unfaithful way, as ever too often done (see the above-mentioned site of the QTEdu project for reports, studies, and articles about these points). Let us better explain our viewpoint in this respect: it is the duty of those who know the details and understand the meaning of a difficult topic to find a way to explain the latter without the former, which is not the same as creating a new, simpler, story, that resembles the original one, but with a different meaning. Whoever is the recipient of our teachings, we must consider that she or he will always have the possibility to get a more detailed education on the taught subject, equipped with more refined formal tools and a richer background. It is hence important that there is no misunderstanding, no confusion, not even the slightest derailment from the conceptual backbone of the subject, in the first encounter with it.

Consider the case of QM: as already mentioned in the first section, attempts of making it more similar to CP thinking that this could help understand the theory were unsuccessful and detrimental. One should rather think about some *layered strategy*, where each layer speaks about QM, but with a different level of detail. Take the above section: there is nothing truly unfaithful to QM and, whenever something is not exactly as we would like it to be, were the recipients equipped with the necessary formal tools, yet there is no lie, in such a way that when the above tools will possibly become available, it will simply be a matter of adding up. Consider the *superposition principle*, or state-postulate: if introduced without mentioning the fundamental fact that states in QM are represented by vectors, there is no way to make the sentence "systems can be in different states at the same time" more precise. We have not mentioned the fact the two coefficients α_0 and α_1 in Eq. (1) are complex numbers, but this is a piece of information that can be easily added to the overall picture if complex numbers will ever become part of the student mathematical background. On the other hand, it is not by chance that we have not mentioned decoherence or the double-slit experiment: there is no way to understand them without referring to the fact that QM uses complex numbers. It is possible to introduce the double-slit experiment to schools in a didactic way, but this implies either evoking a qualitative knowledge about how waves behave or introducing the (complex) mathematical formalism of wave-dynamics.

And yet, the above section alone, without complex numbers, allows one to precisely describe quantum teleportation, quantum cryptography, the Stern-Gerlach experiment, and many other quantum phenomena. Decoherence and the double-slit experiment may come later, when a second layer of knowledge is built upon the solid, despite the simple, basement created. The same is true for the measurement process: we have not mentioned the fact that the process corresponds to a non-unitary dynamic, which is essential and troublesome at the same time. And yet, the idea that observation may alter the state of what is observed is conveyed, Born's rule is provided, and the concept of the right question is introduced. When, and if, the student will meet Hilbert spaces and their orthonormal basis, will see that asking a question means choosing an orthonormal basis and dealing with eigenvectors and eigenvalues of Hermitian operators; the whole formalism will overlap with the previously acquired knowledge and the new information will adhere to the background. Similar considerations can be done for entanglement, where we have only considered two qubits, without speaking about entanglement in more complex systems: this prevents us to comment upon the fact that we do not see a quantum world in front of us, and yet it is enough to define the CNOT quantum gate, an essential element of any universal quantum computer, or explain the Bell inequality (1966) and the reason why experiments demonstrating its violation earned Aspect, Clauder, and Zeilinger the 2022 Nobel prize in physics (2022). Among the existent proposals for introducing QM in high schools, working environments, or to the public in general, which are based on a layered strategy, we like to mention the one designed by *qplayLearn* (https://qplaylearn.com/), the outreach division of the Finnish society Algorithmiq (https://algorithmiq.fi/): on their site one can find multimedia content, such as the *Quantum Pills*, that have been carefully designed in such a way that adding more details become a seamless process.

Let us end this Chapter by underlying that besides the strategy here briefly presented, there are many other proposals for bringing QM into high-school classrooms, and different viewpoints about the goal itself. However, the absolute relevance of creating *"a quantum-ready society, with knowledge about and positive attitudes toward quantum technologies"*, as stated on the homepage of the QTEdu project, is finally emerging, and a dialog between experts in physics, philosophy, sociology, education, and other related disciplines, is the essential tool for moving forward.

References

E. Anderson, in *The Problem of Time: Quantum Mechanics Versus General Relativity,* vol. 190 (Springer International Publishing, 2017). ISBN: 978-3-319-58846-9

J.S. Bell, On the problem of hidden variables in quantum mechanics. Rev. Mod. Phys. **38**, 447 (1966)

G. Carcassi, L. Maccone, C.A. Aidala, Four postulates of quantum mechanics are three. Phys. Rev. Lett. **126**, 110402 (2021)

A. Einstein, B. Podolsky, N. Rosen, Can quantum-mechanical description of physical reality be considered complete? Phys. Rev. **47**, 777 (1935)

A. Einstein, M. Born, H. Born, in *The Born-Einstein Letters: Correspondence Between Albert Einstein and Max and Hedwig Born from 1916–1955, with Commentaries by Max Born* (Macmillan, 1971), p. 240. ISBN-10 0333112679 and ISBN-13 978-0333112670

C. Foti, A. Coppo, G. Barni, A. Cuccoli, P. Verrucchi, Time and classical equations of motion from quantum entanglement via the Page and Wootters mechanism with generalized coherent states. Nat. Commun. **12**, 1787 (2021)

P. Kaye, R. Laflamme, M. Mosca, in *An introduction to quantum computing* (Oxford University Press, 2006). 288 pages. ISBN: 9780198570493

K. Krijtenburg–Lewerissa, Teaching quantum mechanics at the secondary school level. Doctoral thesis, University of Twente, Ipskamp printing–Enschede, The Netherlands (2020)

M.A. Nielsen, I.L. Chuang, in *Quantum Computation and Quantum Information: 10th Anniversary Edition* (Cambridge University Press, 2011). ISBN-10 9781107002173 and ISBN-13 978-1107002173
The Nobel Prize in Physics 2022. https://www.nobelprize.org/prizes/physics/2022/summary/
https://qplaylearn.com/
https://algorithmiq.fi/

History of Physics for Education: The Scientific Contributions of Enrico Fermi

Salvatore Esposito

Abstract Physics teaching often leads students to believe that the content presented is either the result of chance or, rather, is due to geniuses far removed from their world. Telling the story of brilliant individuals who did physics, outlining their biographies and work, and even telling intriguing anecdotes, can then encourage students to follow their example. It can indeed create that spirit of curiosity, of putting themselves to the test, and of feeling the challenge that underlies the success of the learning process at high levels, obviously depending on the enlightened guidance of an educated teacher. We present here the engaging example of Enrico Fermi, who brought back Italian physics to shine with the splendor lost since the time of Galileo (except for the small parenthesis of Alessandro Volta), starting from the small world of his daily life, and arriving to give birth to so many stars wherever he was. A shining example for students and teachers.

Keywords History of physics · Nuclear physics · Particle physics · Enrico Fermi · Physics education · School of physics

1 Introduction

In the last decades, a tradition has been established in incorporating the history (and philosophy) of science in physics teaching (Bevilacqua et al. 2001; Bruneau et al. 2012; Matthews 2014), producing significant contributions (Matthews 2015) in enhancing the understanding of scientific concepts and physical phenomena, as well as allowing teachers to identify and prevent misconceptions of students, revealing the nature of physics as scientific activity and knowledge, and, finally, displaying the elements of physics as a culture (Galili 2008). Although, as pointed out in Viennot

S. Esposito (✉)
Department of Physics "Ettore Pancini", University of Naples "Federico II", Piazzale Vincenzo Tecchio 80, 80125 Naples, Italy
e-mail: salvatore.esposito@na.infn.it; salvatore.esposito12@unina.it

Istituto Nazionale di Fisica Nucleare, Sezione di Napoli, Complesso Universitario di Monte S. Angelo–Via Cinthia, 80126 Naples, Italy

M. Streit-Bianchi et al. (eds.), *New Challenges and Opportunities in Physics Education*, Challenges in Physics Education, https://doi.org/10.1007/978-3-031-37387-9_3

(2008), it is not at all obvious to decide how to stage historical documents for a given public and for a cautiously selected partial goal, several interesting research papers did appear in the literature, showing how the history of physics can be helpful to teaching in multiple ways (Buongiorno and Michelini 2019), and displaying different potentials of using history in physics education.

Indeed, history can be useful in putting physics topics in context, by embedding them in the time in which they were developed, and linking them with other disciplines, especially in the humanities, whose teaching is intrinsically historical. A historically informed educational approach can, e.g., invite students to explore the historical development and philosophical aspects of a given topic, emphasizing the cultural and social relevance of physics. It can also take inspiration from the reading of original texts by the main characters or even founders of the given subject, which can be very useful especially in modern physics, as well as in any other field involving subtle conceptual steps (Mauro et al. 2021). Differently, history can also be used in creating an environment where students can "discover" and learn physics for themselves, in ways similar to how scientists did, according to the fact that the laboratory is a perfect environment to bolster students' authentic learning, allowing them to have their own experiences, augmented by discussions about the development of concepts (and their specific experiments) and supported by original texts (Heering 2000; Gallitto et al. 2021).

Here, however, we focus on a completely different approach concerning an intermediate step in the learning process, and then not directly related to the development of student's skills and attitudes. Rather, it is related to the understanding of the social, political, and scientific contexts, where the given topic was thought and developed, thus helping students to see science as a human endeavor, constantly evolving and fully part of the culture. This is just at variance with students' common belief that scientific achievements are either the result of chance or rather are due to geniuses far removed from their world. To remedy such misunderstanding, simple lectures or seminars devoted to telling (appropriately) the life and work of scholars of the past who did achieve those given results, how they got them, and what was the actual context in which they worked, may serve the purpose. The students may then discover, also through intriguing anecdotes suitably narrated, how the standard situation was just that of "normal" people at work. In such a way, the appropriate storytelling may well induce students to some fruitful "curiosity" that encourages them to follow the example of those scientists, even putting themselves to the test: the wise guidance of an educated teacher can, then, truly lead to the success of the learning process at high levels.

This is exactly what we have been able to see in some activities designed for students from high school to college and university (and even beyond…), dedicated in most cases to relevant characters of modern physics (such as Enrico Fermi and Richard Feynman), closer to our time, although similar results have also been obtained with characters of classical physics(Michael Faraday is an illuminating example, as well as that of Joseph Fourier), often also connected to the local context (in Italy, relevant examples of this kind are those of Alessandro Volta, as well as those of far less known scholars like Giuseppe Saverio Poli (Esposito 2021), Macedonio

Melloni, etc.). In the following, we will report the structure (with relevant information) of one of such seminars, devised for high-school students and teachers and centered on the intriguing figure of Enrico Fermi (Schwartz 2017; Segré and Hoerlin 2016; Cordella et al. 2001; Esposito 2017), whose name is incorrectly associated (in non-educated people) only to the making of the first atomic bomb. What follows can be directly read by students, but it can serve much more usefully to teachers as a model for their seminar or lecture on Fermi, as well as for other possible interventions that can be chosen among the most appropriate ones for the contexts in which they usually operate. We refer the reader to the reported bibliography for further information.

2 From Galilei to Fermi: An Invitation

Leafing through a text of any branch of physics, from nuclear physics to solid state physics, from quantum electrodynamics to elementary particle physics, from mathematical physics to astrophysics, etc., it is impossible not to notice the presence, in many pages, of the name of Enrico Fermi. The extension and recognition of his scientific contributions in practically every field of physics is a fact that reveals the importance that this scientist has held in the history of science of the twentieth century. However, the most remarkable fact can probably be better appreciated by looking at the situation of physics in Italy, his native country, before and after him.

It is well known that modern science was born in Italy, thanks to the irreplaceable work of Galileo Galilei at the beginning of the seventeenth century, but, if we ask students or teachers for another name of an important Italian physicist after Galilei and before Fermi, the answer would probably be slow in coming (even if it would be provided…). Perhaps, the only name provided by those questioned would be that of Alessandro Volta. This is due only to an ignorance of the history of physics since it is well known to scholars that the road that led Italian physics from Galileo to Fermi was by no means deserted or almost deserted, as can be very hastily appreciated in Fig. 1 (left). And as, on the other hand, it must necessarily be, to later allow the appearance of Enrico Fermi's personality. Evangelista Torricelli's name would perhaps be remembered by someone, but probably hardly anyone today would remember the name of Macedonio Melloni, who instead in his day was considered the "Newton of heat", due to his fundamental studies on what today we would call infrared radiation. However, it was precisely through this network of "minor" scientists that modern science was able to spread from Galilei to Fermi (just think of other countries where this did not happen), then allowing it to express all its potential accumulated over the centuries. The change of pace in Italian physics that occurred with the advent of Fermi can be appreciated by looking at Fig. 1 (right), where some illustrious names from the first generation of scientists after Fermi, directly related to his Italian *school of physics* (see below), are shown. The comparison between the two images in Fig. 1

Fig. 1 Comparison of the situation of Italian physics before and after the advent of Enrico Fermi. Left: some relevant Italian physicists between Galilei and Fermi. Right: the first generation of Italian scientists after Fermi

speaks for itself: probably not all would recognize all of the names listed (although many of these names would be recognized), but certainly, the number of notable characters speaks volumes. The aperitif can, then, end and the main actor can enter the scene.

3 The Harbingers of a Brilliant Career

Enrico Fermi was born in Rome on 29 September 1901 to Alberto and Ida de Gattis. His father was a chief inspector of the Italian railways, while his mother was a schoolteacher. From their marriage, three children were born: Maria, in 1899, Giulio, in 1900, and Enrico. Young Enrico's best (and practically only) friend was his brother Giulio, with whom he shared his early school years. At school, Enrico was a brilliant student, with some (apparent) problems only in the Italian language and literature.

In 1915, a tragic event marked the soul of Enrico and his family: following an operation, his brother Giulio died, and Enrico suddenly found himself facing his young life alone. He then tried to overcome this trauma by immersing himself completely in the study of scientific subjects. Fortunately, shortly after the unfortunate event, Enrico established a close friendship with one of his brother Giulio's friends, Enrico Persico, with whom he shared many interests, including physics. The autonomous measurement of the density of *Acqua Marcia* (one mineral water), the acceleration of gravity, and the earth's magnetic field dates back to this period.

This, however, should not suggest that the young Enrico was a "nerd", inclined to spend all his time in study. His habit of being methodical in carrying out the tasks assigned allowed him to practice many sports at the same time, in which he proved to be very competitive, as well as to immerse himself in independent reading, especially of scientific subjects. Indeed, Fermi's interest in mathematics and physics manifested itself very early, guided (1914–1918) by the engineer Adolfo Amidei, a friend of his father who often lent him advanced university-level books on these disciplines.

After having obtained his high school diploma (*licenza liceale*) in July 1918, "skipping" the last year of studies, on the suggestion of the engineer Amidei, Enrico moved to Pisa to study Physics at the *Scuola Normale Superiore* (founded by Napoleon in 1813), after having passed the entrance exam easily and brilliantly ("a boy with extraordinary gifts"). Here, he soon found a fellow student and a friend in Franco Rasetti, a student at the University of Pisa with whom he shared not only an interest in science but also a passion for hiking in the mountains. Meanwhile, his studies at the *Scuola Normale* proceeded brilliantly, and Fermi's exceptional qualities were recognized not only by his fellow students but also by his professors. He always obtained full marks, except in drawing and some chemistry courses.

Fermi and Rasetti were soon "in control" of the university laboratory because the elderly professor in charge (Luigi Puccianti) could no longer keep up with the rapidly expanding field of modern physics. It even happened that Puccianti asked Fermi to teach him the Theory of Relativity, developed by the German theoretical physicist Albert Einstein. By the fall of 1920, at the end of his second year, Fermi had completed all the standard courses and was able to take advantage of a very special opportunity. Once again Fermi and Rasetti had free access to the graduate research laboratory as they once had in the teaching laboratory. Fermi chose to experiment with X-rays and made them the subject of his thesis. In July 1922, Fermi then received the master's degree in physics (*laurea*) from the University of Pisa and, almost simultaneously, a certificate (*diploma di licenza*) from the *Scuola Normale*.

4 Training Abroad and Return to Italy

Soon after graduating, in the winter of 1922–3, Fermi won a scholarship from the Italian Ministry of Education to continue his studies abroad. He then decided to work with the great physicist Max Born at the University of Göttingen, where he was establishing a modern theoretical physics center for research into the new quantum physics, thus attracting many talented students from all over Europe. Fermi, however, didn't profit much from his stay there. Instead, the following months were more fruitful, when in 1924, Fermi moved to Leiden (Holland) with a Rockefeller scholarship to work with the Dutch physicist Paul Ehrenfest. This was one of the greatest Master of Physics and a man of intimate human interests, who immediately appreciated Fermi's great talent, and his advice was an important boost for the young Enrico, still at the beginning of his career.

Returning to Italy in 1924, Fermi began teaching first at the University of Rome and then, in 1925, at that of Florence, where he reunited with his friend Franco Rasetti. Fermi and Rasetti formed again a formidable team here for research into the most current and important topics in physics: Fermi was unsurpassed, especially in mastering the theory underlying the experiments, which were ingeniously conceived and brilliantly conducted mainly by Rasetti.

Fermi's Florentine period also saw one of his greatest contributions to theoretical physics: the discovery of the statistical laws (*Fermi–Dirac statistics*) that govern particles subject to the Pauli exclusion principle (such as electrons, protons and neutrons). Such particles are now generally referred to as *fermions*.

Fermi–Dirac statistics explains how to evaluate the properties of "spin ½" particles. The idea of spin was born in the first models of the atom, which predicted electrons orbiting around the atomic nucleus, just as planets around the Sun and spinning at the same time like a top. Possible values for such a *quantum number* are only 1/2, 3/2, 5/2, etc. in quantum mechanics, and said restrictions were first noticed by Wolfgang Pauli. According to Pauli's "exclusion principle", no two fermions can have the same quantum numbers; in a sense, they cannot be in the same place at the same time in the same state of rotation. This seemingly simple idea explains very well how the physical world works. For example, it explains why lithium is chemically very active, while helium is an inert noble gas, despite differing (chemically) in only one more electron. Fermi's statistics explain, however, much more. For example, it explains how neutron stars and other astronomical objects works.

5 The Right Person for a Tough Job

After Fermi returned from abroad, he began to attend more intensely the Physics Institute of the University of Rome, located in Via Panisperna and directed by Senator Orso Mario Corbino, who was a professor of experimental physics. Completely absorbed in politics and business, Corbino was the most open-minded (in a position of authority) among Italian physicists concerning the revolutions that were taking place abroad in the field of physics. From a theoretical point of view, quantum mechanics was then developing, i.e., a new theory governing the behavior of the smallest units of matter, so different from the classical mechanics of Galilei and Newton that many of the Italian physicists fiercely resisted it. Many of the physicists of the old school, indeed, in Italy and elsewhere, resisted quantum mechanics: the older generation, in fact, often resists the new, the revolutionary. Fermi and Corbino had met several times before Fermi went to Germany, and Corbino had been very impressed, immediately becoming aware of Fermi's exceptional abilities. He understood that Fermi was the right person to bring the revolution in physics, already underway elsewhere in Europe, into Italian classrooms and research laboratories, and, on the other hand, the young Fermi also intended to apply his skills where world interest was concentrated: X-rays, radioactivity and, in general, the physics of the atom and its nucleus.

The turning point came in 1926. Building on his political influence, in 1926 Senator Corbino managed to implement his plan, obtaining the creation of the Chair of Theoretical Physics, the first in Italy, established in Rome. The judging Committee unanimously agreed that Fermi was the best candidate: its members felt that "they could place the best hopes in him for the affirmation and future development of theoretical physics in Italy". And, indeed, the competition for the assignment of the Chair was largely won by Fermi, who in the autumn of the same year moved to Rome

(Enrico Persico replaced him at the University of Florence), beginning one of his most fruitful scientific periods.

Here, however, it was also necessary to create an adequate scientific environment. Fermi and his mentor Corbino then began a triple action. First, Corbino co-opted some students at the University of Rome to study modern physics. Second, Fermi set up a program of experiments and research on new physics topics. Finally, a propaganda action was necessary for the revolution in progress: popular conferences, scientific articles, and another new thing in Italy: writing a textbook on modern atomic physics (Fermi 1928).

To be able to effectively carry out a program of atomic physics experiments, the right person to give impetus to it was required to work alongside Fermi, who for his part knew well who he wanted to start that program; namely, Franco Rasetti, his friend from his time in Pisa. Once again, therefore, Corbino dealt with the matter and, relying on his prestige, at the beginning of 1927 had Rasetti transferred from Florence (where he worked) to Rome: he was appointed Corbino's first assistant. It was a good choice: Fermi would help Rasetti in understanding the new quantum theories of the atom, while Rasetti would help Fermi as an experimenter.

Recruiting students proved a little more difficult. In 1927, Corbino launched a famous appeal to the students of the Faculty of Engineering in Rome to entice the brightest young minds to study physics. Emilio Segré and his friend Edoardo Amaldi accepted the challenge and joined the group of Fermi and Rasetti. With professors Fermi and Rasetti and the students Amaldi and Segré, the first nucleus of the modern physics group—that admirable working group known as "Corbino's boys" or "Via Panisperna's boys"—was formed in Italy. Still, other young people soon joined together, providing exceptional contributions to twentieth century physics: Ettore Majorana, Giovanni Gentile jr, Gian Carlo Wick, Bruno Pontecorvo, and many others. According to Segré, "the speed at which it was possible to form a young physicist at that school was incredible. Naturally, a good deal of success was due to the immense enthusiasm aroused in the young people, never by exhortations or sermons, but by the eloquence of Fermi's personal example" (Segré 1962).

6 The *School of Physics* in Rome

Although at the end of the 1920s, Fermi's success was already guaranteed by his first works (see above), he was not satisfied: he wanted to build a *school of physics*, strengthening first of all the research capacities of Physics Institute. The worldwide affirmation of the newborn physics school in Rome took place rapidly, also thanks to the frequent visits of its members to the research laboratories and to the most advanced study centers abroad, aimed both at mastering the most recent experimental techniques and at weaving a network of international relationships.

The first step taken by Fermi was, indeed, to send his collaborators abroad to learn from more advanced laboratories. Rasetti went to Caltech in Pasadena to work with Robert Millikan and then to Berlin to learn radioactivity techniques with physicist

Lise Meitner. Segré and Amaldi were similarly sent to European laboratories, where they broadened their knowledge of advanced techniques in the study of light and X-rays. Fermi himself spent the summer of 1930 teaching at the University of Michigan in Ann Arbor, his first visit to America. His lectures on quantum electrodynamics became legendary among physicists, including future Nobel laureates.

The process worked. The reputation of Rome and the Roman physicists grew as did their international network of friendships, and, starting in the 1930s, first-rank theoretical physicists found it worthwhile to visit Fermi in Rome. Among these were H. Bethe, G. Placzeck, F. Bloch, R. Peierls, L. Nordheim, F. London, E. Feenberg, etc. In October 1931 Fermi and Corbino also organized a memorable Nuclear Physics Conference which brought together the most authoritative scientists from all over the world. Indeed, after having obtained very important results in the field of atomic physics, in the early 1930s, Fermi and his group shifted their interests to the study of the atomic nucleus, as was happening in other international research centers.

The significant personal events of this period were Fermi's marriage to Laura Capon and his appointment by Prime Minister Benito Mussolini as a member of the Italian Academy (*Accademia d'Italia*). Though well deserved, this honor was unexpected because Fermi's reputation at the time was limited to physicists and, traditionally, at his young age, academic honors were not yet due. The appointment came mainly at the request of Senator Corbino.

7 Nuclear Physics in Italy

Around 1931, Fermi and his group realized that the main interest of atomic physicists shifted increasingly to the study of the inner part of the atom, the nucleus, which is its densest part and has a diameter a hundred thousand times smaller than the whole atom. They, therefore, began to familiarize themselves with the current problems (structure of the nucleus, its disintegrations, etc.).

Many properties of the nucleus were already known at the time. It was clear that most nuclei in nature are stable, but others are radioactive; that is, they spontaneously transform into different elements, usually by changing the value of their electric charge. The radioactive process takes place through the expulsion of an alpha particle (i.e., a helium nucleus) or a beta particle (i.e., an electron). Both phenomena are often accompanied by the emission of electromagnetic radiation in the form of gamma rays.

In 1933, Fermi's activity entered a new phase, when, after Chadwick discovered the neutron and Joliot-Curie's discovery of the production of artificial radioactivity induced by alpha particle bombardment, he began research on neutrons both as experimenter and as a theorist. Fermi and his group thought that neutrons would be much more efficient than alpha particles (due to their lack of electric charge) in inducing artificial radioactivity, and, indeed, they succeeded with this method to create and measure about forty new radioactive elements.

Soon, they also observed unexpected effects of certain substances such as water and paraffin; their mere presence around or near the bombarded element increased its radioactivity. In less than one day, Fermi found the explanation for this phenomenon. The neutrons slow down when they collide with the hydrogen nuclei contained in those substances, thus increasing the likelihood of an effective reaction. Slow neutrons later turned out to be a fundamental key to accessing nuclear energy.

In the autumn of 1933, Fermi also solved a very serious problem concerning the beta decay process: how can a nucleus emit electrons if it does not contain them? Fermi elaborated a complete theory, which immediately gave precise explanations of the experimental facts, assuming that the electrons were created at the same instant in which they were emitted, together with another light, neutral particle, which Fermi later called neutrino: $n \rightarrow p + e + \nu$. Only a few theories in modern physics have been so important: Fermi's theory applies not only to beta decay processes but also to many other transformations observed in unstable particles.

Fermi's scientific contributions at the Physics Institute in Rome in the 1930s earned him the Nobel Prize for Physics in 1938. Unfortunately, however, it was precisely in the autumn of that year that the anti-Semitic restrictions following the Italian-German alliance increased in their intensity, and this put Fermi's wife, Laura Capon, of Jewish birth, in a worrying position. Fermi then decided to accept an offer from Columbia University in New York and, on his family's trip to Stockholm for the Nobel Prize ceremony, in December 1938 he left Italy definitively.

8 Chain Reactions in the U.S.A.

In 1938, three German scientists repeated some of Fermi's early experiments. After bombarding uranium with slow neutrons, Otto Hahn, Lise Meitner and Fritz Strassmann made a careful chemical analysis of the products formed, and on January 6, 1939, they reported that the uranium atom apparently had been split into several parts. Meitner, indeed, secretly fled from Germany to Stockholm where, together with her nephew Otto Frisch, she explained this new phenomenon as a splitting of the nucleus of the uranium atom into barium, krypton, and smaller quantities of other disintegration products. Meitner also realized that this *nuclear fission* induced by neutron bombardment was accompanied by the release of large amounts of energy—by converting part of the mass of uranium into energy—and, apparently, by the emission of more neutrons than used to bombard the uranium nucleus. This last feature made the phenomenon particularly interesting, since it opened the possibility, though only theoretical, of a *chain reaction*, in which many uranium nuclei could be split starting from a single initial bombarding neutron, with the consequent release of a huge total amount of energy.

Informed of this news just after arriving in New York, Fermi saw its implications and immediately set to work to investigate the phenomenon, striving to set up an effective working group at Columbia University as in Rome. The results obtained by him (and by others) were encouraging but, for the effective realization of a chain reaction, the means needed went far beyond those available to individual universities.

Fermi, Leo Szilard, and Eugene Wigner saw the dangers to world peace if Hitler's scientists applied the newly discovered phenomenon to the production of a bomb that released the vast amount of energy produced in a nuclear chain reaction. They, therefore, wrote a letter, subsequently signed by Albert Einstein, which was delivered in the autumn of 1939 to President Franklin D. Roosevelt, warning him of this danger. Roosevelt heeded their warning, and the day before the Japanese attack on Pearl Harbor, on December 6, 1941, finally decided to give the maximum support to the project concerning nuclear chain reaction experimentation.

For this purpose, Fermi was assigned the task of designing the necessary apparatus (which he called an *atomic pile*) to carry out this chain reaction. Thanks to Fermi's unparalleled theoretical and experimental work, the pile went into operation in a laboratory set up in the basement of the University of Chicago Stadium on December 2, 1942, producing a controlled chain nuclear reaction for the first time in history.

The realization of a chain-reacting nuclear pile was scientifically important for the study of the properties of fission, which was necessary, e.g., to understand the internal functioning of an atomic bomb. However, it also served as a pilot plant for the larger reactors later built at Hanford (Washington), which were used to produce plutonium, another fissile element more efficient than uranium for starting a nuclear chain reaction. Over the next two years, Fermi conducted various experiments using this reactor, collaborating as well with the development of an even larger reactor at the nearby Argonne Laboratory.

After a brief period during which Fermi worked primarily with the DuPont Company to produce plutonium on an industrial scale, with his family in August 1944 he moved to Los Alamos (New Mexico), where the U.S. War Department assembled hundreds of scientists to work on the *Manhattan Project* aimed to make an atomic bomb, under the scientific direction of J.R. Oppenheimer and the military leadership of Gen. L. Groves. Fermi was always reluctant to assume administrative responsibilities, so much so that even in the Manhattan Project—at least officially—he never became a prominent personality from an administrative or political point of view. Scientifically, however, by unanimous recognition, his contribution was decisive for this project too. Suffice it to mention that in the Manhattan Project, he was put in charge of the F Division (F standing for Fermi), whose peculiar task was to solve the problems that the other divisions could not solve.

Driven by the critical fear that the Nazis might be the first to make (and then use) the atomic bomb, the scientists assembled at Los Alamos worked with extraordinary determination to complete the project. After continuous experimental tests, which lasted for many months, the time finally came to test the world's first atomic bomb. It was carefully assembled at the Trinity site in the New Mexico desert (Alamogordo) and raised to the top of a 30 m tower: the triggering occurred on Monday, July 16, 1945, at 5:29 in the morning. The explosion of a modest quantity (about 4.5 kg) of plutonium was equivalent to about 20,000 tons of TNT.

The problem of measuring the energy released in the explosion was solved immediately, even if roughly, by Fermi. When the bomb was detonated, he dropped a handful of slips of paper into the air; then, he measured the displacement produced when the front of the shock wave produced in the explosion reached the observation point. From this crude experiment, Fermi deduced an estimate of the energy release within a factor of 2, several days before the detailed estimate later obtained by others.

That explosion meant, in a sense, the end of the initial phase of the Manhattan Project at Los Alamos: the great technical task had been accomplished. Just after the end of the war, Fermi then abandoned the work on atomic bombs, which had no more scientific attraction for him.

9 New Adventures in Chicago

At the end of 1945, Fermi chose to work at the University of Chicago, which planned to form a new institute (the current Enrico Fermi Institute for Nuclear Studies), of which he immediately became a member, carrying out all sorts of research with the new technical means, i.e., nuclear reactors. As the world's foremost authority in this field, Fermi worked on neutron physics for a relatively short time, but long enough to obtain very brilliant results. Later, he became interested in a new branch of physics that was just beginning to develop: high-energy particle physics. He wanted to explore the inner structure of atomic nuclei by working on the experiments (and their theoretical interpretation) conducted with newborn high-energy accelerators. These were indeed built in Chicago, being able to push protons or electrons up to energies a hundred times greater than before the war.

At the end of the 1940s, Fermi also began to develop a keen interest in electronic calculators (their first realization was connected precisely to the work for the Manhattan Project during the war); e.g., he spent the summer of 1953 at Los Alamos, using the MANIAC electronic computer to analyze the results of his experiments on the scattering of the pi meson.

In the enormous calculating potential of these machines, Fermi saw a unique opportunity for the numerical solution of complicated problems that did not admit a mathematical solution. Here too he was, therefore, a pioneer in a completely new branch of science that made the numerical simulation of experiments an irreplaceable tool for investigation, as happened in the following years up to the present day.

Fermi's tireless work, however, did not end with scientific research but was also always aimed at the training of new physicists. At the University of Chicago, he taught physics, gave seminars, and supervised graduate students, both theoreticians, and experimenters. His lessons (which he carefully prepared) and his way of teaching became legendary for their extreme clarity. As evidence of his success also in this activity, it will suffice to recall that seven of his students (E. Segré, M. Gell-Mann, J. Steinberger, C. N. Yang, T. D. Lee, J. Friedman, J. Cronin) have been awarded the Nobel Prize in Physics, while countless relevant results have been obtained by his students thanks to his example and lessons.

One of such Nobel laureates recalled that, while talking with his students "the discussions were kept at an elementary level. The emphasis was always on the essential and practical part of the topic. [...] We learned that *that* was physics" (Segré 1970). And: "Wherever there were new physics problems and young physicists you were likely to see Fermi arrive, always ready with some new and fruitful idea, and also ready to receive information, challenge and inspiration from his younger colleagues" (Segré 1962).

10 Last Years: Remembering His Homeland

After the war, Fermi resumed close relationships with friends and colleagues who remained in Italy; actually, even remaining to work in the United States, his contacts with Italy were never interrupted. Thus, his interest and advice now proved to be invaluable for the rebirth of science in Italy after the war disaster, entrusted only to the work of his former students in Rome, among which Edoardo Amaldi was practically the only one who remained in Italy during the war period.

In 1949, Fermi visited Italy for the first time since the war, and with his memorable lectures in Milan, Rome, and other places he instilled a new impetus in Italian physics research and took an interest in research laboratories. In 1952–53, he contributed to various research projects (discussed mainly with Amaldi and Gilberto Bernardini) on the construction of new facilities in Italy for the study of particle physics at the frontier research level, these projects being later realized in the construction of the electro-synchrotron at the national laboratories of Frascati. Again, when in 1954, Bernardini, Marcello Conversi, and Giorgio Salvini asked Fermi for his advice on how best to invest a certain amount of money for research purposes, and his answer was unequivocal: "build an electronic calculator". From that suggestion, the CEP (Pisa electronic calculator) was realized, and thus computer science research was born in Italy as well.

In the summer of 1954, Fermi came to Italy for the last time, to attend a memorable conference in Varenna, on the Como Lake. After this, when Fermi suddenly returned to the United States, his colleagues noticed that his health was rapidly deteriorating. In September, he underwent exploratory surgery to identify the nature of his illness: unfortunately, it could only recognize a hopeless situation. Fermi survived his surgery for only a few weeks. He went home and tried to revise the notes of his latest course in nuclear physics. This was his last effort: Fermi died on November 29, 1954.

11 Final Remarks

On a day in March 1938, Fermi was in his laboratory in Rome together with a young graduate, Giuseppe Cocconi. Discussing the recent disappearance of one of the "Via Panisperna's boys", to allow the young scholar to understand the loss that this disappearance entailed for the scientific community, Fermi expressed himself in a somewhat unusual way for him. Thirty years later, Cocconi still remembered those words:

> I would like to repeat his words, just as I can still hear them ringing in my memory: "Because, you see, in the world, there are various categories of scientists: people of a secondary or tertiary standing, who do their best but do not go very far. There are also those of high standing, who comes to discoveries of great importance, fundamental for the development of science" (and here I had the impression that he placed himself in that category). "But then there are geniuses like Galileo and Newton". (Amaldi 1966).

Fermi did not consider himself a genius though he had no qualms—and rightly so—in placing Ettore Majorana, one of his "boys", in this category (Esposito 2017). He was a great physicist who managed to arrive at fundamental discoveries for modern science. The echo of such results can only faintly be appreciated from the following list of some of these key contributions where his name enters:

Fermi's theory of beta decay;

Fermi coupling constant, Fermi interaction; Fermi-Dirac statistics; fermions; Fermi gas, Fermi liquid;

Thomas-Fermi model;

Fermi momentum, Fermi energy, Fermi level, Fermi surface, Fermi sea, . . . ;

Fermi's golden ule;

Fermi mechanism, Fermi acceleration, . . . ;

Fermi's nuclear pile;

Fermi paradox;

Fermi questions;

fermi (unit of length);

Fermium;

...

His shining example can be an inspiration to any student interested in scientific matters. And it is precisely for these students that we reserve, at the end of our journey, one of *Fermi's questions* that he asked in his exams at the University of Chicago (Baeyer 1993): *How many piano tuners are there in your town? 1? 10? 100? 1000? 10,000?....* Tomorrow's Fermis have the burden of answering.

References

E. Amaldi, Ettore Majorana: man and scientist, in *Strong and Weak Interactions*, ed. by A. Zichichi (Academic Press, New York, 1966)

F. Bevilacqua, E. Giannetto, M.R. Matthews (eds.), *Science Education and Culture: the Contribution of History and Philosophy of Science* (Springer, Dordrecht, 2001)
O. Bruneau, P. Heering, P. Grapí, M.R. Massa-Esteve, T. de Vittori (eds.) *Innovative Methods for Science Education: History of Science, ICT and Inquiry-Based Science Teaching* (Frank and Timme, Berlin, 2012)
D. Buongiorno and M. Michelini, The conceptual contribution of the history in learning physics: The case of optical spectroscopy, in *Proceedings of the 37th Annual Conference of the Italian Society of Historians of Physics and Astronomy* ed. by B. Campanile, L. De Frenza, A. Garuccio, (Pavia University Press, Pavia, 2019), p. 377
F. Cordella, A. De Gregorio and F. Sebastiani, Enrico Fermi. Gli anni italiani. Editori Riuniti, Rome (2001)
M. Di Mauro, S. Esposito, A. Naddeo, Introducing quantum and statistical physics in the footsteps of Einstein: a proposal. Universe **7**, 184 (2021)
S. Esposito, *Ettore Majorana: Unveiled genius and endless mysteries* (Springer, Heidelberg, 2017)
S. Esposito, From England to Italy: the intriguing story of Poli's engine for the King of Naples. Phys. Persp. **23**, 104 (2021)
E. Fermi, *Introduzione alla fisica atomica* (Zanichelli, Bologna, 1928)
I. Galili, History of physics as a tool of teaching, in *Connecting Research in Physics Education with Teacher Education*, vol. 2, ed. by M. Vicentini, E. Sassi (International Commission on Physics Education, 2008)
A.A. Gallitto, R. Zingales, O.R. Battaglia, C. Fazio, An approach to the Venturi effect by historical instruments. Phys. Educ. **56**, 025007 (2021)
P. Heering, Getting shocks: teaching secondary school physics through history. Sci. Educ. **9**, 363 (2000)
M.R. Matthews, *International handbook of research in history, philosophy and science teaching* (Springer, Dordrecht, 2014)
M.R. Matthews, *Science Teaching: The Contribution of History and Philosophy of Science*, 2nd edn (Routledge, New York, 2015)
D.N. Schwartz, The last man who knew everything, in *The Life and Times of Enrico Fermi, Father of the Nuclear Age* (Basic Books, New York, 2017)
E. Segré, Biographical introduction, in *Enrico Fermi: Collected Papers*, vol. 1. (University of Chicago Press, Accademia Nazionale dei Lincei, Chicago-Rome, 1962), pp. xviixliii
E. Segré, *Enrico Fermi, Physicist* (University of Chicago Press, Chicago, 1970)
G. Segré, B. Hoerlin, The Pope of physics, in *Enrico Fermi and the Birth of the Atomic Age* (Henry Holt and Co., New York, 2016)
L. Viennot, Comments on C2: history of physics as a tool of teaching, in *Connecting Research in Physics Education with Teacher Education*, vol. 2, ed. by M. Vicentini, E. Sassi (International Commission on Physics Education, 2008)
H.C. von Baeyer, *The Fermi Solution* (Dover, New York, 1993)

Three Tales of Gravity

Ugo Moschella

Abstract Science as a factor in human life is a latecomer. Art existed before the last ice age and is at least fifty thousand years old. Modern science, on the other hand, was born only four hundred years ago. Galileo was the father and gravity the midwife; historiography and tradition have assigned the role of founding myth to the tale of Galileo's experiment at the Leaning Tower of Pisa. Two more tales, Newton sitting under the apple tree and Einstein having the happiest thought of his life, articulate the history of our ideas about gravity. These pages tell those tales.

Keywords Gravity · Myth · Geometry

1 Lost Paradise

Gravity stands out among the forces of nature because it acts on all bodies with no exception, there is no way to escape from its attraction. This fact has profound physical consequences that go beyond imagination; it also leads to the mathematical difficulties in solving Einstein's equation—since gravity acts on all bodies, the initial conditions, the initial position of the various bodies, cannot be chosen arbitrarily—and in finding the still elusive quantum version of them, the philosopher's stone of physicists named "quantum gravity."

As regards strength, gravity is the weakest of all forces; its effects on the elementary particles of the microscopic world are negligible.

The situation is quite different for macroscopic bodies. Of course, gravity dominates on the astronomic scale, but in many ways, it dominates on the human scale as well. Electromagnetic and nuclear forces work, in normal situations, in the depth and discreetly; they act on our body in a sensible way only indirectly or occasionally. On the contrary, gravity is so obvious that we tend to forget about it.

U. Moschella (✉)
DISAT Università dell'Insubria, Via Valleggio 11, 21100 Como, Italy
e-mail: ugomoschella@gmail.com

IHES, 11 Route de Chartres, 91440 Bures–sur-Yvette, France

M. Streit-Bianchi et al. (eds.), *New Challenges and Opportunities in Physics Education*, Challenges in Physics Education, https://doi.org/10.1007/978-3-031-37387-9_4

Fig. 1 Painting of the Shaman in the Lascaux cave

We are used to being attached to the ground and feeling our weight. We learnt in our childhood how to exert a muscular reaction with the legs to stay standing and counteract the gravitational attraction of the Earth; we are accustomed to this effort and do not pay attention anymore to *cette Pesanteur dont tous les philosophes ont cherché si long temps la cause en vain, and dans laquelle le vulgaire ne soupçonne pas même de mystère.*[1]

Paradoxically, one might realize that gravity exists exactly when he should not, if he happens to be falling from a roof.

At the human scale, gravity not only acts on the body but also on the psyche of human beings. Since time immemorial, it has engraved its presence in the mythical narrative of the origins, of creation, of life and death, in what Mircea Eliade calls "nostalgia for the origins."[2]

In all archaic cultures, there is indeed a trace of the myth of the lost paradise. Myths of this category have a common trait: all of them put the Earth and the Heavens on the same level before the *fall*. *In illo tempore* (at that time), the Sky was very close to the Earth and men could easily access the Sky through a tree, a liana or a ladder. When the Sky was brutally separated from the Earth, when the Sky became as far from the Earth as is today, when the tree or the ladder between the Earth and the Sky were cut, the Paradisiac epoch was over and humanity *fell* in its present condition; this was the result of a mythical event—every myth has its own—causing the rupture between the Earth and the Heavens (Fig. 1).

---

[1] Voltaire. *Lettres Ecrites de Londres sur les Anglois, et Autres Sujets,* Par M.D.V. Basle 1734.
Gravity, of which all philosophers have sought so long the cause in vain, and in which the vulgar does not even suspect any mystery.

[2] Mircea Eliade. *Mythes, rêves et mystères*, Paris, Gallimard, «Les Essais», 1957.

Myths of a closely related category narrate of the *Magische Flucht*, the magic flight of birdmen and, in particular, of shamans.[3] With his own special techniques, the shaman endeavours to abolish the present human condition to rediscover the primordial condition of the mankind of the heavenly myths before the *fall*. It is the *abolition of the heaviness* which generates an ontological mutation in the human being restoring the proximity of the Earth and the Sky. Myths of the magic flight are ubiquitous in all archaic cultures and were already established in the old stone age, as testified by the famous relief of the man—a shaman—with head of bird fallen into a trance at the Lascaux caves.

2 Gravity, the Ptolemaic Universe and the Copernican Revolution

The archaic separation of the Earth and the Heavens is literally crystallized in the spheres of the Ptolemaic model. Here, the distinction between the Earth and the Sky is more clearly related to the ancient conception of gravity which is in turn rooted in the Aristotelian classification of the movements.

Two types of movements are possible in the sublunary world: the natural movements of falling heavy bodies towards the centre and of rising light bodies away from it are caused by their tendency to proceed to their natural place; violent movements require an external force as a cause.

The Sky over the Moon is divided by concentric crystalline spheres which are made of the fifth element, called aether or quintessence. The aether has no weight and no lightness; it can neither fall towards the centre nor fly away from it; its circular and uniform movement, perfect by nature, has no beginning and no end.

The above distinction explains what gravity is. An apple falls because it goes where heavy bodies naturally go: the motionless centre of the cosmos, where the Earth is located is the place where all the heavy bodies of the sublunary world fall. And how do they fall? The speed of a falling body, Aristotle says, is directly proportional to its weight and inversely proportional to the density of the medium. This law implies also that the vacuum cannot exist; *in vacuo* a body would fall with infinite speed and be simultaneously in many places, which is logically impossible.

The spheres in the Sky, originally invented as a calculation device, were later considered as physically existing. The most external one is fixed and borders the universe which is necessarily finite; indeed, if a (Euclidean) universe has a centre it can only be finite.

The final description of the system of spheres and epicycles was given by Ptolemy in the *Almagest* and has been the basis of the standard vision of the world for centuries; it has also constituted the physical foundation of ethics. That world collapsed under

---

[3] Mircea Eliade, *Le chamanisme et les techniques archaïques de l'extase*, Paris, Payot, «Bibliothèque scientifique», 1950. 2e édition revue et augmentée, 1968; «Payothèque».

the deadly blows of the *De Revolutionibus Orbium Coelestium* by Nicolaus Copernicus. The Copernican revolution "which places the Earth as mobile and the Sun as immobile at the centre of the universe" is based on the same astronomical data of Ptolemy's Almagestus. There was nothing new, no new observation, the skies were still immutable and nothing could change up there. Except that once the Earth is removed from the centre of the universe, a question that was decided and forever closed now is open again: what is gravity?

To find the answer, or rather a way towards the answer, I will now tell three tales of gravity.

3 Tale of Galileo

> Galileo, it must be confessed, was something of a gamin. When still very young he became Professor of Mathematics at Pisa, but as the salary was only 7 ½ d. a day, he does not seem to have thought that a very dignified bearing could be expected of him. [...] He would amuse himself by arranging occasions which would make his colleagues look silly. They asserted, for example, on the basis of Aristotle's Physics, that a body weighing ten pounds would fall through a given distance in one-tenth of the time that would be taken by a body weighing one pound. So, he went up to the top of the Leaning Tower of Pisa one morning with a ten-pound shot and a one-pound shot, and just as the professors were proceeding with leisurely dignity to their respective lecture-rooms in the presence of their pupils, he attracted their attention and dropped the two weights from the top of the tower to their feet. The two weights arrived practically simultaneously. The professors, however, maintained that their eyes must have deceived them, since it was impossible that Aristotle could be in error. (Excerpt from the Scientific Outlook).[4]

Galileo's performance at the Leaning Tower in 1589, as referred by Viviani,[5] was more a social happening than an experiment. If it ever happened, and there is a century-old ongoing controversy about that,[6,7] it had little impact on Galileo thoughts: the answer was surely known to him before climbing the tower and attracting the attention of the academic procession. But, none of the objections that have been raised against this story and that could be raised in future will ever change the status of the Leaning Tower tale. Why is this tale so important? (Fig. 2).

---

[4] Russell (1931).

[5] Viviani (1938), p. 606.

«[...] furono da esso [Galileo] convinte di falsità [...] moltissime conclusioni dell'istesso Aristotele intorno alla materia del moto, [...] come, tra l'altre, che le velocità de' mobili dell'istessa materia, disegualmente gravi, movendosi per un istesso mezzo, non conservano altrimenti la proporzione delle gravità loro, [...] anzi che si muovon tutti con pari velocità, dimostrando ciò con replicate esperienze, fatte dall'altezza del Campanile di Pisa con l'intervento delli altri lettori e filosofi e di tutta la scolaresca».

[6] Segre (1989).

[7] Galileo himself probably told the story to Viviani in 1641, while dictating to him the answer to a letter received from Vincenzio Renieri in which new experiments of fall from the top of the tower

Fig. 2 Galileo Galilei performs the experiment of falling bodies from the Tower of Pisa. Fresco by Luigi Catani, 1816 (Firenze, Palazzo Pitti, sala 15)

Science as a factor in human life is the last comer. The cave paintings recently found in Sulawesi show that art existed before the last glacial epoch and is at least fifty thousand years old. Modern science instead, was born only four hundred years ago: the baby science had a father[8]—Galileo—and a midwife—gravity; historiography and tradition have assigned to Galileo's leaning tower experiment the role of the founding mythical event.

In fact, the mythical narrative of the experiment contains all the major traits of the new-born science:

1. The tale tells of a new procedure, called an experiment, which is the starting point of a new form of knowledge, called science, where the establishment of an exact quantitative law of nature proceeds from the observation of particular phenomena. The law of nature may be then used to predict other yet unseen particular phenomena.
2. The tale contains the reject of every established authority not based on «sensate esperienze» e «certe dimostrazioni».
3. Every statement must be «public», must be presented and demonstrated to others, discussed and subjected to the control by others and to possible rebuttals.[9]

were presented to the Master. That reply letter is lost. It would have proven that the Leaning Tower Experiment really took place.

[8] "Propositions arrived at by purely logical means are completely empty as regards reality. Because Galileo saw this, and particularly because he drummed this into the scientific world, he is the father of modern physics, indeed of modern science altogether". Einstein (1933).

[9] Rossi (2001).

In this new science, the natural laws are not obtained by (more or less) generalizing the empirical observations; rather, they result from an analysis capable of abstractions, having the courage to abandon the conclusions gained through immediate sense perception.

The introduction of abstract thinking was a truly gigantic step of the new science; the collateral phenomena hiding the laws of nature have to be discarded and corrected as much as possible. What features are to be discarded and how to discard them is a matter for the physical intuition and the technical ability of the experimenter; this is indeed a highly sophisticated form of art. For falling bodies, one should make abstraction of the air resistance.[10] Everyday experience suggests that heavier bodies fall faster than light ones. A layman, who probably would agree that the Earth moves around the Sun and not vice versa,[11] will almost surely give the incorrect answer to the question of how bodies fall and observe with surprise what happens when air is removed.[12]

Galileo's Equivalence Principle declaring that all bodies fall in void with the same acceleration, regardless of their mass and composition is, perhaps, the most important result of Galileo's new science. It already contains the germ of the idea that would later play a central role in Einstein's path to General Relativity.

Of course, the discovery is not to be associated with the Leaning Tower Experiment of 1589, it came later. The principle that Galileo was adopting at that time[13] was incorrect: he assumed the speed of the falling body to be proportional to the distance travelled from the starting point rather than to the elapsed time. This wrong hypothesis was abandoned by him in his later years, while Descartes made the same mistake, but he never retracted it. However, it should be said that we are "naturally" inclined to admit that the speed of the falling bodies increases with their distance from starting point while the idea of making their speed depend on the travel time (which is itself a function of the speed) seems unnatural and cumbersome.

---

[10] Galileo Galilei. Discorsi e dimostrazioni matematiche intorno a due nuove scienze - Del violento, o vero de i proietti. Giornata quarta.

[11] Arthur Conan Doyle, A Study in Scarlet. Chapter I. Mr Sherlock Holmes (1891).

[12] Commander David R Scott (2 August 1971, lunar surface): "Well, in my left hand I have a feather; in my right hand, a hammer. And I guess one of the reasons we got here today was because of a gentleman named Galileo, a long time ago, who made a rather significant discovery about falling objects in gravity fields. And we thought: 'Where would be a better place to confirm his findings than on the Moon?

[13] And also a few years later, when he presented the law of free fall in a letter to Paolo Sarpi. There was a mistake in Galileo reasoning that compensated his wrong assumption. Here an excerpt of the letter to Paolo Sarpi where the law of the falling bodies is pronounced for the first time.

GALILEO a PAOLO SARPI in Venezia. Padova, 16 ottobre 1604.

Molto Rev. do Sig. re et Pad. ne Col. mo.

Ripensando circa le cose del moto, nelle quali, per dimostrare li accidenti da me osservati, mi mancava principio totalmente indubitabile da poter porlo per assioma, mi son ridotto ad una proposizione la quale ha molto del naturale et dell'evidente; et questa supposta, dimostro poi il resto, cioè gli spazzii passati dal moto naturale esser in proporzione doppia dei tempi.

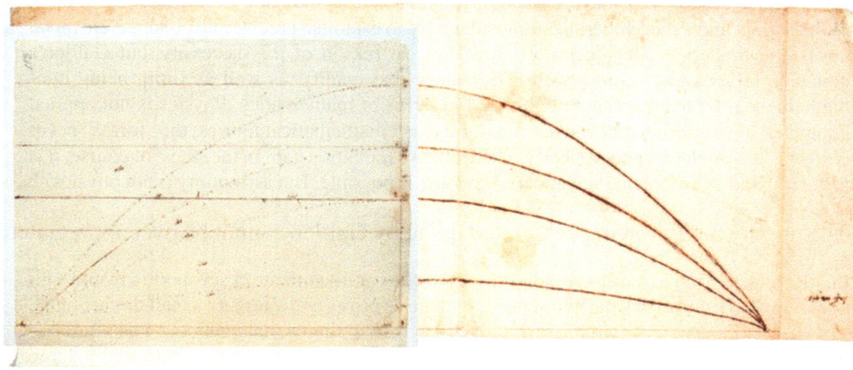

Fig. 3 Trajectories of a projectile drawn with a chainette. Galileo Galilei, Ms Gal. 72, f. 113r and 41v/42r

The correct hypothesis is declared in the *Discorsi e dimostrazioni matematiche intorno a due nuove scienze attenenti alla meccanica e i movimenti locali*, published in Leiden in 1638 to avoid risks with the Roman Inquisition:

> SALV. Possiamo quindi ammettere la seguente definizione del moto di cui tratteremo: Moto equabilmente, ossia uniformemente accelerato, dico quello che, a partire dalla quiete, in tempi eguali acquista eguali momenti di velocità.
>
> SAGR. Io, sì come fuor di ragione mi opporrei a questa o ad altra definizione che da qualsivoglia autore fusse assegnata, essendo tutte arbitrarie [...]
>
> SALV. È bene che V. S. ed il Sig. Simplicio vadano proponendo le difficoltà; le quali mi vo immaginando che siano per essere quelle stesse che a me ancora sovvennero, quando primieramente veddi questo trattato, e che o dall'Autor medesimo, ragionandone seco, mi furon sopite, o tal una ancora da me stesso, co 'l pensarvi, rimosse.

Koyré[14] made an interesting conjecture on the possible origin of Galileo's mistake: *The reason seems to us both very profound and very simple. It lies entirely in the role played in modern science by geometric considerations, by the relative intelligibility of spatial relationships. [...] It is more «natural», more «easy», to imagine in space than to think in time.*[15]

Galileo has always been thinking geometrically about gravity: two examples are the experimental works of his youth, when he was studying movements of projectiles with Guidobaldo del Monte using chainettes, to the *Discorsi* of his maturity, where he demonstrates that the trajectory of a projectile is a parabola resulting from the composition of two movements that do not interfere with each other: a straight horizontal motion and a uniformly accelerated vertical motion whose acceleration does not depend on the mass of the body (Fig. 3).

But one should not forget that geometry had nothing to do with physics in the sublunary world.

---

[14] Koyré (1939).

[15] Alexandre Koyré. Ibidem.

The modern reader [...] finds it difficult to admit that "philosophy" had to content itself with abstract and vague generalization and not try to establish precise and concrete universal laws. The modern reader does not know the real reason of this necessity, but Galileo's contemporaries knew it quite well. They knew that quality, as well as form, being non-mathematical by nature, cannot be treated in terms of mathematics. Physics is not applied geometry. Terrestrial matter can never exhibit exact mathematical figures; the "forms" never "inform" it completely and perfectly. There always remains a gap. In the skies, of course, it is different; and therefore mathematical astronomy is possible. But astronomy is not physics.[16]

This is why the following celebrated quote by Galileo cannot be over appreciated:

La filosofia è scritta in questo grandissimo libro che continuamente ci sta aperto innanzi a gli occhi (io dico l'universo), ma non si può intendere se prima non s'impara a intender la lingua, e conoscer i caratteri, nei quali è scritto. Egli è scritto in lingua matematica, e i caratteri son triangoli, cerchi, ed altre figure geometriche, senza i quali mezzi è impossibile a intenderne umanamente parola; senza questi è un aggirarsi vanamente per un oscuro laberinto.[17]

4 Tale of Newton

On 15 April 1726 I paid a visit to Sir Isaac at his lodgings in Orbels buildings in Kensington, dined with him and spent the whole day with him, alone.... After dinner, the weather being warm, we went into the garden and drank thea, under the shade of some appletrees, only he and myself. Amidst other discourse, he told me, he was just in the same situation, as when formerly, the notion of gravitation came into his mind. It was occasion'd by the fall of an apple, as he sat in a contemplative mood. Why should that apple always descend perpendicularly to the ground, thought he to himself. Why should it not go sideways or upwards, but constantly to the earth's centre? Assuredly, the reason is, that the earth draws it. There must be a drawing power in matter: and the sum of the drawing power in the matter of the earth must be in the earth's center, not in any side of the earth. Therefore, dos this apple fall perpendicularly, or towards the center. If matter thus draws matter, it must be in proportion of its quantity. Therefore the apple draws the earth, as well as the earth draws the apple. That there is a power, like that we here call gravity, which extends its self thro' the universe....[18]

The story of Newton's apple is probably the most famous anecdote ever, related to a scientific discovery. There is only another apple more famous than Newton's: the apple that caused the expulsion of Adam and Eve from the Garden of Eden.[19]

Newton's apple probably truly fell. Several convergent testimonies exist in this regard. The most interesting to us is by Voltaire, who lived in England from 1726 to

---

[16] Koyré (1943).

[17] The philosophy is written in this great book which is continually opened before our eyes (I say the universe) but it cannot be read until we have learnt the language and become familiar with the characters in which it is written. It is written in mathematical language, and the letters are triangles, circles and other geometrical figures, without which means it is humanly impossible to comprehend a single word.

[18] Stukeley (1936).

[19] Like Newton's, the apple tree of the Garden of Eden was a tree of knowledge!

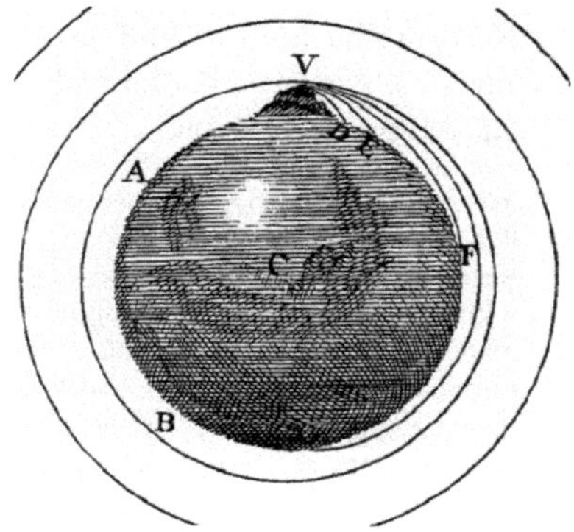

Fig. 4 Newton's diagram of projectile motion (*Principia* 1687)

1728 and was at Newton's funeral in Westminster Abbey on March 28, 1727. Let's hear what Voltaire says in his English letters:

> 'S'étant retiré en 1666, à cause de la peste, a la campagne près de Cambridge, un jour qu'il se promenait dans son jardin, & qu'il voyoit des fruits tomber d'un arbre, il se laissa aller à une méditation profonde sur cette Pesanteur dont tous les philosophes ont cherchéz si long tems la cause en vain, & dans laquelle le vulgaire ne soupçonne pas même de mystère; il se dit à lui-même, de quelque hauteur dans notre hémisphère que tombassent ces corps, leur chute seroit certainement dans la progression découverte par Galilée: & les espaces parcourus par eux sérient come les quarrez des tems. Ce pouvoir qui fait descendre les corps graves, est le même sans aucune diminution sensible a quelque profondeur qu'on soit dans la terre, & sur la plus haute montagne, pourquoi ce pouvoir ne s'etendroit-il pas jusqu'à la lune ?[20]

Why should this power not extend to the moon? This is the question! The answer is that Gravitation is Universal and does not concern exclusively the sublunary world: two bodies attract each other with a force proportional to the product of their masses and inversely proportional to the square of their distance. This law, today known to high schoolers, applies to the apple that falls on the Earth (and the Earth that falls on the apple) as to the Moon that falls around the Earth, and to the planets that turn around the Sun. Newton accomplished Galileo's project of unification and abolished the principle of a substantial difference between Heaven and Earth. Better still: a single law accounts for a wide variety of phenomena. But, contrary to the Shamanic dream in his trance, unity by Earth and the Heavens amounts to bringing the Skies down to Earth! (Fig. 4).

And yet—as everyone knows—Newton declares himself ignorant as to the physical reality of gravitational attraction:

> Hitherto we have explained the phenomena of the heavens and of our sea, by the power of Gravity, but have not yet assigned the cause of this power. [...] I have not been able to discover the cause of those properties of gravity from phenomena, and I frame no hypotheses. For whatever is not deduced from the phenomena, is to be called an hypothesis; and hypotheses, whether metaphysical or physical, whether of occult qualities or mechanical, have no place in experimental philosophy.

[20] Voltaire, Ibidem.

Newton had tried to find an explanation for gravity as a contact force caused by invisible particles, but he realized that explanation could not work. In a letter to Richard Bentley he wrote:

> You sometimes speak of gravity as essential and inherent to matter. Pray do not ascribe that notion to me, for the cause of gravity is what I do not pretend to know and therefore would take more time to consider of it. It is inconceivable that inanimate brute matter should, without mediation of something else which is not material, operate upon and affect other matter without mutual contact, as it must be if gravitation, in the sense of Epicurus, be essential and inherent in it. And this is one reason why I desired you would not ascribe innate gravity to me. That gravity should be innate, inherent, and essential to matter, so that one body may act upon another at a distance through a vacuum, without the mediation of anything else, by and through which their action and force may be conveyed from one to another, is to me so great an absurdity that I believe no man who has in philosophical matters a competent faculty of thinking can ever fall into it.

So, the question "what is gravity" raised by the Copernican Revolution remained unanswered. Neither Galileo nor Newton could answer that question.

5 Tale of Einstein

The New York Times, **Dec. 3, 1919**.

> EINSTEIN EXPOUNDS HIS NEW THEORY: It Discards Absolute Time and Space, Recognizing Them Only as Related to Moving Systems. IMPROVES ON NEWTON Whose Approximations Hold for Most Motions, but Not Those of the Highest Velocity. INSPIRED AS NEWTON WAS but by the Fall of a Man from a Roof Instead of the Fall of an Apple.
>
> When The New York Times correspondent called him at his home to gather from his own lips an interpretation of what to laymen must appear the book with seven seals, Dr. Einstein himself modestly put aside the suggestion that his theory might have the same revolutionary effect on the human mind as Newton's theses. The doctor lives on top floor of a fashionable apartment house on one of the few elevated parts of Berlin—so to say, close to the stars which he studies, not with a telescope, but rather with the mental eye, and so far, only as they come within the range of his mathematical formulae; for he is not an astronomer but a physicist. It was from his lofty library, in which our conversation took place, that he observed years ago a man dropping from a neighbouring roof—luckily on a pile of soft rubbish—and escaping almost without injury. That man told Dr. Einstein that in falling he experienced no sensation commonly considered as the effect of gravity, which, according to Newton's theory, would pull him down violently towards the Earth. This incident, followed by further researches along the same line, started in his mind a complicated chain of thoughts leading finally as he expressed it, "not a disavowal of Newton's theory of gravitation, but to a sublimation or supplement of it."

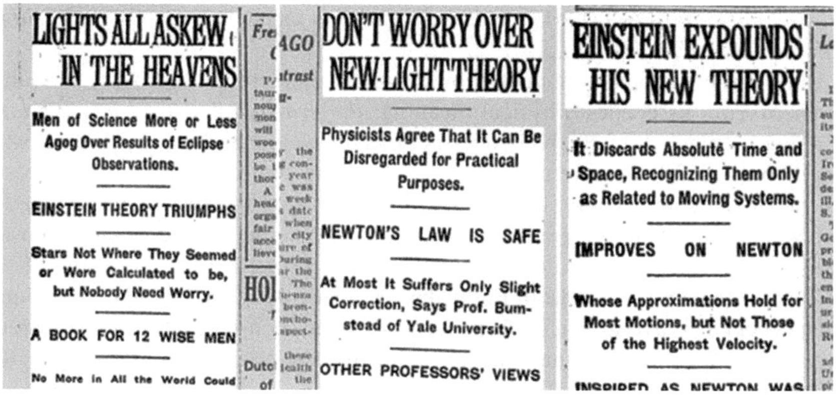

LIGHTS ALL ASKEW IN THE HEAVENS

Men of Science More or Less Agog Over Results of Eclipse Observations.

EINSTEIN THEORY TRIUMPHS

Stars Not Where They Seemed or Were Calculated to be, but Nobody Need Worry.

A BOOK FOR 12 WISE MEN

No More in All the World Could

DON'T WORRY OVER NEW LIGHT THEORY

Physicists Agree That It Can Be Disregarded for Practical Purposes.

NEWTON'S LAW IS SAFE

At Most It Suffers Only Slight Correction, Says Prof. Bumstead of Yale University.

OTHER PROFESSORS' VIEWS

EINSTEIN EXPOUNDS HIS NEW THEORY

It Discards Absolute Time and Space, Recognizing Them Only as Related to Moving Systems.

IMPROVES ON NEWTON

Whose Approximations Hold for Most Motions, but Not Those of the Highest Velocity.

It happened again! This is the third time that someone (Einstein!) observing something falling (actually, someone! Probably the journalist got it wrong: this fall happened in Bern, well before Einstein moved to Berlin), by the pure force of his thoughts, originates a radical change in the history of humanity like Galileo and Newton did before him, a change deeper than the changes generated by men in political or military power. In this case, fortunately, we may listen to the recollections of Einstein himself about «den glücklichsten Gedanken», the happiest thought of his life, it is worth it:

> When I was busy (in 1907) writing a summary of my work on the theory of special relativity for the Jahrbuch für Radioaktivität und Elektronik. I also had to try to modify the Newtonian theory of gravitation such as to fit its laws into the theory. While attempts in this direction showed the practicability of this enterprise, they did not satisfy me because they would have had to be based upon unfounded physical hypotheses. At that moment, I got the happiest thought of my life in the following form: In an example worth considering, the gravitational field has a relative existence only in a manner similar to the electric field generated by magneto-electric induction. Because for an observer in free-fall from the roof of a house there is during the fall—at least in his immediate vicinity—no gravitational field. Namely, if the observer lets go of any bodies, they remain relative to him, in a state of rest or uniform motion, independent of their special chemical or physical nature. The observer, therefore, is justified in interpreting his state as being "at rest." The extremely strange and confirmed experience that all bodies in the same gravitational field fall with the same acceleration immediately attains, through this idea, a deep physical meaning. Because if there were just one single thing to fall in a gravitational field in a manner different from all others, the observer could recognize from it that he is in a gravitational field and that he is falling. But if such a thing does not exist—as experience has shown with high precision—then there is no objective reason for the observer to consider himself as falling in a gravitational field. To the contrary, he has every right to consider himself in a state of rest and his vicinity as free of fields as far as gravitation is concerned.[21]

We see how Einstein begins his reasoning in an ideal dialogue with Galileo when he evokes the (mythical) outcome of the Leaning Tower experiment ...*The extremely*

---

[21] A. Einstein. Collected Works. Volume 7: The Berlin Years: Writings, 1918–1921 (English translation supple-ment) Page 136.

strange and confirmed experience that all bodies in the same gravitational field fall with the same acceleration...

But then, his happiest thought allows him to complete the discovery of Galileo by uncovering its deepest physical meaning: *the gravitational field has a relative existence.* A man who falls freely does not feel his weight and sees his tools floating around him as if there were no gravity. And, on the contrary, a man in a spaceship that accelerates into empty space would feel heavy on his seat; if he dropped an object, it would fall to the ground on a parabolic trajectory. There would be no way to distinguish the effects of gravity and acceleration (locally).

This is Einstein's Principle of Equivalence, the local (i.e. in a small region) equivalence between gravitational field and an accelerated (non-inertial) frame. Einstein's Equivalence Principle continues and accomplishes the discourse initiated by Galileo more than three centuries before.

An immediate consequence of Einstein's principle of equivalence is that gravity must bend the rays of light in the same way as it curves the trajectories of material bodies that would otherwise be rectilinear. Gravity is curving the spacetime!

Wait a moment. Spacetime? What is that? Galileo and Newton taught to us to think of space and of time, but this new word is unheard of. Where does it come from?

There was at the beginning of the twentieth century a strange asymmetry in physics between the laws of mechanics and electromagnetism somehow like the distinction between the sublunary and the celestial worlds. Aether was the name given to the impalpable medium carrying electromagnetic waves. Maxwell's equations were supposed to hold only in the reference frame of where the aether was at rest while all the laws of mechanics were supposed to be the same in every inertial frame. In 1905, the young Albert Einstein abolished that distinction and stated his meta-principle of relativity "*the laws of physics are identical in all inertial reference systems.*"

And since the speed of light appears in Maxwell's equations as a constant of physics, its invariance follows: the speed of light must be the same in every inertial reference frame, that is, it does not depend on the speed of the experimenter who measures it.

The speed of light is therefore a conversion factor: time can be measured in metres and distances in seconds (or light years): "*Accordingly we can express the essence of this postulate very tersely in the mystical formula: 300,000 kilometres* $= \sqrt{-1}$ *s.*" Space and time, thus, are the same thing (up to a minus sign): "*Henceforth, space by itself, and time by itself, are doomed to fade away into mere shadows, and only a kind of union of the two will preserve an independent reality*" *(H. Minkowski, 1908).*

Without having taken this fundamental step, that is to say without having merged space and time into the Minkowski spacetime, the question of what gravity is could not be posed. Newton's law of universal gravitation cannot find its place within Einstein's relativity framework of 1905 also because it presupposes an instantaneous action at a distance—this is also exactly what shocked Newton.

A new conception of gravity therefore emerges: gravity is not to be thought of as a force that acts at a distance but rather as a geometric property of spacetime; the gravitational attraction is a manifestation of the spacetime curvature. The spacetime around the Sun is curved and the planets follow geodesics, i.e. fall freely like an apple, in a curved non-Euclidean geometry. The resulting orbits are ellipses, with the notable exception of Mercury, whose trajectory is not closed. The light grazing the Sun is deflected for the same reason. This is Einstein's answer to the question *a quo moventur planetae.*

And what is the cause of the spacetime curvature? The matter, the radiation and all the forms of energy contained in the universe (more precisely, its energy–momentum content). Einstein's equations of 1915 tell exactly how this happens. They mathematically translate the following idea into a set of equations:

$$\text{Spacetime Curvature} = \text{Energy content of the universe}$$

$$R_{ab} - \frac{1}{2} g_{ab} R = \frac{8\pi G}{c^4} T_{ab}$$

6 Epilogue

If you meet three people on the street who throw objects on the ground, look at those objects carefully, argue animatedly, take notes and do it again, and again and again, you will think that they are some crackpots. What's there to look at? If one lets a body go, it falls what else? Everybody knows[22] it, what the hell!

But if the three men are Galileo, Newton, and Einstein, observing the falling bodies and strongly thinking about them, they will change the world.

The last word is to Wittgenstein: his famous.

Die Welt ist alles, was der Fall ist

Usually translated "The world is everything, that happens to be the case," might be also read

The world is everything that falls!

[22] *Cette Pesanteur [...] dans laquelle le vulgaire ne soupçonne pas même de mystère.*

Someone who throws a cigarette butt on the ground, might in fact be not a rude knucklehead who dirties the city, but a peripatetic student who let go the butt to its natural place.

References

A. Einstein, On the method of theoretical physics. Philos. Sci. **1**(2), 163–169 (1933)
A. Koyré, *Études galiléennes* (Hermann, Paris, 1939)
A. Koyré, Galileo and Plato. J. Hist. Ideas **4**(4), 400–428 (1943)
P. Rossi, La nascita della scienza moderna in Europa. Laterza (2001)
B. Russell, *The Scientific Outlook* (George Allen & Unwin, London, 1931)
M. Segre, Galileo, Viviani and the tower of Pisa. Stud. Hist. Philos. Sci. Part A **20**(4), 435–451 (1989)
W. Stukeley, *Memoirs of Sir Isaac Newton's Life, etc. White* (London, 1936), pp. 19–21
V. Viviani, Racconto istorico della vita di Galileo Galilei, in *Le opere di Galileo Galilei*, XIX, 1938 [1654], p. 606.

Gravity Between Physics and Philosophy

Mariano Cadoni and Mauro Dorato

Abstract We present a dialog between a philosopher and a theoretical physicist on some aspects of the historical and conceptual foundations of gravitational theories, in particular the equivalence principle, the relation between gravity/geometry, the ontological nature of space–time, and some relevant aspects of the historical development of the concept of action at distance and locality.

Keywords Gravitational theories · Newtonian gravity · XVII century gravity and geometry · Einstein's general relativity

1 Prolog

Despite the huge progress on the experimental side (gravitational wave detection, black hole imaging, cosmology) (Abbott et al. 2016; Event Horizon Telescope Collaboration et al. 2022; Planck Collaboration: Aghanim et al. 2018) achieved in recent years, gravity remains the most enigmatic of the four fundamental interactions of the physical world. Gravity appears to be intrinsically different from electromagnetic, weak, and strong interactions. It is only attractive, is characterized by a tiny coupling strength (Newton's constant G), and extends to infinite distances. These features imply that gravity determines the behavior of physical systems at large scales (astrophysical and cosmological) but, unlike the other interactions, is completely irrelevant in the regimes, where the other three interactions play an essential role, at least for the energy tested by present particle accelerators like LHC. On

M. Cadoni (✉)
Dipartimento di Fisica, Università di Cagliari, Cittadella Universitaria, 09042 Monserrato, Italy
e-mail: mariano.cadoni@ca.infn.it

I.N.F.N, Sezione di Cagliari Cittadella Universitaria, 09042 Monserrato, Italy

M. Dorato
Dipartimento di Filosofia, Comunicazione e Spettacolo, Università Degli Studi Roma Tre, Via Ostiense 234, 00146 Rome, Italy
e-mail: mauro.dorato@uniroma3.it

M. Streit-Bianchi et al. (eds.), *New Challenges and Opportunities in Physics Education*, Challenges in Physics Education, https://doi.org/10.1007/978-3-031-37387-9_5

the other hand, gravity is believed to become relevant again at extremely short scales of the order of Planck's length $\sim 10^{-33}$ cm. But the description of gravity and space-time at these short scales remains a big puzzle because we still do not have a satisfactory theory of quantum gravity. Moreover, also the description of gravity at extremely large distances presents many unresolved enigmas. To explain the rotational trajectories of galaxies, the structure formation, and the accelerated expansion of the universe, we must postulate the existence of exotic forms of matter called *Dark Matter* and *Dark Energy* (ESA 2017), which, despite important theoretical and experimental efforts, are still not well understood.

As is well known, progress in our understanding of gravitational interaction has marked two of the most important scientific revolutions, implying radical changes in the fundamental categories with which we describe and explain the physical world. The first, most important revolution characterized the birth of modern science (Butterfield 1959) and was triggered by the investigations of the kinematics and dynamics of bodies falling toward the Earth. Newton's law of universal gravitation unified two kinds of motions that until then were regarded as distinct, namely the terrestrial motion, which makes bodies fall toward the earth (the famous apple), and the centripetal force, which keeps the planets in closed orbits around the Sun, both subject to the same gravitational force. Via this *synthesis*, as Koyré called it (Koyré 1965), the Earth could not be regarded any longer as ruled by "parochial", anthropomorphic, "unique" laws independent of, and different from, those governing the rest of the universe. During the seventeenth century, the teleologically based, Aristotelian distinction between the vertical terrestrial motion and the circular celestial motion together with the related general view of nature, was abandoned forever.

Given the close conceptual relationship between space, time, matter, and motion, the modern scientific revolution also brought about a radical change in our view of *space and time*. Analogously, Einstein's General Relativity (GR) marked another paradigmatic shift in our understanding of the physical world, thanks to the formulation of a revolutionary theory of space and time (space-time) that brought about important ontological consequences. The present search for the *Holy Grail* of theoretical physics, namely a quantum theory of gravity, is expected to lead to another deep reshaping of our understanding of the nature of space-time at extremely short distances and to radically new relations between quantum features such as entanglement and the geometric properties of space-time.

In this contribution, we present a dialog between a philosopher of science (Mauro Dorato) and a theoretical physicist (Mariano Cadoni) on some foundational aspects of gravitational theories: very briefly and synthetically, (i) the landmarks of the historical development of the concept of action at distance and locality, (ii) the equivalence principle, (iii) the relation gravity/geometry, and (iv) the ontology of space-time. These four points will be discussed in the form of answers to four questions raising these issues.

2 Gravity from Physics to Philosophy

2.1 How Did the Seventeenth-Century Historical Debate on the Concepts of Action at Distance and Locality Affect the Explanation of the Gravitational Interaction?

2.1.1 MD

Given the lack of space, we will focus our discussion of Newton's still *approximately valid* theory of gravitation with an eye, however, to the subsequent, twentieth's century development. For the same reason, we will not be able to discuss or refer to the immense secondary literature.

The opposition between the so-called continental physics (Descartes, Huygens, Leibniz, and Bernoulli) on one hand, and Newton and his English on the other, was brilliantly summarized by Voltaire in his *Lettres Anglaises* (Coward et al. 1733).

A Frenchman who arrives in London will find philosophy, like everything else, very much changed there. He had left the world a *plenum*, and he now finds it a *vacuum*. At Paris, the universe is seen composed of *vortices* of subtle matter; but nothing like it is seen in London. According to Cartesians, everything is performed by an *impulsion*, of which we have very little notion; and according to Sir Isaac Newton, it is by an *attraction*, the cause of which is as much unknown to us.

Let us comment very briefly on the italicized words considering the subsequent development of the relationship between space-time and gravitation. As we will see in the section dealing with the ontological status of space-time in general relativity, the opposition between space regarded as a "plenum" by continental natural philosophers and space considered as an infinite "vacuum" containing point particles (Newton) characterizes the history of ancient and modern natural philosophy well *before* and *after* the seventeenth century. The Greek atomists (Leucippus, Democritus, Lucretius) had to postulate empty space to justify the motion of the atoms, which follow a vertical motion. Aristotle and Descartes, for different reasons, argued that the world is a *plenum*. With various arguments, and against the mechanistic worldview of the atomists, the former argued that "Nature abhors a vacuum". According to the latter, 'empty space' is an oxymoron, since space is essentially identical to matter: absence of matter (i.e., empty space) also entails the absence of space and therefore no vacuum, a logical impossibility! Nevertheless, the Cartesian identification of matter and space was very important for the progress of physics, since it justified the application of geometry, the science of space, to the physical world. It is of the utmost importance to recall that this divide on the nature of space is logically connected to the understanding of gravity. Not only did Descartes and the other mentioned continental philosophers attack Newton's introduction of a force acting at a distance since it was not understandable and "unknown" (see Voltaire). Relatedly, they also argued that action at a distance reintroduced *animistic suggestions* that the early modern mechanical view of nature had been overthrown.

According to Descartes, any type of motion is locally due to an impulse. Based on this hypothesis, he argued that the planets are carried around the Sun by a subtle, fluid matter rotating around it. With their impulse, these vortices prevented the tendency of bodies to continue their inertial motion along the tangent to the circular trajectory. Voltaire, however, correctly claimed that also the vortices introduced by Descartes and defended also by Leibniz are, despite their local action, "unknown".

In addition to their empirical inaccessibility, in his *De Gravitatione et equipondium fluidorum* (Newton 1962), Newton argued that Descartes' theory was contradictory and therefore insufficient to explain gravity. Pure relative rotation, conjoined with the absence of the notion of force in Descartes' physics, could not explain the inertial forces (the centripetal force) or the tendency of the planets to resist the impulse of the ether. Relative to the vortices, the planets are at rest, but at the same time they accelerate, due to their tendency to move toward the Sun. And the planets cannot be at rest and accelerating, since in the latter case there is a force acting on them, and a force cannot be "relativized away" (see Huggett 2000). It is important to note that also Newton believed that a gravitational force acting at a distance could not truly explain the physical relationship between any two massive bodies. He tried, without success, to find a mechanical model that could explain by acting at contact the motion of an attracted body. An influential theory of the nature of scientific explanation due to Hempel (Hempel and Oppenheim 1948) and endorsed by other influential twentieth-century philosophers of science like Popper, defends the view that *explaining* any physical phenomenon means deducing it from physical laws plus initial conditions. In this view, Newton's gravitation law would explain the motion of massive bodies. Newton, however, would have rejected this deductive nomological model, since he did not believe in action at a distance and did not forge hypotheses (hypotheses non fingo) on its nature. The following quotation reveals Newton's true attitude toward the nature of gravity (Newton 2020):

> That gravity should be innate, inherent, and essential to matter, so that one body may act upon another at a distance through a vacuum, without the mediation of anything else, by and through which their action and force may be conveyed from one to another, is to me so great an absurdity that I believe no man who has in philosophical matters a competent faculty of thinking can ever fall into it. Gravity must be caused by an agent acting constantly according to certain laws; but whether this agent be material or immaterial, I have left open to the consideration of my readers.

The solution to this conundrum could only come from a theory that denied tout court the existence of a gravitational force, a radical conceptual change brought about by Einstein's general theory of relativity, in which gravity is regarded as a geometrical feature of a curved space-time acting *locally* on material bodies. The general relativistic identification between space-time and the metric field is the fruit of the long 19th development of the electromagnetic *field* theory, in which the connection between any two charged bodies is carried *locally* by electromagnetic waves oscillating in a new version of a physical plenum at absolute rest, the ether, filling the physical space. Einstein was deeply influenced by the field paradigm based on locality, in such a way that he refused to accept quantum mechanics due to its non-local effects, therefore opting for its incompleteness.[1]

2.1.2 MC

Not only did the discoveries of physicists like Faraday and Maxwell imply the unification of optical, electric, and magnetic phenomena, but they also provided evidence for the existence, independently of the material sources (charges), of a *propagating electromagnetic (EM) field*. In the absence of a material medium, this field could, however, be seen either as an electromagnetic perturbation propagating in a vacuum or as a perturbation in an EM medium (*the ether*). In the latter case, the propagation of EM waves was seen in complete analogy with the propagation of sound waves in an elastic medium, with the dielectric constant and magnetic permeability playing the role of the "elastic" parameters of the ether.

The negative outcome of the famous Michelson–Morley experiment (1887), designed to show the existence of a relative motion between the Earth and the ether, created additional problems for the supporters of the existence of the ether and paved the way for the formulation of Einstein's Special Relativity (1905). However, the first solution of the plenum-vacuum dichotomy had to wait until the formulation of General Relativity (1915) and of quantum field theory (QFT) in the middle of the twentieth century. Quantum field theory extends the conceptual framework of quantum mechanics from point particles to *fields*, i.e., to systems with an infinite number of degrees of freedom. This extension led to the formulation of quantum theories describing the electromagnetic, the weak, and the strong interactions, which found their final systematization in the so-called *Standard model of particle physics* (CERN 1999). A very important conceptual outcome of QFT is that the vacuum of quantum fields is far from being "empty" and unique but has a rather complex structure. For instance, the quantum vacuum of the EM field is not "empty space" like in the atomistic doctrine, but a quantum state full of particle-antiparticle virtual excitations. In this sense, "nothing" is relative. On the other hand, GR shows us that the geometry of space-time is not fixed but is a dynamic entity determined by matter and dark energy. Thus, quantum field theory and general relativity teach us that the vacuum of the quantum field and the space-time geometry, being determined dynamically by matter, can be compared to the elastic medium whose structure is not static or inert but changes.

Despite the enormous conceptual progress achieved by QFT and GR, one cannot consider the above view as a full synthesis of the continental/English opposition in physics discussed by Mauro. The main open problem left is that the vacuum for quantum matter fields and the gravitational geometric background (thought as a differentiable manifold endowed with a geometry) is conceptually very different entities. QFTs can be defined only in a fixed geometric background. Moreover, the

[1] For a very informative review, see Berkovitz (2016).

geometric gravitational background is an object of classical physics, whereas the vacuum in QFT is a quantum object.

Einstein was aware of this asymmetry and in his later years tried, without success, to unify the gravitational with the electromagnetic interaction. Nowadays, it seems quite evident that a complete resolution of the Newton/Descartes opposition will require the formulation of a consistent quantum theory of gravity, such that all the four fundamental interactions of nature can be described on the same footing. The two main proposals for a quantum theory of gravity advanced so far, namely String Theory (Green et al. 1987) and Loop quantum gravity (Rovelli 1998) present difficulties precisely for what concerns the determination of the vacuum and the backgrounds that should describe the physical world at the (relatively low) energies that we can test experimentally. String theory allows for a huge number of vacua, the so-called *landscape* (Kachru et al. 2003), preventing a unique determination of the world where we live. Conversely, in Loop Quantum, gravity is very difficult to obtain the smooth space-time description of general relativity in the low-energy regimes. Thus, after 250 years of progress in our scientific understanding of the physical world, fundamental questions about action at a distance and locality in physics still wait for an answer.

2.2 *What is the Meaning of the Equivalence Principle?*

2.2.1 MC

The equivalence principle (EP) is the conceptual formulation of the universality of gravitational interaction and is the essential pillar of Einstein's General relativity. It states that, *locally*, one can always eliminate or mimic a gravitational field by choosing an accelerated reference frame. It has its roots in Newtonian and Galilean physics, where it takes the form of the equality between the inertial mass m_i, characterizing the inertia of a material body and the gravitational mass m_g, i.e., the source of gravitational forces. The equality between inertial and gravitational mass implies that in absence of friction bodies of different mass, form, or internal structure near the surface of the earth fall with the same constant acceleration.

In Newton's theory, the equality $m_g = m_i$ was considered a mere coincidence. In the general theory of relativity, Einstein took a step forward by promoting it as a fundamental principle of physics. Nowadays, this equality $m_i = m_g$ is experimentally tested with extremely high precision (1 part over 10^{15}) (Schlamminger et al. 2008). This equality can be used both to eliminate the gravitational field on the surface of the earth by choosing as a reference frame a freely falling lift and to generate a fictitious gravitational field in empty space (far from any gravitational source) by choosing as a reference frame a spacecraft that accelerates in the opposite direction. Notice that the equivalence principle holds only *locally*, that is, in a neighborhood of a given point of space-time. For an extended region of space, tidal forces prevent the possibility of eliminating a gravitational field by choosing the appropriate reference frame.

In its relativistic form, the equivalence principle (EP) states that in the presence of a gravitational field, for every space-time point, we can always find a reference frame for which the laws of physics take the form of those of special relativity, a theory that can be applied only in the absence of gravitational fields (*Einstein's equivalence principle*). In GR, the equivalence principle allows for the full geometrization of the gravitational interaction, given that, as we will see, the motion of a point-like test mass particle in a gravitational field generated by a source mass is fully independent of the mass of the test particle. The trajectory of the test mass is described by a geodesic of the space-time obtained by connecting the local freely falling reference frames. Owing to the use of accelerated frames, the EP is also deeply intertwined with the principle of general covariance, stating that the laws of physics must be written in a form that is independent of the choice of the reference frame.

At the epistemological level, the EP is strongly related to the debate on the existence of an absolute reference frame and the origin of inertia in Newtonian mechanics. Einstein was deeply influenced by Mach's ideas about the non-existence of absolute reference frames and the physical origin of inertia in terms of the overall distribution of masses in the universe. Nevertheless, the fact that the EP, together with the fact that in the absence of matter, Einstein's equations have flat space-time solution, makes GR intrinsically a non-Machian theory. Through the geometrization of the trajectories of the point particles in a gravitational field, the inertial mass of a test particle moving in a gravitational field does not appear in the equations for the dynamics. Inertia is not determined by the overall distribution of mass in the universe but is simply removed from the dynamics.

Present experiments test the validity of the EP from distances ranging from *mm* to solar distances. Thus, the EP (and hence GR) could be falsified, at least in principle, at short scales and/or large scales. For instance, this could occur if the gravitational interaction loses its universality when describing the dynamics at galactic scales (see, e.g., Cadoni 2004). Although presently there is no indication that this could occur, the possibility has not been ruled out. Another controversial point is the meaning of the EP in a quantum mechanical framework. In general, in the quantum realm, the non-local character of quantum mechanics makes the status of the EP rather unclear. Research on its possible quantum extension is ongoing (Giacomini et al. 2019; Cadoni et al. 2020).

2.2.2 MD

The attempt to unify different physical concepts characterized Einstein's scientific heuristics throughout his career. In this respect, not surprisingly, his results were amazingly successful: space and time, energy and mass in the special theory, inertial and gravitational mass, space-time, and the metric field in the general theory of relativity. The same effort toward unification is exemplified by his failed attempt to find a unified field theory that could bring together gravitation and electromagnetism. From a conceptual viewpoint, the philosophical significance of the equivalence principle as formulated in GR consists in the fact that Einstein succeeded in "transforming a problem into a postulate". This expression, originally coming from Goethe, has been used by Cassirer in regard to the origin of the special theory of relativity (Cassirer 1910). In our context, "postulate" refers to a fundamental assumption of GR. The problem with Newton's theory of gravity was making sense of the notion of a gravitational force acting at a distance. However, this force is unknown, as Voltaire had it is not directly observable: we can observe *directly* only the acceleration impressed on a massive body. Einstein solved the problem that characterized the dispute between the continental and the English natural philosophers presented above by considering free fall as "the natural state" of bodies, and, in this sense, as a postulate of the new gravitational theory. As such, free fall does not require any explanation, because the gravitation force is no longer part of the ontology of the theory, for the simple reason that it does not exist!

In GR, what needs to be explained is rather the "*deviation*" from this natural state, which holds only locally. The question now becomes: why does the free fall cease holding? The explanation brings into play the tidal force. There is an important Aristotelian distinction that runs across important stages of the history of physics. Aristotle divided all motion into two categories: *natural* (up and down on Earth, circular for the Heavens) and *violent*. The former kind is represented by the behavior of the four elements: Earth and water move toward the surface of our planet (down), while fire and air move toward the sky (up). The latter kind of motion, non-natural, is exemplified by objects thrown by human beings: stones or arrows, for instance, during part of their flight do not follow a downward direction. Their parabolic trajectory is a *deviation* from the vertical motion of all bodies (their natural motion) and as such requires an explanation. In Aristotelian physics, violent motions cried out for an explanation because, Aristotle and his followers, of course, did not know the law of inertia, which is the natural tendency of all bodies to continue in a straight line with the same velocity. This law lies at the foundations of the new dynamics. In Newton's physics, unlike the Aristotelian one, we don't have to explain why bodies keep on moving in the same direction with the same velocity. Rather, we must explain why they fail to do that, and the explanation, of course, brings into play gravity, friction, and in general *forces*. Free fall in GR is the natural state of motion in the Aristotelian sense: the natural motions in GR correspond to free fall that, qua natural or postulational, does not need any explanation. The analogy is limited by the fact that Aristotle *had* some explanation for vertical motion, given by a teleological tendency of all bodies to their natural places.

In a word, problems that in the pre-revolutionary theory seemed to call for an explanation, in the post-revolutionary became *"natural"*, that is, regarded as not needing an explanation. It is remarkable that this switch marks also the origin of the special theory of relativity: the postulation of the light principle (one of the two axioms of the theory) explained away the need to give a *dynamical* explanation of the null result of Michelson and Morley's experiments. It has also been suggested that the same transformation may also characterize the existence of entangled states in quantum mechanics, namely states whose parts are not separable from the whole to which they belong (Fine 1984): if such entangled states are the natural states of quantum systems, what needs to be explained is the breaking of entangled states, that is the appearance of separate and well-defined measurement outcomes.

2.3 *What is the Relationship Between Gravity and Geometry?*

2.3.1 MC

The paradigm of classical physics developed in the seventeenth, eighteenth, and nineteenth centuries describes the physical world as an ensemble of objects interacting through forces in a fixed stage. The arena in which the "performance" of the physical world takes place is measured by rigid rods (to record the spatial position of objects, e.g., their Cartesian coordinates *x, y, z*) and by clocks, recording the absolute time *t* needed to describe their dynamics, i.e., the change of their positions in time. The properties of this arena are essentially described by the spatial three-dimensional, Euclidean, i.e., homogeneous, and isotropic, geometry and by a separated, one-dimensional, timeline measured by a clock. During the nineteenth century, mathematicians like e.g., Gauss, Lobachevsky, and Riemann discovered curved, non-Euclidean geometries differing from the flat Euclidean spaces. Around 1915, physicists like Emmy Noether realized that conserved dynamical quantities of physical objects could be derived from properties of geometry (symmetries) (Kosmann-Schwarzbach 2011). However, her identification of conservation laws with geometrical symmetries did not presuppose that the interactions between objects and their dynamics could influence the geometry defined by rods and clocks.

At the beginning of the twentieth century, this well-established paradigm was completely overthrown by a scientific revolution that changed the very foundations of physics. Einstein's special relativity challenged the existence of rods that have the same length in every reference frame and clocks ticking in an absolute way. The existence of a maximum speed for information flow (the speed of light) implies that space and time become intimately intertwined in a four-dimensional flat space-time described by equal-footing inertial coordinates (*t, x, y, z*) and that observers moving at different (constant) speed measure different values for the lengths of the rods and for time intervals. But it was Einstein's General Relativity theory of gravity to *"put in motion" the inert geometric stage of Newtonian physics*. Space-time itself participates in the drama and is not an inert arena. Einstein's equations of GR tell us that the space-time geometry is not fixed and given by a flat geometry but is dynamically determined by the physical matter sources. Einstein's field equations brilliantly related the curvature of the space-time, which in turn determines the gravitational metric field, with the distribution of mass, energy, and momentum in the space-time itself. This intimate matter-gravitation-geometry relation is then completed by Einstein with his equivalence principle from which follows that matter must move in the space-time along a geodesic (the analog in curved geometries of a straight line in Euclidean geometry) of the space-time determined by Einstein's equations. Rephrasing the famous sentence of the physicist John Wheeler: Matter tells space-time which geometry must be used, and geometry tells matter how to move (Wheeler and Ford 2000).

This interplay between matter-gravity-geometry has several astonishing consequences, (most of them already tested), when gravity becomes extremely strong or when the gravitational matter source becomes "exotic". This happens for instance in the case of the solutions of Einstein's equations describing black holes, gravitational waves, cosmological models, and wormholes.

Black holes are configurations resulting from collapses of ultra-massive stars in which the gravitational field becomes so strong that the resulting geometry develops an *event horizon* (a mathematical surface separating the black hole interior, i.e., the region out of which any physical object cannot escape, from the outside world) preventing any kind of signal to escape from the hole. The image of a supermassive black hole sitting in the center of our galaxy has been recently observed by the EHT collaboration (Event Horizon Telescope Collaboration et al. 2022). Gravitational waves are tiny deformations of the space-time geometry which propagate at the speed of light. They can be generated by the merging of two black holes (the cosmic dance) and have been recently first detected by the LIGO-VIRGO collaboration (Abbott et al. 2016). We have also recently discovered that our universe expands in an accelerated way, leading physicists to conjecture the existence of an exotic form of matter called *dark energy* which produces a repulsive gravity field instead of the usual attractive one. Finally, wormholes are hypothetical gravitational configurations created by extremely exotic forms of matter. They could connect parallel universes and generate closed time-like curves: starting from any point of the curve, one can come back in time to the original point of departure. In this way, the distinction between past and future loses any meaning.

One of the most difficult questions with which theoretical physicists are faced today is to understand how gravity and geometry talk to each other when quantum theory is brought into play. This is believed to be necessary and unavoidable at least at a very short distance of the order of the Planck length $\sim 10^{-33}$ cm. Despite more than 50 years of effort, the formulation of a quantum theory of gravity remains elusive. Therefore, questions like *How does space-time geometry look like at very short distances? Is the space-time geometry at very short distances still smooth or becomes discrete? Is the beautiful relation of gravity/geometry of GR preserved in some way in a quantum theory of gravity?* Remain still unanswered.

2.3.2 MD

From a conceptual viewpoint, the relationship between space-time and geometry enters the stage only with Einstein's GR. As anticipated by Mariano's answer to this question, before 1916, space and time were regarded as an *inert* arena where the actors (the physical world) recited their drama (the physical world evolved). With GR, the arena becomes an active actor, and space-time itself becomes dynamic, that is, it changes its structure in different regions. In this sense, change is not eliminated from the theories of relativity because, as Meyerson claimed, time keeps its difference from space despite its inseparability. From a little-known letter of Einstein to Emile Meyerson, we know that Einstein agreed with this reading (Einstein 1928) (See also Capek 1961).

2.4 In What Sense Does Space–Time Exist?

2.4.1 MD

This vexed question was posed in an explicit and philosophically mature way in the famous epistolary Leibniz-Clarke (Alexander 1956). Clarke was a theologian follower of Newton, and the added importance of this exchange lies in the fact that, most probably, Newton himself supervised the text of Clarke's letters. In this sense, it was a confrontation between giants, certainly the best mathematicians of their time. Newton was a better physicist, Leibniz, an equally talented mathematician a better-learned scholar and philosopher. This distinction is reasonable only considering the current division of subjects since Newton's masterpiece was titled *Philosophiae Naturalis Principia Mathematica*, and the name 'Natural Philosophy' refers to what today we call physics. The opposition between Newton and Leibniz was on the ontological and epistemic nature of space and time. But before going to discuss what divided them, we should remind the reader that before Einstein's special theory of relativity space and time were separated and represented as the Cartesian product between three-dimensional Euclidean space and time, three coordinates for three-dimensional space and one for time, a separation grounded in the frame-independence of the relational of simultaneity. For both Newton and Leibniz, the simultaneity between any two events is absolute, so that the relation 'before then' is total: either *a* is before *b* or *b* is before *a* and this holds for any pair of events in time. In the special theory of relativity instead, it is not true that for any pair of events either *a* is before *b* or *a* is simultaneous with *b*, a fact that is due to the limiting character of the speed of light. As to the ontological divide, Newton is traditionally presented as a *substantivalist* and Leibniz as a *relationalist*. This opposition is not wholly justified from a historical viewpoint. There is/are various sense/s of the philosophical word 'substance', and the one that applies to Newton's definition of time is "*existence independent from anything else*". According to Newton, space and time exist independently of physical events and objects. In his famous Scholium, he writes (Rynasiewicz 2022).

"Absolute, true, and mathematical time, of itself, and from its own nature, flows equably *without relation to anything external,* my emphasis".

Time "flows equably" independently of anything external means that it ticks independently of the most uniform motion of the clocks (physical systems) that we must use to measure its duration. Unlike the Aristotelian conception, according to Newton time exists without change. Just below the definition of time, we read (Rynasiewicz 2022).

"Absolute space, in its own nature, *without relation to anything external*, remains always similar and immovable, my emphasis".

In this restricted sense, both space and time *are* substances because they would exist independently of the physical world. According to Leibniz, instead, as is evident from his letters to Clarke, space and time are *relations*: the former is the order of coexistence, the latter the order of succession. This is equivalent to saying that space and time would not exist without physical events and objects. They are like genealogical trees: the relation of being a father, a mother, or an uncle would not exist without physical persons. Leibniz's criticism of Newton's absolutist conception (to designate the latter's position on space and time, absolutism is philologically more appropriate than substantivalism) proposes two famous arguments, based on two axioms: the (controversial) principle of the identity of the indiscernible (PII) and the principle of sufficient reason (applied in this case to God). The first principle claims that any pair of spatial points and instants are homogeneous, and therefore indistinguishable. Based on the first principle, they are identical. If we shift the physical universe westward or eastward or rotate it 180 degrees via static shifts, or change the velocity of everything via dynamic shifts, based on the Galilean principle of relativity (which was known to Newton of course), both shifts do not result in physical changes. By presupposing Leibniz's relationalism, the spatial and temporal relations of the original world and the shifted ones produce indistinguishable, and therefore identical worlds. At this point, the second principle intervenes: why did he not create the world a year before or a year later? Why did he decide to place the world W where it is rather than translate it to the right (say W_1)? Based on Newton's physics, as Leibniz interprets it, these two worlds would be *different*, since its point and instants have an intrinsic *identity*. Thanks to the PII, it follows that there would be *no sufficient reason* to explain why God decided to create our actual world W rather than W_1 obtained via some sort of shift. Since according to Leibniz, the two worlds are indistinguishable and therefore identical, the above questions raised by Newton's physics would collapse: if the two worlds are identical, we don't have to explain God's choice. Note that Leibniz's second principle could be reformulated without the theological ingredient: based on coherent and strict empiricism, Galilei's principle of relativity implies that there is no experiment thanks to which we could distinguish between the two worlds: a difference that does not make any empirical difference, not a difference. By a careful study of Newton's unpublished manuscript quoted above, we now know that Newton claimed that space 'is neither a substance nor an accident but has an existence of its own.

The dispute between Leibniz and Newton received a new twist after the formulation of Einstein's GR. The main question has now become: which element of the new theory plays the role of Newton's empty space and time? Consider a set of points with a purely differentiable structure: this is 'the matter-empty' manifold on which the metric, gravitational field 'sits'. The latter, however, is a *physical* field and therefore cannot have the same role played by the *empty* space and time postulated by Newton, given that, according to Newton, space and time are *non-physical* and yet exist independently of the physical world. It would then seem that we should identify

Newton's empty space-time with *the manifold* which, however, has no metric structure and therefore cannot tell objects "where to go" (see above Wheeler's slogan). For this reason, it would seem plausible to identify space-time with the metric field, which however, by carrying energy and momentum, is a substance in a different sense. On the other hand, however, it cannot be a substance *in Newton's sense*, precisely because it is physical and non-empty. Since in GR, there is no clear distinction between what is empty and what is physical, or between 'content' and 'container' as in the Newton-Leibniz debate, authoritative philosophers of physics have concluded that the Newton-Leibniz dispute cannot be defined in the new relativistic context and is therefore 'outdated' (Rynasiewicz 1996). To overcome this impasse, one must grant Rynasiewicz that neither manifold nor metric field substantivalism is suitable to reconstruct not arbitrarily choose the debate and the Newtonian views. Dorato has therefore advanced a third option which he calls *structural space-time realism* (Dorato 2000). In essence, this option is a synthesis between the two traditional positions in the same sense in which the metric field is both matter and space-time. He defends at the same time the relational, structural character of space and time (i.e., the identity of space-time points depends on their structure), while claiming that the geometrical structure used to represent them is "really", mind-independently exemplified by the physical world.

2.4.2 MC

As pointed out by Mauro, we must discuss separately the situation before and after GR. Before GR, in the context of Newtonian mechanics (and in Einstein's special theory of relativity), the concept of absolute space and time was perfectly meaningful, and the epistemic discussion about their ontological nature was well-grounded. Newton was aware that a logically consistent formulation of the three fundamental laws of mechanics required an absolute reference frame, which he identified with the *fixed stars*. He also proposed an experiment (the famous bucket experiment described in Newton's *Principia*) to test the physical existence of this absolute reference frame (Newton 1687). Newton's argument was criticized by Mach, who argued that the notion of an absolute reference frame was metaphysical and that only relative motions are measurable.

It is interesting to notice that the ontological opposition between substantivalism and relationalism becomes much milder in a hypothetical world of non-interacting particles, which owing to the absence of forces must move with uniform velocity. In this case, we can identify absolute space and time with Euclidean space plus a timeline or, in the context of Special Relativity, with Minkowski space-time. The symmetries of these space-times (invariance under translations, rotations, and Lorentz boosts) are determined by the properties of free particles and vice versa. This follows from the above-mentioned Noether's theorem and from F. Klein's Erlangen Program, defining geometry in terms of invariance under a symmetry group. In fact, for free particles impulse, angular momentum, and energy must be conserved, which using Noether's theorem implies that space must be homogeneous and isotropic, whereas time must be homogeneous, i.e., invariant under the translation and rotation group (plus Lorentz boosts in the case of special relativity). According to Klein's definition of geometry, this implies that the geometry of space must be Euclidean in the Newtonian case and Minkowskian in Special Relativity. Absolute space-time can be, therefore, completely reconstructed from the properties of matter.

The situation changes completely when we turn on interactions (forces) between matter and particles. In the case of electromagnetic, weak, and strong forces, space-time geometry cannot be derived from the properties of matter particles, particle motion is in general accelerated motion and forces cannot be universally, even locally eliminated by choosing an accelerated reference frame. The only fundamental interaction for which the geometry of space-time follows from properties of matter and for which forces can be universally, and locally eliminated by choosing an accelerated reference frame is the gravitational one. The way this happens is completely different from the free particle case and it is explained by Einstein's GR.

As argued by Mauro, after the formulation of GR, the debate between Newton and Leibniz became somehow outdated because of the difficulty of defining the notion of absolute space-time in the Newtonian sense. On the other hand, for what concerns recent developments in canonical gravity and Loop Quantum Gravity, GR has given a new impulse to the relational point of view. It has been shown that, in the GR context, if one wants to describe gravitational systems using quantities that do not depend on a specific reference frame, there are no dynamics, no "time", i.e., these quantities do not evolve dynamically (Bergmann 1961). A possible solution to this "problem of time" is to define *relational observables*, that is, to look for the evolution of the system not with respect to time but, instead, with respect to other fields, like e.g., a scalar field (see for instance Rovelli 2002). Within this approach, a complete set of relational observables can be defined also in the GR context. There are attempts to generalize this relational approach to quantum gravity, e.g., in Loop Quantum Gravity (Rovelli 1996).[2]

3 Epilog

In this contribution, we have presented a theoretical physicist's and a philosopher's viewpoint, on (i) the landmarks of the historical development of the concept of action at distance and locality, (ii) the equivalence principle, (iii) the relation gravity/geometry, and (iv) the ontology of space-time. We have seen that the seventeenth-century historical debates about the meaning of action at a distance and the ontological interpretation of space and time are still alive since they play an important role also in contemporary scientific research about gravitational interaction. Several conceptual aspects of Newtonian gravity and of Einstein's General Relativity can be fully understood only if framed in the more general philosophical debate about the

[2] For a recent evaluation of the ontological consequences of this approach to quantum mechanics, see Dorato and Morganti (2022).

nature of both gravitational interaction and space and time. On the other hand, difficult challenges of the contemporary scientific research on gravity, such as, e.g., the question about the fate of smooth space-time geometry and locality at short distances, are often modern reformulations of old conceptual problems. It is therefore quite possible that future scientific advances in our understanding of gravitational interaction will also shed light on century-old philosophical questions.

References

B.P. Abbott et al., Observation of gravitational waves from a binary black hole merger. Phys. Rev. Lett. **116**(6), 061102 (2016). https://doi.org/10.1103/PhyRevLett.116.061102

H.G. Alexander (ed.), *The Leibniz-Clarke Correspondence, Together with Extracts from Newton's Principia and Optics* (Manchester University Press, 1956)

P.G. Bergmann, Gauge-invariant variables in general relativity. Phys. Rev. **124**(1), 274–278 (1961). https://doi.org/10.1103/PhysRev.124.274

J. Berkovitz, Action at a distance in quantum mechanics, in *The Stanford Encyclopedia of Philosophy*, ed. E. N. Zalta (Spring, 2016). https://plato.stanford.edu/archives/spr2016/entries/qm-action-distance/

H. Butterfield, *The Origins of Modern Science 1300–1800*. (The MacMillan Company, New York, 1959), p. 242. First published 1949 https://archive.org/details/originofmoderns007291mbp/page/n7/mode/2up

M. Cadoni, An Einstein-like theory of gravity with a non-Newtonian weak field limit. Gen. Rel. Grav. **36**, 2681–2688 (2004)

M. Cadoni, M. Tuveri, A.P. Sanna, Long-range quantum gravity. Symmetry **12**(9), 1396 (2020)

M. Capek, *The Philosophical Impact of Contemporary Physics* (Publisher D. Van Nostrand, Princeton, New Jersey, 1961)

E. Cassirer, *Substanzbegriff und Funktionsbegriff: Untersuchungen über die Grundfragen der Erkenntniskritik* (Verlag von Bruno Cassirer, Berlin, 1910), p.459

CERN, *The Standard Model of Particles Physics*. https://press.cern/science/physics/standard-model

D. Coward, Voltaire, Letters Concerning the English Nation. Printed for C. Davis in Pater-Noster-Row, and A. Lyon in Russel-Street, Covent-Garden. 1733 Wikipedia or https://monadnock.net," *Letters Concerning the English Nation by Voltaire*

M. Dorato, Substantivalism, relationism, and structural spacetime realism. Found. Phys. **30**(10), 1605–1628 (2000). https://doi.org/10.1023/A:1026442015519

M. Dorato, M. Morganti, What ontology for relational quantum mechanics? Found. Phys. **52**, 52–56 (2022)

A. Einstein, A propos de la "Déduction Relativiste" de M. Émile Meyerson. *Revue philosophique de la France et de létranger* 105, 046005 (1928)

ESA, The cosmic energy budget (2017)

Event Horizon Telescope Collaboration et al., First Sagittarius A* event horizon telescope results. I. The shadow of the supermassive black hole in the center of the milky way. Astrophys. J. Lett. **930**(2), L12 (2022). https://doi.org/10.3847/2041-8213/ac6674

A. Fine, *The Shaky Game. Einstein, Realism, and the Quantum Theory*, 2nd edition. (The University of Chicago Press, 1984)

F. Giacomini, E. Castro-Ruiz, V. Brukner, Quantum mechanics and the covariance of physical laws in quantum reference frames. Nat. Commun. **10**(1), 494 (2019)

M. Green, J. Schwarz, E. Witten, *Superstring Theory* (Cambridge University Press, New York, 1987)

C.G. Hempel, P. Oppenheim, Studies in the logic of explanation. *Philos. Sci.* **15**(2), 135–175 (1948). Published by: The University of Chicago Press
N. Huggett (ed.), *Space from Zeno to Einstein* (The Mit Press, 2000)
S. Kachru et al., De Sitter Vacua in string theory. Phys. Rev. D **68**, 046005 (2003). https://doi.org/10.1103/PhysRevD.68.046005
Y. Kosmann-Schwarzbach, in *The Noether Theorems. Invariance and Conservation Laws in the Twentieth Century.* Series: Sources and Studies in the History of Mathematics and Physical Sciences (Springer-Verlag, New York, 2011)
A. Koyré, *Newtonian Studies.* (Ltd, London Chapman and Hall, 1965)
I. Newton, *Philosophiae Naturalis Principia Mathematica, Iussu Societatis Regiae ac typis Josephi Streater, Londini, anno MDCLXXXVII* (1687)
I. Newton, *Unpublished Papers of Isaac Newton: A Selection from the Portsmouth Collection in the University Library, Cambridge*, ed. By A. R. Hall (Cambridge University Press, Cambridge, 1962)
I. Newton, Web reference:www.sophiararebooks.com. *Four letters from Sir Isaac Newton to Doctor Bentley* (2020)
Planck Collaboration: N. Aghanim et al., Planck 2018 results. VI. Cosmological parameters. *Astron. Astrophys.* **641**, A6 (2020). [Erratum: *Astron. Astrophys.* **652**, C4 (2021). https://doi.org/10.1051/0004-6361/201833910e
C. Rovelli, Relational quantum mechanics. Int. J. Theor. Phys. **35**, 1637–1678 (1996). https://doi.org/10.1007/BF02302261
C. Rovelli, Loop quantum gravity. Living Rev. Rel. **1**, 1 (1998)
C. Rovelli, Partial observables. Phys. Rev. D **65**, 124013 (2002). https://doi.org/10.1103/PhysRevD.65.124013
R. Rynasiewicz, Absolute versus relational space-time: an outmoded debate? J. Phil. **93**(6), 279–306 (1996)
R. Rynasiewicz, in *Newton's Views on space, Time, and Motion*, ed. by E.N. Zalta, The Stanford Enyclopedia of Phylosophy (Springer, New York, 2022)
S. Schlamminger et al., Test of the equivalence principle using a rotating torsion balance. Phys. Rev. Lett. **100**(4), 041101 (2008). https://doi.org/10.1103/PhysRevLett.100.041101
J.A. Wheeler, K.N. Ford, *Geons, Black Holes, and Quantum Foam: A Life in Physics* (W.W. Norton & Company, New York, 2000)

Einstein's Theory at the Extremes: Gravitational Waves and Black Holes

Mariafelicia De Laurentis, Paolo Pani, and Michele Punturo

Abstract Einstein's theory of gravity, General Relativity, is one of humankind's greatest intellectual achievements. Over the last century, its interpretation of the gravitational interaction in terms of spacetime curvature has been brilliantly confirmed by many different experiments. Two astonishing predictions of General Relativity are gravitational waves—spacetime perturbations that travel at the speed of light—and black holes—mysterious objects made solely from pure spacetime fabric which challenge the laws of physics. Once considered only as bizarre solutions to Einstein's equations, black holes have now acquired a central role in astrophysics, cosmology, and even in foundational physics and philosophy. The recent radio images of supermassive black holes are providing us with invaluable information about the dynamics of matter in the vicinity of an event horizon. Gravitational waves carry information about the most dramatic astrophysical processes in the Universe, such as black-hole collisions. Their detection has inaugurated the era of gravitational-wave astronomy, which is reshaping astrophysics, cosmology, and fundamental physics. Thanks to future experiments, black holes, and gravitational waves will shed new light on the unknown Universe. In this chapter, we will explore the theory of these fascinating objects and the observations that have confirmed their existence.

Keywords Gravity · Gravitational waves · Black holes · Spacetime · General relativity · Interferometers

---

M. De Laurentis
Department of Physics "Ettore Pancini", University of Naples, Monte Sant'Angelo Campus, Via Cinthia 21, Building 6, 80126 Naples, Italy
e-mail: mariafelicia.delaurentis@unina.it

P. Pani (✉)
Physics Department, La Sapienza University of Rome, Piazzale Aldo Moro 5, 00185 Rome, Italy
e-mail: paolo.pani@uniroma1.it

M. Punturo
INFN Sezione Perugia, Via A. Pascoli, 06123 Perugia, Italy
e-mail: michele.punturo@pg.infn.it

M. Streit-Bianchi et al. (eds.), *New Challenges and Opportunities in Physics Education*, Challenges in Physics Education, https://doi.org/10.1007/978-3-031-37387-9_6

1 Introduction: Gravity as Spacetime Curvature

Why do objects fall? Since the dawn of humanity, this (only apparently trivial!) question has kept the brightest minds busy, pushing them to formulate increasingly sophisticated theories of gravitation (from the Latin *gravitas*, weight) whose predictions are as accurate as they are incredible. It is, in fact, amazing how a phenomenon that we have known well since our infancy (gravitational attraction) is the same as the basis of the Big Bang, the evolution of the universe, black holes, gravitational waves, and some of the most important open problems of fundamental physics.

But let us take a step back. For the early Greek philosophers, gravity was caused by the attraction between "like matters". However, this forerunner concept was shelved for centuries due to the influential but rather unscientific vision of Aristotle, according to which the motion of bodies was attributable to their tendency to move toward their "natural" place. This vision was supplanted only thanks to the studies of Galileo Galilei and the introduction of the scientific method. Galileo identified a property of all objects (what we now call "*inertial mass*") that determines their resistance to changing their state of motion or rest. A generation later, Isaac Newton formalized these concepts in his theory of mechanics, according to which inertial mass is the constant of proportionality between the "force" acting on a body and its acceleration. Newton went further and identified in the "gravitational force" the cause of the terrestrial attraction on all bodies, from the famous apple to the Moon, as well as of the mutual universal attraction between all celestial bodies. According to Newton, any two objects attract each other proportionally to their masses (what we now call "*gravitational masses*") and inversely proportional to the square of their distance. It was a revolution. This simple and elegant theory, combined with mechanics, provided an incredibly precise description of all gravitational phenomena: the motion of the Moon around the Earth, eclipses, tides, the laws empirically discovered by Kepler for the Solar System, and the motion of bullets and satellites. The predictions were so precise that small anomalies in the motion of the planet Uranus, instead of suggesting corrections to Newton's theory, led to the hypothesis of the existence of a new planet (later actually discovered and named Neptune), which perturbed its orbit. Not bad for a theory born to explain the fall of an apple!

However, there were two major loose ends in Newton's theory of gravitation. One was clear from the very beginning: As Galileo's famous experiments had demonstrated, from an experimental point of view, the "inertial" and "gravitational" masses are identical. Why two such different concepts (inertia and gravity) were related to each other remained a mystery. The second problem was only understood much later by Albert Einstein when, in his annus mirabilis (1905), he formulated the theory of Special Relativity, whose cornerstone is the constancy of the speed of light in a vacuum and the fact that this is the maximum speed attainable in nature. Special Relativity immediately found various experimental confirmations, but it starkly contrasted Newton's theory, in which the gravitational interaction propagates at infinite speed. Einstein took these problems (for many at the time considered only as marginal) extremely seriously. It took him ten years of intense work to arrive at his masterpiece: the general theory of relativity (now universally known as General Relativity). As often in Einstein's works, General Relativity is based on some principles, i.e., theoretical axioms based on empirical observations. Starting from Galileo's observations, Einstein formulates the "Principle of Equivalence", according to which inertial and gravitational mass are not accidentally equal or very similar, they are precisely the same physical entity. This simple principle has very remarkable consequences, such as the fact (cleverly explained by Einstein with the famous *gedankenexperiment* of the elevator) that it is impossible to locally distinguish whether a system is subject to gravitational attraction or if it is moving with accelerated motion. Einstein's new theory was also based on another principle: The laws of physics must be the same regardless of the motion of the observer (i.e., for any reference frame). Therefore, the fact that gravitational attraction can be locally eliminated with an appropriate choice of the reference system is a general and unavoidable fact of physical systems. This is a revolutionary concept that does not apply to any other of the fundamental interactions. Furthermore, the theory went beyond the concept of inertial systems, which Galileo and Newton saw as privileged compared to a generic observer. Einstein spent the following years looking for an appropriate mathematical formalism for his theory. He found it in the differential geometry developed in the previous century, mainly by Gauss and Riemann. Einstein's ingenious intuition was to describe the gravitational interaction as a purely geometric effect: The fact that gravity can be eliminated locally is analogous to the well-known fact that, on small scales, the Earth appears flat to us despite being spherical. Einstein identified in the *curvature of spacetime* the fundamental physical quantity to describe gravity. Locally spacetime is flat, and the laws of Special Relativity apply, but globally spacetime is a *dynamic entity*: It can bend, twist, and deform. The source of this distortion is the energy of the system in all its forms (or, more precisely, what is technically called the energy–momentum tensor). Everything is summarized by Einstein's famous equations:

$$R_{\mathrm{ab}} - \frac{1}{2} g_{\mathrm{ab}} R = \frac{8\pi G}{c^4} T_{\mathrm{ab}}$$

The left-hand side of these equations has to do with the curvature of spacetime (the so-called curvature tensor and the metric tensor), while the right-hand side (suitably multiplied by some fundamental constants: the speed of light c and gravitational constant G) has to do with the energy content of the system, including, of course, its kinetic energy. Physicist John Archibald Wheeler (one of the pioneers of General Relativity as a modern field of research) summed up the situation in a very effective way: "*Spacetime tells matter how to move, and matter tells spacetime how to curve*". This sentence underlies both the beauty and the beast of General Relativity. Such elegant interplay makes it extremely challenging to solve Einstein's equations, as the motion of bodies and curvature of spacetime are indissolubly connected. From such subtle exchange, two entities emerge which are solely made of spacetime fabric: black holes and gravitational waves.

Before discussing these concepts in the next sections, let us take a moment to reflect on the fact that Einstein spent ten years of his life pondering seemingly trivial topics such as elevators and free-falling apples until he arrived at one of the highest theoretical constructs ever formulated by the human mind, what physicist Carlo Rovelli defines "the most beautiful of all theories". The definition is apt because the implications of General Relativity are disparate and surprising. The new theory of gravity was, in fact, difficult to accept even for the best scientists of Einstein's time. For this reason, some phenomena predicted by Einstein, and which would have been inexplicable with Newton's theory, the so-called classic tests of General Relativity, were looked at with great interest. The first and perhaps most famous is the deflection of light. If gravity is ultimately motion in curved spacetime, this curvature must influence the motion of every object, including light which, therefore, in the vicinity of large masses (i.e., large energy) must deflect with respect to the rectilinear trajectory expected in flat spacetime. This unthinkable phenomenon was observed by Arthur Eddington during the famous solar eclipse of 1921 and helped to form Einstein's image in public opinion as a global scientific star. Furthermore, Einstein's theory predicts small variations in the Solar System with respect to Keplerian orbits. It was able to brilliantly explain a small discrepancy in the advancement of Mercury's perihelion that has puzzled astronomers for decades. Finally, it predicts the so-called gravitational redshift, namely that the frequency of a wave emitted near a star is higher than that observed at a great distance. This phenomenon, also readily verified experimentally, is linked to the fact that time passes more slowly near large masses and is a fundamental ingredient for the functioning of the GPS, which must consider the influence of the Earth's gravitational field on the signals sent to satellites for geolocation.

2 Hic Sunt Leones

However extraordinary, the effects listed above are small and require precision observations. In the Solar System, Newtonian gravity is a very good approximation because the curvature produced by a star like the Sun is modest, and the spacetime of the solar system is almost flat. However, there are situations in which the curvature of spacetime can be extreme; in that case, General Relativity shows all its spectacular beauty. The first area is cosmology. Before Einstein's theory, the study and evolution of the universe were the sole prerogatives of philosophy and theology. But the Universe as a whole can be described as spacetime that evolves in relation to (visible and invisible) matter and to the radiation contained in it, starting from an initial configuration of very high density and spacetime curvature (the so-called Big Bang). Furthermore, the current accelerated expansion of the universe (the observation of which was awarded the Nobel Prize in Physics to Saul Perlmutter, Brian Schmidt, and Adam Riess in 1999) is explained by General Relativity through a term linked to the infamous cosmological constant. Initially introduced by Einstein, then denied by himself as one of his greatest mistakes, this constant is currently a fundamental ingredient of the cosmological standard model which, despite the enormous observational successes, is today at the center of an important debate precisely linked to velocity expansion of the universe and the origin of the cosmological constant.

2.1 Black Holes: The Unbearable Lightness of Spacetime

Another context in which General Relativity predicts incredible effects is compact stars. If it is true that the Sun distorts spacetime just a little, what if all its mass were compressed into a much smaller and denser star? In that case, the curvature of spacetime would be much greater, and phenomena such as the deflection of light and gravitational time dilation would be greatly amplified. For example, the light might be deflected to such an extent that it becomes trapped in the star's gravitational field. Such a star could not emit any kind of radiation, and it would appear as a *black hole*[1] in the sky. Furthermore, since the speed of light is the maximum possible, anything that enters this black hole (beyond that region of no return known as the event horizon) would be swallowed up forever and could no longer communicate with the external world. In other words, the interior of a black hole is beyond our "horizon", we can never be affected by anything happening inside a black hole.

The existence of black holes is another great theoretical prediction of General Relativity. Regarded for decades as mere mathematical curiosities, we now know that black holes are everywhere in our universe. Virtually every galaxy (including our Milky Way) is thought to host a black hole of millions to billions of solar masses with which it evolves in symbiosis. Around these cosmic monsters, matter and light are strongly deflected, and time is strongly distorted.

For example, the complex stellar orbits around the supermassive black hole at the center of our galaxy allow us to indirectly reveal its presence and estimate its mass (about 4 million solar masses), and their observations were awarded the Nobel Prize in Physics to Reinhard Genzel and Andrea Ghez in 2020.

Supermassive black holes energize the luminous centers of most galaxies, where they convert the gravitational potential energy of in-falling matter to radiant power and jets of charged particles that stretch for tens of thousands of light years. Despite their commanding presence, black holes remain "unseeable" as they consume the light that could illuminate them and remain hidden behind a superheated haze of stellar matter. It is possible to deduce their presence in the core of a galaxy by measuring the velocities of stars that orbit the black hole. More massive black holes will accelerate nearby stars to greater speeds, so the velocities of stars can reveal not only the presence of a black hole but also its mass.

In addition, many black holes are encircled by disks of superhot gas called *accretion disks*. In the case of a star-sized black hole, the gas usually comes from a nearby companion star; the black hole's powerful gravity pulls gas off the star's surface. In

[1] The name was allegedly coined by John Archibald Wheeler, although he probably heard it from an unknown participant in a conference in 1967.

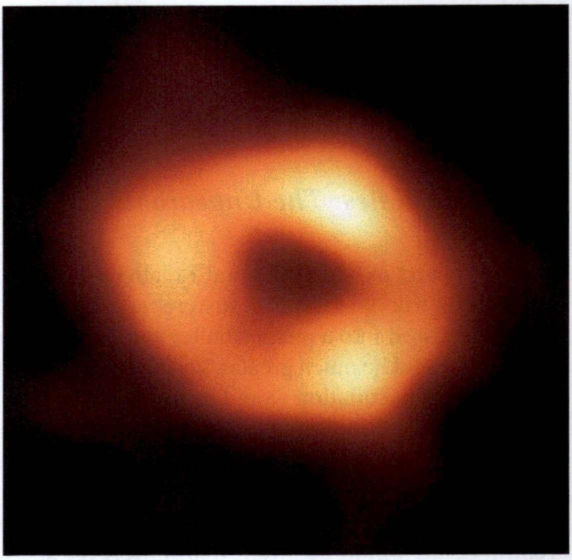

Fig. 1 Image of the Black Hole SgrA* at the hearth of the Milky Way galaxy. *Credit* Event Horizon Telescope (EHT) collaboration

the case of a supermassive black hole, the source is large clouds of gas in the crowded core of a galaxy or stars that pass close to the black hole and are torn apart by its gravitational pull. As the gas spirals into the black hole, it forms a wide, flat accretion disk. The gas moves faster and faster as it spirals closer, so it is heated to millions of degrees. The glowing gas reveals a tell-tale signature: a dark central region, called a *shadow*, surrounded by a bright ring-like structure. The gas radiates bright light at many frequencies, including radio waves which can be detected by radio telescopes.

This phenomenon is the basis of the famous images of the supermassive black hole at the center of the Milky Way (Sagittarius A*) and that in the giant elliptical galaxy Messier 87, "taken" by the Event Horizon Telescope in 2022 and 2019, respectively (Fig. 1). The Event Horizon Telescope captured the images of the shadows of these monsters using a technique called very long baseline interferometry (VLBI), which combines radio observatories on multiple continents to form a virtual Earth-size telescope, an instrument with the highest resolution in all of astronomy. These images opened a new window into intensive investigations of spacetime within such an extreme environment.

The black hole concept was introduced only in the 1960s, but already in the late eighteen century, "dark stars" were explored using Newton's theory. Indeed, every high-school student can repeat Michell's 1783 original argument and compute the escape velocity of an object at the surface R of a star with mass M. This velocity is

$$v = \sqrt{\frac{2GM}{R}}$$

Because light moves at a speed $c \approx 300000$ km/s, the condition that not even light escapes from the surface of a "dark star" imposes a limit on the star's radius, which must be smaller than

$$R_S = \frac{2GM}{c^2}$$

Curiously, this is precisely the radius of a black hole in Einstein's theory, known as Schwarzschild[2] radius!

Thus, spacetime can be deformed so much that not even light can escape. But what can cause such enormous deformation? The Schwarzschild radius gives us the answer: our Earth (whose mass is roughly 6×10^{24} kg, should be compressed in one centimeter to become a black hole! It seems, therefore, impossible that such compact objects can ever be formed. However, gravity has a unique feature that makes it possible—and, in fact, likely—to squeeze matter in such small regions. Namely, gravitational interactions are always *attractive*. Starting from the pioneering works of Subrahmanyan Chandrasekhar, we know that black holes form as the final stage of the evolution of very massive stars, which collapse under the action of their gravitational attraction. These black holes have a mass that varies from a few to tens of times the solar mass, even if recent gravitational-wave observations (see next section) are providing us with some surprises.

When matter starts accumulating toward the center of a big star, it becomes more compressed as more matter is attracted. Sometimes (as in the case of the Sun), the pressure at the surface is sufficient to balance the effect of gravity, and the object remains in equilibrium. However, in some other situations, gravity can be so strong that no material—no matter how solid and incompressible—can sustain its gravitational weight. The situation is like a house of cards: If a critical mass is reached, the house cannot sustain its weight and collapses. In stars, this process is called gravitational collapse, and in many cases, the leftover (after a huge amount of energy is released in what is called a "supernova explosion") is precisely a black hole.

Black holes show very special and counterintuitive features. Imagine a far-away observer watching a spacecraft falling into a black hole. According to Einstein's gravity, the observer will see the spacecraft getting closer and closer to the black hole's event horizon but, in fact, never crossing it, like in the famous Zeno's paradox, in which Achilles never reaches the tortoise. At the same time, the light emitted by the spacecraft gets infinitely redshifted as it approaches the horizon, so the signal that reaches the far-away observer will become fainter and fainter. In practice, the spacecraft will appear to the far-away observer as if it "dissolves" while approaching the horizon, until it disappears forever. If this sounds crazy, imagine that nothing special happens for the crew inside the spacecraft as it crosses the event horizon. However, once inside the black hole, not only they cannot escape outside, but time

[2] Karl Schwarzschild's discovery has a fascinating and sad history. Schwarzschild found his famous solution while serving in the German army during World War I, only a few months after Einstein had published his theory. Unfortunately, Schwarzschild died the year following his discovery from a disease he developed while at the Russian front.

(as measured by the crew) freezes, and the spacecraft is forced toward the center of the black hole, where a *singularity* is lurking. A singularity is a region of spacetime in which the curvature becomes so large that Einstein's theory and physics laws break down. When approaching the singularity, tidal forces become so large that they rip the spacecraft apart, stretching it (and its crew!) along the radial direction in a process known as "spaghettification" (fortunately, the external world would not be able to watch this gravitational carnage!).

Some fundamental theorems proved by Stephen Hawking and Roger Penrose show that a singularity must necessarily exist inside a black hole (Penrose shares the Nobel for Physics 2020 precisely for his theoretical studies on the formation of black holes and singularities). What happens near the singularity is hidden by the event horizon, and it is, therefore, impossible to observe it from outside the black hole. However, as discussed in another chapter of this book, we expect that when the curvature near the singularity reaches the Planck scale, the quantum effects of gravity become crucial and drastically modify General Relativity. The search for the quantum theory of gravity, which should solve the paradox of singularities, is the holy grail of current theoretical physics. Undoubtedly, black holes will play a decisive role in understanding the elusive quantum properties of gravitational interaction. As demonstrated by Hawking, quantum mechanics drastically modifies the event horizon, causing a black hole to emit a feeble radiation named after the great British physicist. The temperature of Hawking radiation is described by this beautiful equation

$$T_H = \frac{hc^3}{16\pi^2 G k_B M}$$

which contains several of the main fundamental constants of nature: Newton's constant G, Planck's constant h, the speed of light c, and Boltzmann's constant k_B, thus suggesting that Hawking emission is a bridge connecting gravity, quantum theory, relativity, and thermodynamics. Due to Hawking radiation, black holes slowly evaporate until they disappear. Where the information originally contained inside the black hole ends up is another big paradox that quantum gravity should resolve.

2.2 *Gravitational Waves: A New Cosmic Messenger*

How can we "observe" an object like a black hole that, by definition, swallows everything, even light? Apart from the study of the orbits around a black hole (as in the case of Genzel and Ghez) or of the hot matter that falls into it (as in the case of the radio images taken by the Event Horizon Telescope), the possibility of "seeing" a black hole is provided by another amazing prediction of Einstein's theory: gravitational waves. Indeed, although black holes do not shine, they can talk very loudly. While any other object can be seen by the electromagnetic radiation it emits, black holes can be detected by the gravitational radiation they produce.

Gravitational waves are ripples in spacetime that are generated by large moving masses and propagate at the speed of light, like how a motorboat generates waves on a calm lake or how moving electric charges generate electromagnetic waves. Any moving mass generates gravitational waves, but these are typically too weak to observe. The power released by two colliding black holes, instead, is enormous.[3] Because they interact very feebly with the matter, the gravitational waves generated by these cosmic collisions propagate undisturbed in the universe for billions of light years until they reach us on Earth. However, because of the large distance of the most powerful gravitational wave sources and because of the smallness of the spacetime curvature produced even by such energetic events, by the time they reach us the amplitude of the gravitational waves is tiny: Its relative effect is typically about 10^{-21}, that is, a rod as long as the Sun–Earth distance would shrink by only one atomic size during the passage of such a weak gravitational wave!

This explains why the decade-long quest for the detection of gravitational waves has been so challenging. Impossible as it might seem, some instruments can achieve such unbelievable sensitivity. These detectors are called *gravitational-wave interferometers* because they use the principles of interferometry to measure tiny differences in the light propagation induced by the passage of a gravitational wave. The huge experimental and technological effort that led to the development of gravitational-wave interferometers has paid off: as discussed elsewhere in this book, on September 14, 2015, the LIGO-Virgo Collaboration detected for the first time the gravitational waves produced by the merger of two black holes. This historical event (dubbed GW150914) marked the beginning of a new astronomy that is currently in full blossom. In the last 7 years, the LIGO and Virgo laser interferometers for the detection of gravitational waves have discovered almost 100 signals produced by the collisions of black holes and neutron stars. At least for one of them (the merger of two neutron stars, named GW170817), it was possible to observe a series of coincident electromagnetic signals, thus giving rise to *multimessenger* astronomy, that is, the possibility of observing an astrophysical object with different "messengers". These observations require highly advanced instruments and perfect theoretical knowledge of the expected signal, which can be obtained with a sophisticated combination of numerical simulations and theoretical calculations, both based on Einstein's equations. The detection of gravitational waves is thus a further astonishing confirmation of General Relativity and was awarded the Nobel Prize in Physics to Rainer Weiss, Barry Barish, and Kip Thorne in 2017.

At the time of writing, the U.S. interferometer (LIGO) and its European counterpart (Virgo, operated by the European Science Observatory in Italy) are terminating a recent stage of improvements to start their fourth observational run in Spring 2023, this time also together with the new detector KAGRA operating underground in Japan. The enhanced sensitivity of these detectors would allow them to detect binary inspirals almost daily.

The huge potential of gravitational-wave astronomy is still far from being fully explored. While current interferometers will keep searching for gravitational waves

---

[3] It is about 10^{52} W. For comparison, the Sun luminosity is "only" about 10^{26} W!

in the next few years, the scientific community is working hard to develop the next generation of gravitational-wave detectors. In particular, the Einstein Telescope is a European project for an underground gravitational-wave interferometer, whose size is more than double the size of the LIGO detector and that is planned to be over ten times more sensitive, thanks to the optimal conditions of its underground location and the use of novel technologies such as cryogenic facilities. The Einstein Telescope will be a unique discovery machine: Recent studies have shown that it will detect more than 10 thousand mergers per year and will detect black-hole mergers at very high redshift when the universe was still in its infancy. Furthermore, its low-frequency sensitivity (in the 1–10 Hz range) will allow the observation of the coalescence of intermediate-mass black-hole binaries (black holes having masses of 10^2–10^4 solar masses), unveiling a missing tile of the Universe puzzle.

Gravitational-wave science is moving forward not only on (and under) ground but also in space. In recent years there has been a flurry of activity related to the Laser Interferometer Space Antenna (LISA), the first-ever gravitational-wave detector in space, expected to be launched by ESA and NASA around 2035. LISA will be sensitive to low-frequency (millihertz) gravitational waves as those emitted by supermassive black holes, like the one at the center of our galaxy. Finally, complementary to laser interferometers, the prospect of detecting gravitational waves using the signal from various pulsars as a huge interferometer on a cosmic scale (so-called pulsar timing arrays) has been recently explored, and a gravitational-wave signal may be already lurking in data astronomers are currently analyzing.

This global effort is motivated by the huge scientific potential of gravitational-wave astronomy and by its disruptive impact on astrophysics, cosmology, and even fundamental physics. First, gravitational waves allow studying events (like black-hole mergers) that would otherwise be invisible to us. This helps understand much better the conditions in which black holes form and the astrophysical environment where they live. For example, the LIGO-Virgo-KAGRA Collaboration has observed black holes in certain mass ranges (as light as a few solar masses and as heavy as one hundred solar masses), which challenge the standard astrophysical formation scenarios, forcing astronomers to improve their models or maybe signaling new and exotic formation channels. The sensitivity of the detectors limits current investigations, but the future Einstein Telescope will allow us to answer these open questions.

Furthermore, precisely because gravitational waves interact so feebly with matter, they can travel the entire Universe without being perturbed or quenched. This is very important for astrophysics because the signal they carry is uncontaminated. For example, it does not suffer from obstruction typical to electromagnetic signals and can carry direct information from the interior of neutron stars, where the density of matter is higher than that of a nucleus and physical processes are not well understood. Even a single detection of the binary neutron star merger GW170817 has constrained the behavior of nuclear matter to an unprecedented level, and its coincident electromagnetic signal has shed light on heavy-element nucleosynthesis and the origin of short gamma-ray bursts. We can only wonder about the wealth of information that will come with the future Einstein Telescope, which will detect more than 10,000 neutron star mergers per year (about 1 signal every 5 min!).

Not being attenuated during cosmic travel, gravitational waves can also carry invaluable information about the early universe. For example, future detectors such as the Einstein Telescope and LISA will shed light on how the first stars formed and can help us explore the existence of novel families of black holes of primordial origin. They can also be used to measure the rate of acceleration of our universe, shedding light on the dark energy and the cosmological constant, which are believed to be responsible for such acceleration. Finally, they can detect the imprint of cosmic inflation right after the Big Bang.

Finally, by detecting gravitational waves, we are testing Einstein's theory of gravity in its most extreme regimes. The events detected in the last few years have already placed some of the most stringent constraints on theories beyond General Relativity, by confirming that gravitational waves have two polarizations propagating at the speed of light, like electromagnetic waves. Furthermore, they have confirmed one of the cornerstones of General Relativity, namely that all black holes in the universe—regardless of how they have formed—are uniquely defined by only two quantities: their mass and angular momentum.

3 New Horizons

The observations of black holes and gravitational waves are opening hitherto unexplored boundaries of the universe and are reshaping entire areas of astrophysics, cosmology, and fundamental physics. Future experiments such as the third-generation interferometer Einstein Telescope, the LISA space mission, and accurate observations with radio pulsars will make it possible to capture gravitational waves in frequency bands never observed before. It is a revolution that equals the discovery of radio waves or X-rays and whose consequences will amaze us for decades to come. In parallel, the next generation of the Event Horizon Telescope will be able to resolve the region near the event horizon of the supermassive black holes in our Milky Way and in other galaxies with unprecedented resolution, testing once more Einstein's theory at the extremes.

In addition to reaching a deeper understanding of the universe, the hope is that these observations will lead to the discovery of unexpected new signals, perhaps shedding light on some open questions in astrophysics and cosmology, for example, the fate of black holes and the nature of dark matter and dark energy.

Christopher Nolan's movie *Interstellar* was about a future human civilization able to undertake cosmic travels to black holes using "wormholes", which are special shortcuts in spacetime also predicted by General Relativity (but never observed!). Published in 2014, Interstellar serendipitously anticipated the gravitational-wave and black-hole revolutions that the scientific community has witnessed in the last few years. Science-fiction as it might have seemed, Interstellar screen-players worked side by side with Kip Thorne to ensure that the movie did not contain scenes that would make Albert Einstein cringe. Some of the spectacular effects that appear in the movie went beyond science fiction and have been observed in real experiments! Does this mean that travel agencies are about to sell (roundtrip?) tickets to a black hole? Not quite, but in the next few years, physicists and astronomers will be able to study them as never before.

Other Sources of Information

Book: The evolution of physics, by A. Einstein and L. Infeld available also as paperback or Kindle
Book: The Irresistible Attraction of Gravity: A Journey to Discover Black Holes, by L. Rezzolla, Publisher Cambridge University Press
Book: The Science of Interstellar, by K. Thorne. Publisher W.W. Northon & Company
LIGO educational resource: a large database with information, games, and the latest news on gravitational waves
https://eventhorizontelescope.org
https://eso.org/public
https://supernova.eso.org
https://webbtelescope.org
www.asimetrie.it (in Italian): the INFN outreach magazine has published an entire issue on "Gravity" and various articles on black holes and gravitational waves
Youtube: "How to understand the black hole images"
Youtube: Gravitational waves explained by PhD Comics
Youtube (in Italian): INFN "Dialoghi sui massimi sistemi" (interview with the authors of this chapter)
Youtube (in Italian): Viaggio ai confini dello spaziotempo

The Dark Universe

Something New Exists But We Do Not Know Its Nature

Riccardo Murgia, Walter M. Bonivento, and Cristiano Galbiati

Abstract Cosmic Microwave Background (CMB) observations, interpreted within the context of General Relativity, indicate that visible matter, i.e., stars and interstellar gasses, makes up $< 5\%$ of the total energy density of the Universe today. If Einstein's theory of gravity holds up to the largest observable distances, the remaining 95% of the current energy budget of the Universe is made of two distinct unknown entities, a quite elusive form of *Dark Matter* ($\sim 24\%$), and a very exotic form of *Dark Energy* ($\sim 71\%$). The presence of Dark Matter is necessary to explain the birth, growth, and dynamics of all the cosmic structures existing today, such as galaxies and galaxy clusters. The existence of Dark Energy is required to explain the observed accelerated expansion of the Universe. Whereas proofs of the existence of these dark entities are considered extremely solid from astronomical observations, their very nature remains mysterious. This fact poses profound issues in physics, as well as in the philosophy of science, related to the problem of unobserved entities. In this chapter, the authors develop an informal dialogue on some of the most advanced topics in Cosmology and Astroparticle Physics. The conversation is between two fictional characters: a *Boomer* (B) and a *Kid* (K).

Keywords Dark matter · Dark energy · Cosmology · Astroparticle physics · General relativity · Gravitation · Cosmic microwave background · CMB · Universe

R. Murgia (✉) · C. Galbiati
Astroparticle Physics Department, Gran Sasso Science Institute, 67100 L'Aquila, AQ, Italy
e-mail: riccardo.murgia@gssi.it

C. Galbiati
e-mail: galbiati@princeton.edu; cristian.galbiati@gssi.it

W. M. Bonivento
Istituto Nazionale di Fisica Nazionale, Sezione di Cagliari, Complesso Universitario di Monserrato, S.P. per Sestu – Km 0,700, 09042 Monserrato, CA, Italy
e-mail: walter.bonivento@ca.infn.it

C. Galbiati
Physics Department, Princeton University, Princeton, NJ 08544, USA

M. Streit-Bianchi et al. (eds.), *New Challenges and Opportunities in Physics Education*, Challenges in Physics Education, https://doi.org/10.1007/978-3-031-37387-9_7

1 Introduction

Humankind is the result of about 13.7 billion years of cosmic evolution. That is the age of the Universe where we belong. We live on Earth, one of the eight planets orbiting around our closest star—the Sun—roughly 150 million km from us. It takes light about 8 min to travel such a distance, and less than a day to cross the entire Solar System.

We call a light-year the distance traveled by light in one year. Proxima Centauri, the nearest star outside of our Solar System, is 4.2 light-years away from us.

We all are part of a Galaxy—the Milky Way—containing about 100 billion stars spread over a disc with a radius of ~ 100,000 light-years.

Given the finiteness of both the age of the Universe and the speed of light, the portion of the Universe that we can observe is finite as well. It is determined by the distance traveled by light in a 13.7-billion-year-old Universe that is expanding like an inflating balloon. The so-called observable Universe has a size of about 46 billion light-years, and it contains roughly 100 billion galaxies. The Andromeda Galaxy, the nearest to us, is about 2 million light-years away. On even larger scales, galaxies are not randomly displaced. They form the so-called cosmic web: a filamentary structure interconnecting over dense regions—clusters and superclusters of galaxies—separated by giant cosmic voids (https://www.illustris-project.org/, https://www mpa.mpa-garching.mpg.de/galform/virgo/millennium/, https://www.nottingham.ac.uk/astronomy/sherwood/media.php).

Since the Universe is expanding (Hubble 1929), at very early times it was much hotter and denser: an opaque cosmic plasma, where extremely high interaction rates keep all components in thermal equilibrium. Different particle species fall out of thermal equilibrium when their interaction rate drops below the expansion rate of the Universe. At that instant, the given species decouples from the rest of the plasma: the particles stop interacting and their relic abundance freezes. About 370,000 years after the beginning of our 13.7-billion-year journey, it is time for the light to decouple. The Universe becomes transparent to light, which starts propagating freely everywhere. Having progressively decreased its energy due to the Universe's expansion, this primordial relic is now observed as a Cosmic Microwave Background (CMB) radiation component (Penzias and Wilson 1965) (Fig. 1). More than any other currently available cosmological probe, the CMB provides us with an extraordinary amount of information about the babyhood of our Universe (Akrami et al. 2020). It is a bit like a cosmic "selfie" taken just 370,000 years after the Big Bang. By studying patterns and correlations in the CMB as well as in several other observational probes, the goal of Cosmology is to reconstruct a complete and consistent history of our Universe (Ryden 2016; Baumann 2022). One major thing that we know is that most of the leading characters in such a long and fascinating history are unknown.

We live in a Universe largely dominated by two invisible species: Dark Matter (Zwicky 1933) and Dark Energy (Zel'dovich 2020). The existence of Dark Matter (de Swart et al. 2017; Bertone and Hooper 2018) is required to explain the birth, growth, and dynamics of all the cosmological structures existing today, such as galaxies (Rubin and Ford 1970) and galaxy clusters (Clowe 2006). Dark Energy is required to explain the observed accelerated expansion of the Universe (Perlmutter et al. 1999; Riess et al. 1998). Normal, visible matter—all the stuff that we know—accounts for less than 5% of the total matter and energy content of the Universe. The rest is dark (https://play.google.com/store/apps/details?id=com.thecerealkiller.darkmaster&hl=it&gl=US; https://apps.apple.com/it/app/dark-master/id1489301577).

The rest of the chapter is developed as an informal dialogue on the advanced topics in Cosmology and Astroparticle Physics that we have just briefly sketched. The conversation is between two fictional characters: a *Boomer* (B) and a *Kid* (K).

2 More Than 95% of the Current Content of the Universe is Invisible to Us

B: What we call light is electromagnetic radiation. What we call visible light is made by electromagnetic waves in the very narrow range of wavelengths that human eyes are sensitive to.

K: What if we were butterflies? They can see ultraviolet radiation, a range of wavelengths invisible to humans! Is more than 5% of the universe visible to butterflies?

B: Ultraviolet light, as well as all its other wavelength bands, are not invisible to us as humans either, since we have got telescopes. Astronomers nowadays make use of several different instruments, both on Earth and in space, to take incredibly accurate pictures of the sky. It does indeed look very different when observed in different wavelengths. Astronomers today can explore and combine all the wavelengths of the electromagnetic spectrum, from radio waves to gamma rays. The results are stunning, as anyone can appreciate thanks to several open-source softwares (https://worldwidetelescope.org). Our universe is incredibly dynamic and colorful. Billions of galaxies, each one with billions of stars being born, living, and dying through majestic astrophysical phenomena that absorb and emit light of all wavelengths. Stars and galaxies, as well as us humans and everything else we know, are all made of visible matter. Matter that can emit and absorb light, electromagnetic waves.

K: And that is pretty much all that any human, butterfly, or telescope could ever see?

B: Sure if you mean directly. Via electromagnetic interaction. That is why all the astonishing variety of visible matter that we know, from the hydrogen nuclei burning in the sun to the oxygen molecules that you are breathing right now… is pretty much all that we know.

K: What else do we know?

B: We also know that even if all kinds of visible matter were packed together, they would still represent < 5% of the total content of our universe. The remaining 95% is unknown.

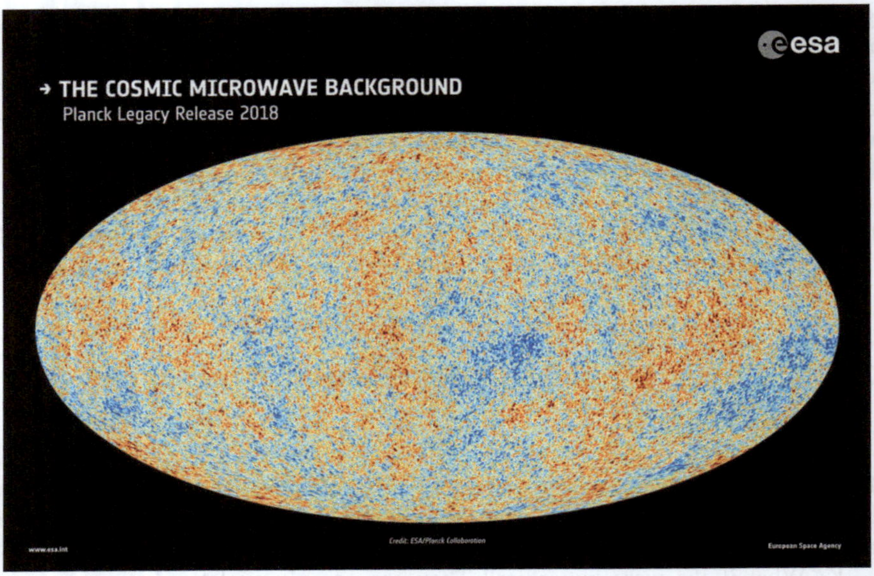

Fig. 1 *Planck*'s view of the cosmic microwave background. *Credits* ESA/Planck collaboration

3 How Can We Know that Something Unknown Exists?

B: Visible matter is made of particles that can interact through electromagnetic interaction, which is one of the four fundamental interactions of nature. Each particle has a mass. The mass of an astrophysical object is given by the sum of the masses of the particles composing it. Massive objects interact with each other through another fundamental interaction: gravitational interaction.

K: The gravitational force?

B: That is what Isaac Newton would have called it, more than two centuries ago. Nowadays, thanks to Albert Einstein's theory of general relativity, we interpret gravity as an intrinsic property of each point of the Universe, rather than as a force between objects. Our Universe is described as a 4-dimensional system. The so-called spacetime, where any event is fully characterized by 4 coordinates: the 3 spatial ones, plus one identifying the time when each event occurs. The curvature of spacetime reflects the mass distribution. An empty universe would mean a flat spacetime: zero curvature everywhere. The presence of masses—inhomogeneously localized—curves the spacetime region in their vicinity.

K: Wait, what?

B: You can visualize it (https://www.youtube.com/watch?v=wrwgIjBUYVc). Imagine an elastic carpet, anchored on its sides. What happens if you place a big lead ball on top of it?

K: The mass of the lead ball curves the portion of carpet around it…

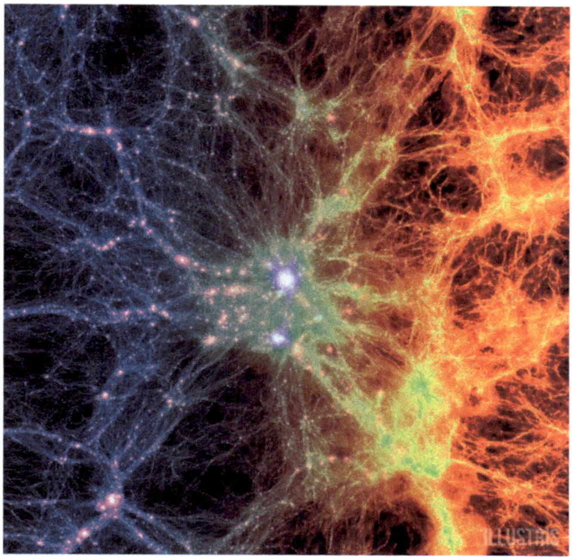

Fig. 2 The cosmic web: dark matter density (*left*) transitioning to ordinary matter density in form of gas (*right*). *Credits* Illustris simulation project

B: Analogously, the mass of a planet like Earth induces a non-zero curvature in the region of spacetime around it. Being about 300,000 times more massive, a star like our sun induces an even larger non-zero curvature, and so on. The intensity and direction of the gravitational attraction between different astrophysical objects are governed by the mass distribution of the objects themselves (Fig. 2) (Wheeler 1998).

4 "Spacetime Tells Matter How to Move; Matter Tells Spacetime How to Curve"

K: Cool. How does that affect my imaginary lead ball and elastic carpet?

B: What happens if you drop on the same carpet a less massive lead ball?

K: It still curves the carpet but to a lesser extent.

B: Lower mass implies lower gravitational attraction. What if you now throw a marble toward the largest lead ball?

K: It orbits around the lead ball, following the curvature of the carpet…

B: Pretty much as the moon orbiting around Earth, or the Earth orbiting around the Sun. In all cases, there is no attractive force among them. Planets, stars…

K: and imaginary marbles…

B: …simply follow the local curvature of spacetime/carpet. Technically, they move along the so-called geodesics: curves representing the shortest path between two points.

K: Straight lines?

B: That is true in a flat space. It is not always so on your imaginary elastic carpet, as well as in our 4-dimensional Universe (https://www.youtube.com/watch?v=NfqrCdAjiks), where even light is forced to move along geodesics, as it is deflected by the presence of very massive structures, such as galaxy clusters. The importance of this effect—called gravitational lensing (https://esahubble.org/science/gravitational_lensing/)—depends on the total mass of the cluster playing the role of the lens. So that gravitational lensing can be used to "weigh" clusters.

K: What does it mean?

B: As light travels through the Universe, it gets "lensed" by the intervening distribution of matter. Measuring the amount of light deflection allows us to "weigh" the mass of the gravitational lens. That is one of the reasons why we know that most of the mass in the Universe… is unknown.

K: What about the other reasons?

B: If we rely on General Relativity theory as the exact theory of gravity on cosmological scales, almost 90% of the total mass in the Universe must be dark. Some exotic form of matter that we cannot see—it neither absorbs nor emits light—yet its existence is required to explain cosmic structure growth and stability.

K: Structures such as…?

B: Galaxies, as well as galaxy clusters. The existence of large Dark Matter halos around galaxies is required to explain the high rotation speed of gas in their outer regions. Also, the amount of visible matter is far from enough to keep different galaxies gravitationally bound together in a cluster. The presence of a large amount of Dark Matter is required to curve spacetime enough to hold the cluster together (Mo et al. 2010).

K: Since what we "weigh" is almost 10 times larger than what we see, are we sure that we are "weighing" correctly the ordinary matter? Are we sure that using General Relativity is the right way to go?

B: We are quite sure that now there is no alternative gravity theory providing a more consistent explanation for our observations on all testable scales (Fig. 3).

5 Does Einstein's Theory Hold Everywhere, Under Any Condition?

B: The theory of General Relativity has brilliantly passed all its experimental tests for more than a century. However, the largest distances involved in all these experiments are of the order of the solar system.

K: So, what happens when we apply the same theory to study the evolution of the universe as a whole, i.e., trillion times larger distances?

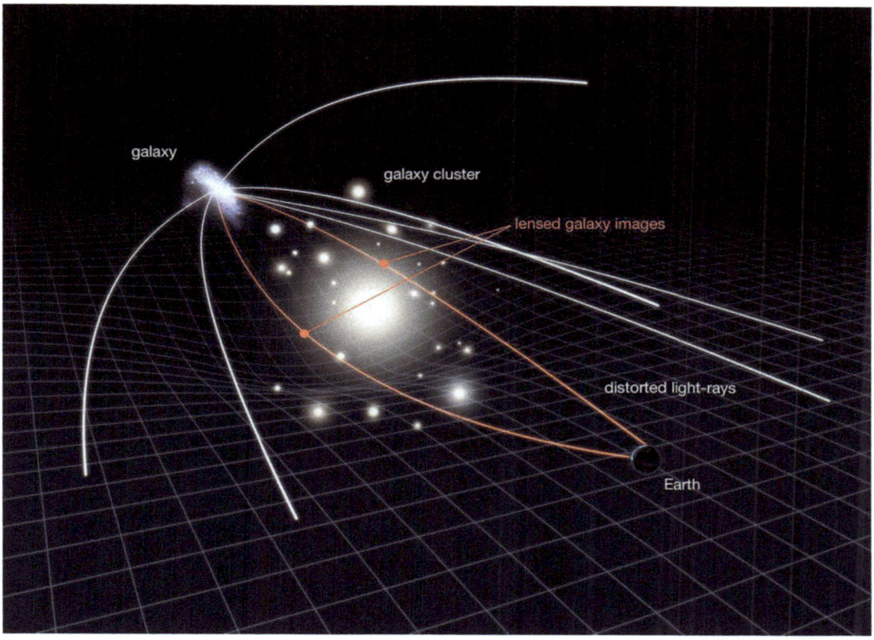

Fig. 3 Gravitational lensing in action. *Credits* NASA, ESA and L. Calçada

B: Besides Dark Matter, an even more exotic species must be introduced. To interpret the current accelerated expansion of the Universe, as observed by measurements of type I-a Supernovae and other astrophysical tracers. A new form of invisible energy which exerts some sort of repulsive effect that balances the gravitational attraction of (both visible and dark) matter, pushing our Universe to expand faster and faster.

K: Some sort of dark… energy?

B: Exactly. Dark Energy is the catch-all nickname for the origin of cosmic accelerated expansion, regardless of whether due to an unknown form of energy or a modification of General Relativity. Within General Relativity, one can view Dark Energy as a cosmological constant, associated with the constant energy density of the vacuum. However, cosmic acceleration could arise even from a more complex species, as an exotic fluid with negative pressure, mimicking to some extent a cosmological constant behavior. One of the candidate explanations is a new quantum scalar field, named *quintessence*, whose effect could be made manifest by observing violations of Einstein's equivalence principle and variation of the fundamental constants in space or time, never observed so far.

6 Can We Observe What We Cannot See?

B: There are several ongoing and proposed experiments to characterize in depth Dark Energy, either based on observations of galaxies in the sky, such as the Dark Energy Survey (DES) (https://www.darkenergysurvey.org/), which uses a 570-megapixel camera to map one-eighth of the sky aiming to measure the expansion of the universe using supernovae, weak gravitational lensing, and other observables; and the Euclid space telescope (https://sci.esa.int/web/euclid), set to launch in 2023, which will study the large-scale structure of the universe and probe the nature of Dark Energy through weak lensing and galaxy clustering. The CMB also provides very important information on Dark Energy. Indeed, the measurements of CMB anisotropies have indicated that visible and dark matter together make up only ~ 29% of the total energy density of the Universe today (https://www.esa.int/Enabling_Support/Operations/Planck). Since the overall geometry of the Universe appears close to flat, its total energy density must be close to the so-called critical density. Dark Energy must thereby account for the remaining ~ 71%. Next-generation gravitational wave observatories, such as the Einstein Telescope (ET) (https://www.et-gw.eu/) and the Laser Interferometer Space Antenna (LISA) (https://lisa.nasa.gov/), will also significantly contribute to shedding light on the nature of Dark Energy. They will enable us to test Einstein's theory with unprecedented accuracy.

K: What about Dark Matter?

B: There are several popular Dark Matter particle candidates, not included in the Standard Model of particle physics, the theory which includes all particles and interactions known today in a coherent framework, each with their unique properties and characteristics (Bertone and Tait 2018). Some of the most widely studied theoretical models include Weakly Interacting Massive Particles (WIMPs), ultra-light scalar particles such as axions, and sterile neutrinos.

K: Well, If Dark Matter is made up of invisible particles, how should we be able to observe them?

B: WIMPs are particles that interact only very weakly with normal matter, only through the third fundamental interaction of nature, i.e., the so-called weak interaction. This fact makes them quite difficult to detect directly. WIMPs (Roszkowski 2004) are one of the most promising candidates for Dark Matter as they could account for the observed Dark Matter density in the universe. WIMPs might be Supersymmetric particles (Baer et al. 2015), a theory that assumes the existence of a partner particle, with the spin of the partners differing by half a unit, for each particle of the Standard Model; in particular, the lightest neutralino may serve as a suitable candidate for Dark Matter. Axions (Weinberg 1978) are light particles that were first proposed to solve the so-called strong CP problem in quantum chromodynamics, i.e., the predicted yet unobserved violation of the charge-conjugation (C) and parity (P) symmetries in experiments involving only the strong interaction—the fourth fundamental interaction of nature (Peccei and Quinn 1977). Axions are a strong candidate for Dark Matter as they can be produced in large quantities in the early universe, and they have properties that are consistent with the observed Dark Matter density.

Sterile massive right-handed neutrinos (Giunti and Kim 2007; Lesgourgues, et al. 2013) are a type of neutrino that do not interact via weak interaction. They are a Dark Matter candidate because they are feebly interacting and can be produced in the early universe. Searching for the existence of these new phenomena is one of the main goals of Astroparticle physics. Once some new particle is detected by, e.g., a laboratory experiment, it will then be crucial to confront the corresponding cosmological implications with complementary astrophysical data. Hopefully proving that the existence of such a Dark Matter particle can solve the longstanding cosmic puzzle we have been chatting about. Or a part of it, at least.

K: What are the current experiments looking for WIMPs?

B: Experimental efforts to detect WIMPs include: the search for products of WIMP annihilation, i.e., gamma rays, neutrinos, and cosmic rays in nearby galaxies and galaxy clusters; direct detection experiments designed to detect the collision of WIMPs with nuclei in the laboratory, and attempts to directly produce WIMPs in colliders, such at the highest energy proton-proton collisions at the Large Hadron Collider (LHC) (https://home.cern/). Direct detection experiments are usually located deep underground in laboratories such as LNGS (https://www.lngs.infn.it/en) in Italy or SNOLAB (https://www.snolab.ca/) in Canada, to shield them from cosmic rays. The most used target nuclei are argon, xenon, germanium, silicon, calcium tungstate, and sodium iodide. Nuclear recoils are detected from their energy transfer to the media, either to nearby electrons, causing scintillation or ionization, or to the bulk material as heat, such as with bolometers. In some cases, more than one method is used.

K: What are the current experiments looking for axions?

B: Several different types of experiments have been proposed or are currently being conducted to search for axions (Choi et al. 2021). Some of them aim at detecting the axions supposedly present in the Milky Way Dark Matter halo, e.g., the so-called haloscope experiments, which use a resonant cavity in the presence of a strong magnetic field to convert axions into microwave photons, based on the inverse Primakoff effect, that can be detected by a sensitive receiver, such as the ADMX experiment (ADMX Collaboration 2021). These are high-technology experiments using state-of-the-art instrumentation, such as an 8 T superconducting solenoid magnet and sensitive photon detectors cooled at 4 K. Other experiments looking for axions, independently of their Dark Matter nature, are light-shining-through-a-wall experiments, like QSQAR (OSQAR Collaboration 2015), which use a strong magnetic field to convert axions into photons that can pass through a wall and be detected on the other side; and helioscope experiments aiming at detecting axions produced in the sun, such as the pioneering CAST experiment that started data taking at CERN in 2003 (CAST Collaboration 2017) and the proposed International AXion Observatory IAXO (IAXO Collaboration 2019). Of high relevance are also astrophysical observations looking for the effects of axions on astronomical phenomena such as the cooling of neutron stars (Sedrakian 2016).

K: What are the current experiments looking for sterile neutrinos?

B: Sterile neutrinos may be detected through their decay products. The main decay mode to search for sterile neutrino dark matter is the radiative one, i.e., the decay into a photon and a light neutrino. The radiative decay mode implies that, despite its small rate (much smaller than the inverse age of the Universe), sterile neutrino Dark Matter leads to potentially observable, nearly monoenergetic O(keV) X-ray emission in regions of high Dark Matter density (Galactic Centre, galaxy clusters, etc.).

7 Can We Hear What We Cannot See?

K: Is it possible that Dark Matter is made of many Black Holes, given that they are invisible?

B: A Black Hole is a region of the Universe where the gravitational attraction is so strong that nothing, not even light, can escape it. Astrophysical Black Holes originate from the death and subsequent gravitational collapse of very massive stars. However, another class of Black Holes might exist, the so-called primordial Black Holes (PBHs), which could have formed in the very early universe, much before the birth of the first stars. Since Dark Matter was present in the Universe since very early times, at least part of the current Dark Matter abundance might in principle be constituted by PBHs. If so, even though we cannot *see* them, we might hope to *hear* them.

K: Wait, what? Do you hope to hear them?

B: On September 14th, 2015, the first historic direct detection of gravitational waves from a binary black hole coalescence (LIGO and Virgo collaboration 2016), has opened a new revolutionary window for astrophysics, cosmology, and fundamental physics. Two years later, the first detection of a neutron star (NS) binary coalescence (LIGO and Virgo collaboration 2017), together with the subsequent observation of the electromagnetic counterpart in all bands of the electromagnetic spectrum, confirmed the beginning of the era of multi-messenger astronomy. Electromagnetic instruments have been allowing us for decades to *look*—in all wavelengths—at the Universe. Nowadays, thanks to gravitational wave observations, we are finally able to *listen* to it! We can now *hear* the spacetime perturbations produced by very massive astrophysical objects orbiting around each other and merging.

K: Is it possible to detect PBHs from gravitational waves?

B: Yes! PBHs that are in binary systems, or two PBHs orbiting each other, are supposed to emit gravitational waves that could be detected by ongoing gravitational wave experiments such as LIGO/Virgo/KAGRA (Bird et al. 2016) or next-generation observatories (Kovetz 2017). The method to distinguish astrophysical black holes from PBHs relies on studying the black hole's intrinsic properties, such as their mass and spin, and spacetime properties such as their redshift, i.e., the increase in the wavelength of emitted light from sources in relative motion to us. Black Holes with masses smaller than ~ 3 solar masses, as well as black hole mergers occurring earlier than the first stars formed, cannot originate from astrophysical objects.

K: Are there other ways to detect PBHs?

Fig. 4 Computer simulation image showing the merger of two black holes. *Credits* SXS, the simulating eXtreme spacetimes (SXS) project

B: One is by looking at gravitational microlensing effects (Niikura et al. 2019). When a PBH passes in front of a star bends the light from the star, resulting in a temporary brightening of the star. Another way is thanks to Hawking evaporation (https://www.inverse.com/science/stephen-hawking-radiation-information-paradox): PBHs with a mass smaller than a few solar masses are predicted to emit thermal radiation via the Hawking mechanism, which could be detected by telescopes in the form of X-ray or gamma-ray emissions. Constraints on the abundance of PBHs as Dark Matter (Carr et al. 2021) are also provided by CMB (Ricotti et al. 2008) and structure formation observations (Afshordi et al. 2003; Murgia et al. 2019).

K: Despite all these huge efforts, both Dark Matter and Dark Energy remain unobserved. That is puzzling (Fig. 4).

8 The Problem of Unobserved Entities

K: By postulating the existence of unobserved entities such as Dark Matter and Dark Energy, are we undermining the reliability of our theories?

B: Physics is a discipline in continuous evolution thanks also to technological developments. Over time, new experiments allow scientists to observe new phenomena, and it is often necessary to postulate the existence of new entities to interpret them. Newer experiments will confirm whether such theoretical forecasted entities exist or not. As in the case of the theory of celestial spheres in the Ptolemaic Universe. Or the phlogiston theory (seventeenth century) to explain combustion and rusting (Mauskop 2002), or the luminiferous ether, which was postulated in the nineteenth century to be the substrate for the propagation of electromagnetic waves. All these ideas were eventually proven to be wrong, as they were disproved by experimental verification. However, other visionary hypotheses turned out instead to be correct, such as the existence of neutrinos (Brown 1978), or the quantum interpretation of atomic spectral lines as electron transitions between different energy levels inside the atom (Bohr 1918).

K: It looks like visionary ideas have played a very important role in the history of physics.

B: Indeed. A good scientific approach is always visionary. Always ready to look at things differently. Physicists are constantly thinking of newer, more refined, visions of the cosmos to encompass all the astonishing phenomena that we observe on Earth and in the sky. From cosmic rays to tropical cyclones. From the afterglow of the Big Bang to the motion of water molecules. The presence of unknown entities should not surprise. Until a few decades ago, neutrinos and radio-waves were unknown too, yet we now know that they are everywhere in the Universe.

K: So, it is good for a scientist to be a bit of a visionary.

B: Vision, creativity, and imagination are crucial ingredients of a good scientific approach. Yet, the creative impulse is instinctive to each person. In science, as in arts, philosophy, or in any other aspects of human life, it is very important to be able to think "out of the box". How to describe the Dark and mysterious Universe where we live, without using metaphors and concrete examples (elastic carpets, marbles, inflatable balloons, springs, etc.)? Moreover, nowadays to visualize what is invisible we cannot neglect the importance of digitalization (https://www.youtube.com/@CosmologyTalks, https://www.youtube.com/@CosmologyfromHome). Many publicly available online outreach resources, such as the ones included among the references of this chapter, can be used to make these topics more accessible for classroom purposes (https://www.jwst.nasa.gov/content/forEducators/formal.html, https://www.jwst.nasa.gov/content/forEducators/informal.html) and the learning process more interactive (https://www.youtube.com/@CupOfCosmology, https://www.youtube.com/@EuropeanSpaceAgency, https://www.youtube.com/@NASA).

The dialogue is now ending. We leave the reader with examples and references to be looked at to deepen the topics and quests that have been discussed throughout this chapter.

References

ADMX Collaboration, Search for invisible axion dark matter in the 3.3–4.2 μeV mass range. Phys. Rev. Lett. **127**(26), 261803 (2021). https://doi.org/10.1103/PhysRevLett.127.261803

N. Afshordi et al., Primordial black holes as dark matter: the power spectrum and evaporation of early structures. Astrophys. J. Lett. **594**, L71–L74 (2003). https://doi.org/10.1086/378763

Y. Akrami et al., Planck 2018 results i overview and the cosmological legacy of Planck. Astron. Astrophys. **641**, A1 (2020). https://doi.org/10.1051/0004-6361/201833880

H. Baer et al., Dark matter production in the early Universe: beyond the thermal WIMP paradigm. Phys. Rept. **555**, 1 (2015). https://doi.org/10.1016/j.physrep.2014.10.002
D. Baumann, *Cosmology* (Cambridge University Press, Cambridge, 2022)
G. Bertone, D. Hooper, History of dark matter. Rev. Mod. Phys. **90**(4), 045002 (2018). https://doi.org/10.1103/RevModPhys.90.045002
G. Bertone, T. Tait, A new era in the search for dark matter. Nature **562**(7725), 51–56 (2018). https://doi.org/10.1038/s41586-018-0542-z
S. Bird et al., Did LIGO detect dark matter? Phys. Rev. Lett. **116**(20), 201301 (2016). https://doi.org/10.1103/PhysRevLett.116.201301
N. Bohr, On the quantum theory of line spectra, part I: on the general theory. KDM **4**(1), 1–36 (1918)
L. Brown, The idea of neutrino. Phys. Today **31**(9), 23 (1978). https://doi.org/10.1063/1.2995181
B. Carr et al., Constraints on primordial black holes. Rept. Prog. Phys. **84**(11), 116902 (2021). https://doi.org/10.1088/1361-6633/ac1e31
CAST Collaboration, New CAST limit on the axion-photon interaction. Nat. Phys. **13**, 584–590 (2017). https://doi.org/10.1038/nphys4109
K. Choi et al., Recent progress in the physics of axions and axion-like particles. Annu. Rev. Nucl. Part. Sci. **71**, 225–252 (2021). https://doi.org/10.1146/annurev-nucl-120720-031147
D. Clowe, A direct empirical proof of the existence of dark matter. Astrophys. J. **648**(2), L109–L113 (2006). https://doi.org/10.1086/508162
J. de Swart et al., How dark matter came to matter. Nat. Astron. **1**, 0059 (2017). https://doi.org/10.1038/s41550-017-0059
C. Giunti, C. Kim, *Fundamentals of Neutrino Physics and Astrophysics* (Oxford University Press, Oxford, 2007). https://doi.org/10.1093/acprof:oso/9780198508717.001.0001
https://esahubble.org/science/gravitational_lensing/
https://home.cern/
https://lisa.nasa.gov/
https://play.google.com/store/apps/details?id=com.thecerealkiller.darkmaster&hl=it&gl=US; https://apps.apple.com/it/app/dark-master/id1489301577
https://sci.esa.int/web/euclid
https://worldwidetelescope.org
https://www.darkenergysurvey.org/
https://www.esa.int/Enabling_Support/Operations/Planck
https://www.et-gw.eu/
https://www.illustris-project.org/
https://www.inverse.com/science/stephen-hawking-radiation-information-paradox
https://www.jwst.nasa.gov/content/forEducators/formal.html
https://www.jwst.nasa.gov/content/forEducators/informal.html
https://www.lngs.infn.it/en
https://www.nottingham.ac.uk/astronomy/sherwood/media.php
https://www.snolab.ca/
https://www.youtube.com/@CosmologyfromHome
https://www.youtube.com/@CosmologyTalks
https://www.youtube.com/@CupOfCosmology
https://www.youtube.com/@EuropeanSpaceAgency
https://www.youtube.com/@NASA
https://www.youtube.com/watch?v=NfqrCdAjiks
https://www.youtube.com/watch?v=wrwgIjBUYVc
https://wwwmpa.mpa-garching.mpg.de/galform/virgo/millennium/
E. Hubble, A relation between distance and radial velocity among extra-galactic nebulae. Proc. Natl. Acad. Sci. **15**(3), 168–173 (1929). https://doi.org/10.1073/pnas.15.3.168
IAXO Collaboration, Physics potential of the international axion observatory (IAXO). JCAP **06**, 047 (2019). https://doi.org/10.1088/1475-7516/2019/06/047

E. Kovetz, Probing primordial-black-hole dark matter with gravitational waves. Phys. Rev. Lett. **119**(13), 131301 (2017). https://doi.org/10.1103/PhysRevLett.119.131301

J. Lesgourgues et al., *Neutrino Cosmology* (Cambridge University Press, Cambridge, 2013). https://doi.org/10.1017/CBO9781139012874

LIGO and Virgo collaboration, Observation of gravitational waves from a binary black hole merger. Phys. Rev. Lett. **116**(6), 061102 (2016). https://doi.org/10.1103/PhysRevLett.116.061102

LIGO and Virgo collaboration, GW170817: observation of gravitational waves from a binary neutron star inspiral. Phys. Rev. Lett. **119**(16), 161101 (2017). https://doi.org/10.1103/PhysRevLett.119.161101

S. Mauskop, Richard Kirwan's phlogiston theory: its success and fate. Ambix **49**(3), 185–205 (2002). https://doi.org/10.1179/amb.2002.49.3.185

H.J. Mo, F.C. van den Bosch, S.D.M. White, *Galaxy Formation and Evolution* (Cambridge University Press, Cambridge, 2010). https://doi.org/10.1017/CBO9780511807244

R. Murgia et al., Lyman-α forest constraints on primordial black holes as dark matter. Phys. Rev. Lett. **123**(7), 071102 (2019). https://doi.org/10.1103/PhysRevLett.123.071102

H. Niikura et al., Microlensing constraints on primordial black holes with Subaru/HSC Andromeda observations. Nat. Astron. **3**(6), 524–534 (2019). https://doi.org/10.1038/s41550-019-0723-1

OSQAR Collaboration, New exclusion limits on scalar and pseudoscalar axionlike particles from light shining through a wall. Phys. Rev. D **92**, 092002 (2015). https://doi.org/10.1103/PhysRevD.92.092002

R. Peccei, H. Quinn, CP conservation in the presence of instantons. Phys. Rev. Lett. **38**, 1440–1443 (1977). https://doi.org/10.1103/PhysRevLett.38.1440

A. Penzias, R. Wilson, A measurement of excess antenna temperature at 4080 Mc/s. Astrophys. J. **142**, 419–421 (1965). https://doi.org/10.1086/148307

S. Perlmutter et al., Measurements of Ω and Λ from 42 high-redshift supernovae. Astrophys. J. **517**(2), 565–586 (1999). https://doi.org/10.1086/307221

M. Ricotti et al., Effect of primordial black holes on the cosmic microwave background and cosmological parameter estimates. Astrophys. J. **680**, 829 (2008). https://doi.org/10.1086/587831

A. Riess et al., Observational evidence from supernovae for an accelerating universe and a cosmological constant. Astron. J. **116**(3), 1009–1038 (1998). https://doi.org/10.1086/300499

L. Roszkowski, Particle dark matter: a theorist's perspective. Pramana **62**, 389 (2004). https://doi.org/10.1007/BF02705097

V. Rubin, W. Ford, Rotation of the Andromeda nebula from a spectroscopic survey of emission regions. Astrophys. J. **159**, 379 (1970). https://doi.org/10.1086/150317

B. Ryden, *Introduction to Cosmology* (Cambridge University Press, Cambridge, 2016)

A. Sedrakian, Axion cooling of neutron stars. Phys. Rev. D **93**, 065044 (2016). https://doi.org/10.1103/PhysRevD.93.065044

S. Weinberg, A new light boson? Phys. Rev. Lett. **40**, 223–226 (1978). https://doi.org/10.1103/PhysRevLett.40.223

J.A. Wheeler, *Geons, Black Holes and Quantum Foam: A Life in Physics* (W. W. Norton Company, New York, 1998)

Y.B. Zel'dovich, Soviet Phys. Uspekhi **11**, 3 (2020). https://doi.org/10.1070/PU1968v011n03ABEH003927

F. Zwicky, Die Rotverschiebung von extragalaktischen Nebeln. Helvetica Phys. Acta **6**, 110–127 (1933)

Gravitational Waves: An Historical Perspective

Adele La Rana

Abstract On September 14, 2015, the Earth was hit by a very brief and extremely weak signal, which was the only trace of a catastrophic cosmic event that took place 1.3 billion light years from our planet. That tiny signal recounted the last whirling moments of the life of a binary system of black holes, before the two bodies—of masses 30 times greater than the mass of the Sun—merged into each other at speeds comparable to the speed of light. Captured by two special experimental devices—the interferometric detectors LIGO in the United States—the radiation of September 14 represents the first gravitational signal ever observed by man and the first confirmation that binary systems of black holes exist. Einstein had predicted the existence of gravitational waves as early as 1916. Nevertheless, their reality as physical entities—and not just mathematical solutions of Einstein's field equations—were still the subject of theoretical discussions in the 1950s, when the first ideas of how to detect them started to develop. Since the first detection in September 2015, an extraordinary new field of cosmic investigation is born: gravitational wave astronomy.

Keywords Gravitational waves · Cosmic event · Binary system · LIGO · Virgo

1 Introduction: An Extremely Ambitious Scientific Challenge

> The following example may help us grasp the magnitude of the task. Imagine this distance: travel around the world 100 billion times (a total of 2400 trillion miles, or one million times the distance to Neptune). Take two points separated by this total distance. Then a strong gravitational wave will briefly change that distance by less than the thickness of a human hair. We have perhaps less than a few tenths of a second to perform this measurement. And

A. La Rana (✉)
Department of Education, Cultural Heritage and Tourism - University of Macerata, Via Luigi Bertelli N° 1, 62100 Macerata, Italy
e-mail: adele.larana@unimc.it

INFN Section Rome 1, Department of Physics "Guglielmo Marconi", Sapienza University of Rome, Piazzale Aldo Moro N° 5, 00185 Rome, Italy

M. Streit-Bianchi et al. (eds.), *New Challenges and Opportunities in Physics Education*, Challenges in Physics Education, https://doi.org/10.1007/978-3-031-37387-9_8

we don't know if this infinitesimal event will come next month, next year, or perhaps in 30 years. (Tyson 1991)

With these words, on March 13, 1991, the Bell Laboratories physicist J. Anthony Tyson tried to dissuade the members of the House of Representatives of the United States of America from investing 200 million dollars in the construction of two large new generation detectors for gravitational waves: a pair of twin interferometers with 4 km long arms, called Laser Interferometer Gravitational Observatory (*LIGO*).

Twenty-four years later, on September 14, 2015, the two US interferometric antennae—completed despite their initial detractors—captured a gravitational signal for the first time in history. The signal lasted less than two-tenths of a second and accounted for the measurement of an amazingly small distance variation: 10^{-18} m, i.e. a thousandth part of the diameter of a proton (Abbott et al. 2016).

Skepticism about the possibility of observing gravitational waves was a *fil rouge* running through the entire history of their experimental research. A substantial part of the international scientific community shared a critical attitude already starting from the 1960s, when Joe Weber devised and built the first-generation detectors, the *resonant bars*, at the University of Maryland.

When Weber first proposed the idea of an apparatus capable of vibrating to the passage of gravitational radiation and of measuring the magnitude of the vibration, the very existence of such waves was a theoretical question still debated. Were they purely mathematical entities—an artificial consequence arising from the choice of the coordinates in the field equations of General Relativity—or did they possess a physical and, as such, measurable reality? Not only were the effects of the passage of a gravitational wave so small as to make their measurement seem impossible: it was even doubted that such effects existed.

A Weber *resonant bar* is a metal cylinder suitably suspended to minimize the seismic noise acting on it and isolated as well as possible from environmental disturbances. The cylinder connects to a transducer, which transforms the mechanical vibrations of the resonant bar into amplified and measurable electrical signals. The basic idea of a bar detector recalls a common phenomenon related to sound waves, called "resonance". When a sound of a certain frequency strikes a tambourine, its membrane begins to vibrate. To produce the vibration of the membrane, the frequency of the incoming sound wave must equal the characteristic frequency of the tambourine, i.e. its "resonant frequency".

If a gravitational wave hitting the bar detector spans frequencies including the cylinder resonant frequency, it can cause it to vibrate. If the cylinder is sufficiently isolated from external and spurious disturbances, it should be possible, in principle, to distinguish the vibration induced by the gravitational wave from those due to other causes. It is worthwhile noticing that, however optimized the resonant bar isolation can realistically be, the sought signal is much tinier than external disturbances, and extremely sophisticated data analysis techniques are required to extract the useful signal from the noise.

To identify any gravitational signal drowned in noise, it is necessary to formulate models as accurate as possible of both the various disturbances acting on the detector and the waveforms that we are looking for. At the time of Weber's ambitious proposal, the modeling of gravitational radiation emitted by astrophysical sources was in its infancy. The very definition of the energy carried by the wave was still finding a coherent formulation from a theoretical point of view. These considerations highlight the courage and scientific creativity of Joe Weber, an experimental physicist of singular ability, he brought the scientific debate on the existence of gravitational waves from a theoretical ground to the empirical one.

2 Early Hypothesis About Gravitational Waves

In 1916, just a year after completing his theory of gravitation, Einstein linearized the field equations of General Relativity in the approximation of weak gravitational field. He showed that they admit solutions that propagate through space at the speed of light (Einstein 1916). The result seemed to overcome a controversial assumption underlying Newton's theory of universal gravitation: the fact that gravitational force was considered to act instantaneously between two massive bodies.

Before Einstein, at least two illustrious precursors discussed the hypothesis of the non-instantaneous propagation of the phenomenon of gravity. The first was the French mathematician and astronomer Pierre Simon Laplace (1749–1827). In his treatise *Sur le Principe de la Gravitation Universelle*, published in 1776, among various astronomical issues he addressed the problem of the gradual decrease in the orbital period of the Moon, which was observed over time by taking into consideration the data relating to well-documented eclipses (Laplace 1776).

Laplace pointed out that the slow acceleration of the lunar motion did not seem to follow the law of Newton's universal gravitation. Interestingly, he hypothesized *ad absurdum* that gravity was due to a fluid emanating from the center of gravity at a certain finite speed. The friction between the fluid and the Moon should have caused a gradual decrease of the satellite's orbit (and therefore of its period). However, Laplace calculated that the decrease would occur very quickly and, since this did not correspond to the observations, he argued that the speed of propagation of gravity should have been very large, well exceeding the maximum speed ever measured, i.e. the speed of light. In the limit of infinite speed, in fact, one returns to Newton's assumption, which does not predict a decrease of the orbital radius and period.[1]

[1] It is worthwhile recalling that Newton himself was unsatisfied by his own assumption, which we may consider motivated by a scientific pragmatism. In a letter addressed to the British philologist and theologian Richard Bentley on February 25, 1962, Newton argued: «It is unconceivable that inanimate brute matter should (without the mediation of something else which is not material) operate upon and affect other matter without mutual contact [...]. That gravity should be innate, inherent, and essential to matter so that one body may act upon another at a distance through a vacuum without the mediation of anything else by and through which their action or force may be conveyed from one to another, is to me so great an absurdity that I believe no man who has in

As underlined by the American physicist Bernard Schutz, one of the leading experts in the field of gravitational wave detection, Laplace was ahead of his time, relating the finite speed of propagation of gravity to the decay of the orbital period, a link that Relativity General would develop successfully only a century and a half later (Schutz 2012).

French was also a second mathematician precursor of Einstein, Henri Poincaré, who took up Laplace's argument in the context of what would become the special theory of relativity and supported the existence of what he called '*ondes gravifiques*' ("ondes" is the French word for "waves"). In a 1905 article Poincaré wrote that for Laplace "the introduction of a finite speed of propagation was the only modification to Newton's law he took into consideration"; this had led him to erroneously conclude that this speed must be, if not infinite, at least much higher than that of light (Poincaré 1905).

According to Poincaré, the classical theory of gravitation needed a much more radical revision, in the light of the recent studies of the Dutch physicist Hendrik Lorentz[2] and his transformation laws (1904), which generalized the principle of Galilean relativity and extended it to electromagnetic phenomena. Poincaré argued that to make gravitational force transform according to the Lorentz transformations in the same way as the electromagnetic forces do (*invariance under Lorentz transformations*), it was necessary to introduce modifications into the law of universal gravitation. One should assume for gravity a finite propagation speed, lower than the speed of light in a vacuum. The latter was in fact postulated in the Lorentz transformations as the maximum achievable by a physical entity.

In analogy to electrodynamics—where accelerated electrically charged particles emit electromagnetic waves—Poincaré hypothesized that massive bodies generate gravitational waves when their distribution of matter in space varies. He suggested that the gravitational radiation takes place at the expense of the energy of the source. Like Laplace, he considered a system made up of two celestial bodies orbiting each other. As the motion proceeds, one can observe the planetary orbit and its period gradually decrease, due to a dissipative process (Poincaré 1908). According to Poincaré, a model of this phenomenon should account for the energy loss related to the emission of gravitational waves.

---

philosophical matters any competent faculty of thinking can ever fall into it. Gravity must be caused by an agent constantly according to certain laws, but whether this agent be material or immaterial is a question I have left to the consideration of my readers».

[2] In 1902, Lorentz shared the Nobel Prize with Pieter Zeeman, "in recognition of the extraordinary service they rendered by their research into the influence of magnetism upon radiation phenomena".

3 Einstein's Theory of Gravitation and the Weakness of Gravitational Radiation

Less than ten years later, in November 1915, Einstein presented the General Theory of Relativity—a new theory of gravitation—to the Royal Prussian Academy of Sciences. A few months later, as anticipated, he also argued how his field equations—under special approximation conditions—could have solutions in the form of waves (Einstein 1916).

In his second paper on gravitational radiation, published in 1918, Einstein corrected relevant mistakes made in his first work and derived the so-called *quadrupole formula* (Einstein 1918). For large distances from the emitting source, gravitational radiation depends to the first order on variations in the quadrupole moment of the source, unlike electromagnetic radiation, which depends on variations in the dipole moment.[3]

The *quadrupolar nature* of gravitational waves is a consequence of the conservation of the total mass and angular momentum of the system. In practice, the quadrupole dependence implies that a spherically symmetrical body moving or contracting does not emit gravitational waves as far as it maintains its symmetry during the process, since its quadrupole momentum is constant. Conversely, an accelerating electrically charged sphere produces electromagnetic waves. It is a profound difference between gravitational waves and electromagnetic radiation. Looking for a source of gravitational radiation, one should only consider dynamical systems deviating from spherical symmetry, such as the case of two stars orbiting each other.

The quadrupole formula expresses the amplitude of gravitational radiation as a function of the quantities that characterize the source. It shows that the amplitude of the radiation is larger the greater are the mass-energy and the velocity of the radiant system. The amplitude is also inversely proportional to the distance from the source.

Summarizing: gravitational wave sources of greater intensity are to be found among astrophysical systems having relativistic speeds, high mass-energy densities, and a high degree of asymmetry.

Nevertheless, from the first years of their prediction in the framework of General Relativity, it was clear that even considering astrophysical sources of this type, the amplitude of gravitational radiation would be extremely small, as well as the possible effects caused by the passage of such radiation through matter. Measuring such effects was long thought impossible.

The quadrupole term, in fact, contains the factor (G/c^4), which has an extremely modest value, given that the universal gravitational constant G is very small, and the speed of light c is very high. This very small multiplying factor implies that gravitational radiation is extremely weak and interacts very poorly with matter, i.e.

[3] The amplitude of the gravitational wave is proportional to the second derivative with respect to time of the quadrupole moment Q. G is the universal gravitational constant, c the speed of light, and r the distance from the emitting source. In its turn, the quadrupole moment depends on the geometry and mass distribution of the system.

with nearly no energy dissipation. A gravitational wave can travel through matter keeping its energy virtually intact.

Gravitational radiation interacts very little with matter present in space, as well as with matter specially devised to detect it, such as Weber's resonant bars. Gravitational waves transfer a minimal part of their energy to the detector: while crossing it, they pass practically undisturbed. The effect of their passage is so tiny that it has been unmeasurable for more than half a century.

4 "Do Gravitational Waves Exist?"

The existence and measurability of gravitational waves were for several decades the subject of theoretical debate. Einstein himself questioned the physical reality of waves, in a paper he co-authored with his colleague Nathan Rosen (Einstein and Rosen 1937). The original work had an emblematic title, "Do gravitational waves exist?", and the authors concluded the article giving a negative answer to the question. Indeed, Einstein and Rosen had tried to find exact solutions of the field equations in the form of plane wave, but they had come up against the impossibility of proceeding without introducing singularities in describing the amplitude of the radiation. This result led them to deduce that the field equations did not admit periodic and regular solutions in the form of waves.

This conclusion contrasted with Einstein's early work on the subject and may surprise. Einstein attributed the novel result to the nonlinearity of the field equations. In a coeval letter to the physicist Max Born, he stated:

> Together with a young collaborator, I arrived at the interesting result that gravitational waves do not exist, though they had been assumed a certainty to the first approximation. This shows that the non-linear general relativistic field equations can tell us more or, rather, limit us more than we have believed up to now. (Kennefick 1996, 2005)

In other words, Einstein argued that the highly nonlinear equations of General Relativity could hide surprises when solutions are calculated using different orders of approximation.

Einstein and Rosen's paper was sent for publication in the American scientific journal *Physical Review*. According to the peer review system—a practice, which, in its modern form, was gradually establishing itself internationally—the editor sent a copy to a secret referee (an expert in the field able to evaluate and comment on the correctness of the scientific content). The latter found an error that invalidated the result and reported it to the editor, who sent the work back to the illustrious sender with the request to consider the observations of the anonymous referee. Einstein reacted very negatively. In his response letter, he addressed the editor:

> Dear Sir,
>
> We (Mr. Rosen and I) had sent you our manuscript for publication and had not authorized you to show it to specialists before it is printed. I see no reason to address the—in any case

> erroneous—comments of your anonymous expert. On the basis of this incident I prefer to publish the paper elsewhere. (Kennefick 1996, 2005)

Later, Einstein re-thought his work in collaboration with his Polish colleague Leopold Infeld. Together, they corrected the calculations, changing the conclusions of the article and publishing the new version in the *Journal of the Franklin Institute* in Philadelphia (Kennefick 2007). Although he acknowledged the error, Einstein never again approached *Physical Review* for publication of his work. As unveiled later, the anonymous referee was the well-known American cosmologist and relativist Howard Percy Robertson.

The debate on the existence of gravitational waves, on the calculation of the energy they carry, and on the effects of their passage through matter was greatly enriched during that lively scientific period that physics historians call the Renaissance of General Relativity.[4] Between the 1950s and 1960s, thanks to the contributions of theorists such as Herman Bondi, Felix Pirani, Richard Feynman, Subrahmanyan Chandrasekar, and John Wheeler, these issues were approached with new vigor (Weber 1969). However, it was certainly the experimental work by Joe Weber that gave the greatest impetus to research on gravitational radiation.

5 Claimed Evidence for Discovery

On June 16, 1969, the scientific journal *Physical Review Letters* published an article with the triumphal title "Evidence for discovery of gravitational radiation", signed by Weber (1969). He reported the analysis of coincident signals from six room-temperature bar detectors, one located in the Argonne National Laboratory (near Chicago) and five in the physics department of the University of Maryland, where Weber was a professor. The devices showed some coincident peaks in the data collected during the first months of the year, which exceeded the threshold value established for detection. The fact that the Argonne bar was located about 1000 km away from the others appeared to support the idea that the observed coincidences had a common cosmic cause. The article began with the following words:

> The probability that all of these coincidences were accidental is incredibly small. Experiments imply that electromagnetic and seismic effects can be ruled out with a high level of confidence. These data are consistent with the conclusion that the detectors are being excited by gravitational radiation.

Having several detectors placed at a certain distance from each other and operating in coincidence is a fundamental requirement in the search for gravitational waves. If two or more detectors are far from each other, the noises acting on the bars will be largely uncorrelated. A gravitational wave from an astrophysical source will hit the detectors almost simultaneously, inducing a characteristic vibration. The probability

[4] For a description of the Renaissance of General Relativity, see Blum et al. (2015), Blum et al. (2020).

of having spurious coincidences decreases as the number of detectors increases. In principle, the coincidence analysis of the data allows identifying by comparison the spurious and uncontrollable local noises affecting the individual devices.

Following Weber's 1969 paper, several researchers around the world set up similar experiments in their laboratories. Nevertheless, none of them confirmed the results obtained by the father of the bar detectors. Even experiments carried out with devices that were more sensitive and using more sophisticated data analyzes were unable to highlight significant coincidences.

One of these first-generation experiments (resonant bars at room temperature) was set up in Italy, in Frascati, in the laboratories of the newborn European Space Research Institute.[5] In 1969, the German electronics engineer Karl Maischberger and the Italian physicist Donato Bramanti began to build a Weber resonant bar, stimulated by the theoretical physicist and relativist Bruno Bertotti (La Rana and Milano 2016). They soon came into contact with another research group that had begun a similar experiment in Munich, under the leadership of Heinz Billing, one of the pioneers of computing and data archiving systems. A collaboration took place to operate the two detectors in coincidence; the Frascati-Munich experiment was the first international collaboration for the detection of gravitational waves. At the time of room-temperature resonant bars, it provided the strongest evidence against the validity of Weber's results. The sensitivity achieved by the devices of the two groups was higher than that of the Weber bars, but after about 350 days of coincident operation, no significant event was observed.

Despite the lack of confirmation from all the experiments of the following decades, Weber remained steadfast in the belief that he had detected the first gravitational signal and continued to support his scientific reasons until his death, in 2000.

The march toward the first direct detection of gravitational waves has been long, arduous, and studded with failures, false alarms, illusions, and clashes within the international scientific community.[6] Like Weber, other pioneers in the field risked or even dented their scientific credibility along the way, claiming in good faith that they had detected a signal, which, however, never found confirmation in the results of subsequent experiments, including those obtained by second-generation detectors: *cryogenic resonant bars*, bars cooled down to a temperature of a few kelvins, to remove thermal noise.

---

[5] ESRIN was born in 1968 as an offshoot of the *European Space Research Organization* (ESRO), directed by Herman Bondi, one of the theoretical physicists contributing to the lively scientific debate on gravitational waves started during the so-called Renaissance of General Relativity. ESRO was a brand-new institution, founded just four years earlier. In the 1970s, it would merge with the European Launcher Development Organization to form ESA, the *European Space Agency*. The idea of ESRO, the first European space research centre, had originally been stimulated by Edoardo Amaldi, one of the founding fathers of the Conseil Européen pour la Recherche Nucléaire (CERN), the first European laboratory born in Geneva in the early 1950s. In Amaldi's view, ESRO should have been modeled on the principles of CERN.

[6] This bumpy and emblematic path has been studied and analyzed for several decades by the English science sociologist Harry Collins, who has been collecting a huge number of documents and testimonies from researchers involved in one of the most ambitious scientific enterprises of all time. See Collins (2004).

Fig. 1 Edoardo Amaldi and Guido Pizzella, in a picture taken in 1983 by Emilio Segrè (Courtesy of Guido Pizzella)

6 On the Italian Side

Also the second international collaboration for the detection of gravitational waves was born in Italy, around the same time as the Frascati-Munich experiment and a short distance away, at the University of Rome.

Between 1970 and 1971, Edoardo Amaldi[7] and his young collaborator Guido Pizzella (Fig. 1) established agreements with two teams in the United States: the Stanford University team, led by William Fairbank and the one led by William Hamilton at Louisiana State University. The collaboration involved the construction of three cryogenic gravitational antennae, to be operated in coincidence in the three different laboratories.

During the 1960s, Edoardo Amaldi had attempted more than once at establishing a research activity on gravitational waves in Rome. His interest in experiments on gravity had begun in the late 1950s, in the wake of the renaissance of General Relativity and had grown with the flourishing of relativistic astrophysics in the following years (La Rana and Bonolis 2018; Bonolis et al. 2018; La Rana 2022). His program could only materialize thanks to the return from the United States of the young Guido Pizzella, who became to all intents and purposes the leader of the new research activity.

[7] At the time, Edoardo Amaldi was about 60 years old and represented the greatest scientific personality on the Italian scene. A pupil and collaborator of Enrico Fermi, he had been part of the glorious group of boys from via Panisperna, contributing to the discovery of radioactivity induced by slow neutrons. The only researcher of that group to remain in Italy during and after the war years, he became one of the most active promoters of the reconstruction of science in Italy and in Europe.

The development of the technologies needed to bring an aluminum bar weighing several tons to temperatures close to absolute zero took some time. The construction of the Roman antenna was completed during the 1980s at CERN, after an initial development period at the SNAM-Progetti laboratories in Monterotondo.[8] *Explorer*, as the detector came to be called, was the first cryogenic antenna to achieve the design sensitivity and stability needed to operate over long periods. Explorer's aluminum cylinder weighed 2300 kg, was 3 m long, with a diameter of 60 cm, and was cooled by liquid helium to a temperature of 2.6 K. As for all bar detectors, its resonant frequency hovered around 1 kHz.

The decision to build the bars resonating around 1 kHz was motivated on one side by practical reasons, on the other by the type of gravitational radiation source one aimed at detecting. For several years, the best candidate had seemed to be a supernova explosion. The gravitational signals expected from an astrophysical event of this type had peak frequencies of the order of the kilohertz. The bars were then optimized to resonate at the frequencies expected for gravitational radiation emitted by a supernova event.

Initial hopes dashed, supernova explosions have given way to more promising sources: binary systems made up of very dense and compact objects such as neutron stars or black holes, which spiral around a common center of mass, until they merge into one body. During the last instants preceding the collision (*coalescence phase of the binary system*) and during the collision itself (*merger phase*), the system radiates an enormous amount of energy in the form of gravitational waves, rapidly changing in frequency and amplitude. The frequencies of the radiation vary according to the value of the masses involved and can go from mHz to tens of Hz.

Since a bar detector is sensitive in a very narrow band of frequencies, centered around its resonance frequency, this kind of device is not suitable for detecting and following the signals emitted by coalescent binary systems, whose amplitude and frequency grow over time. This is where interferometric detectors come into play. Interferometric antennae are third-generation detectors and can achieve high sensitivity over a wide frequency range.

7 Indirect Evidences of the Existence of Gravitational Waves

After the many unfulfilled hopes of the 1970s, a remarkable discovery in astrophysics gave new impetus to the search for gravitational waves.

In 1974, the astronomers Russell Hulse and Joseph Taylor discovered the first binary system composed of a neutron star and a radio pulsar (the system PSR B1913 + 16) (Hulse and Taylor 1975). The radio signal emitted by the pulsar and observed

[8] The SNAM-Progetti laboratories in Monterotondo were part of the ENI Group and were expressly dedicated to fundamental research. Its director was Giorgio Careri, a low-temperature physicist at the University of Rome.

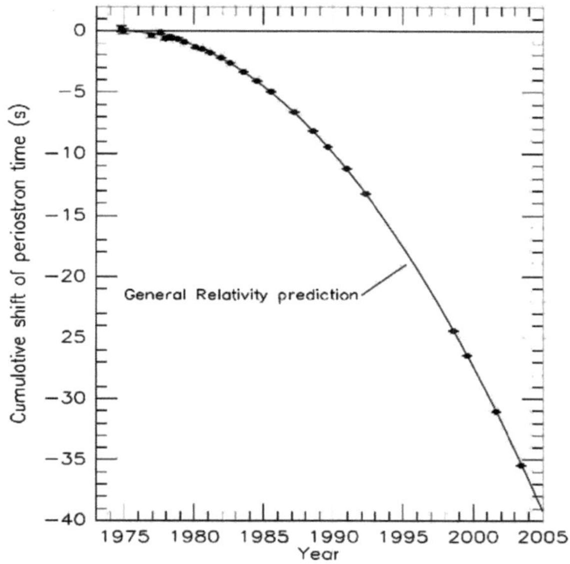

Fig. 2 Decrease in the orbital period of the binary system PSR B1913 + 16 during about 30 years of observations, starting from the discovery of the system in 1974. The dots represent the results of the measurements, while the solid line indicates the theoretical orbital decay curve predicted by General Relativity for the emission of gravitational waves by a binary system (Weisberg and Taylor 2005)

at regular time intervals from the Earth made it possible to precisely calculate the orbital period of the system and measure its evolution over the years. Such a binary constituted, as Hulse argumented in his Nobel Lecture, an extraordinary astrophysical laboratory for testing the theory of General Relativity (Hulse 1993).

In 1982, after a few years of observations, Taylor and Joel Weisberg were able to verify that the orbital period decreased progressively over time, following with very high precision the curve predicted by gravitational wave emission models in the framework of General Relativity (Fig. 2).

The issue, which had been addressed prematurely by Laplace and Poincaré, had finally found a coherent explanation supported by accurate astrophysical observations. While the two stars of PSR B1913 + 16 rotate around their common center of mass, the system loses energy by the emission of gravitational waves. The latter propagate at the speed of light, while the orbits of the two celestial bodies shrink more and more, and their orbital period decreases.[9]

In receiving the Nobel Prize together with Hulse in 1993, Taylor explained:

> The clock-comparison experiment for PSR 1913 + 16 thus provides direct experimental proof that changes in gravity propagate at the speed of light, thereby creating a dissipative mechanism in an orbiting system. It necessarily follows that gravitational radiation exists and has a quadrupolar nature. (Taylor 1993)

However, this wonderful scientific achievement was still an indirect confirmation of the existence of the waves, made by observing the orbital motion of the stars. It was not a direct measurement of their effects, made through a detection device.

[9] For the conservation of angular momentum.

The results of Hulse, Taylor, and Weisberg greatly contributed to convincing the funding agencies to finance very expensive projects for large third-generation detectors. In the early 1990s, the US National Science Foundation gave the green light to the construction of the twin interferometers LIGO, while the Italian *National Institute of Nuclear Physics* (INFN) and the French *Center National de la Recherche Scientifique* (CNRS) signed the agreement to build Virgo (Fig. 3) (La Rana 2020).

Fig. 3 Virgo interferometer (3 km arm length), located in Cascina, close to Pisa (Italy)(Courtesy of the Virgo Collaboration). The working principle of interferometric gravitational wave detectors is based on a special physical effect. A gravitational wave that hits two small test masses floating in space—i.e. subject to gravity alone or, in other terms, in *free fall*—produces a periodic contraction and expansion of their distance. To detect a gravitational wave, it is thus necessary to measure a distance variation between two test masses. This distance variation is extremely small and evolves over time with a characteristic trend, depending on the type of source that radiated the wave. In an interferometric detector, the test masses consist of mirrors located at each end of the orthogonal arms of the antenna. Along each arm of Virgo, the mirrors are 3 km apart, while in the two LIGO they are 4 km apart. Each mirror is suspended in a vacuum chamber through a very sophisticated noise attenuation system, capable of isolating the mirror from external disturbances, so to achieve a condition as close as possible to free fall along the two perpendicular directions of the interferometer. A laser beam is split in two orthogonal beams by a beam-splitter; the beams then travel along the two arms of the antenna, and are reflected multiple times between the suspended mirrors of each arm. After these multiple reflections, the two beams finally recombine at the output in a single beam, which is detected by a sensor. Variations in the distance between the mirrors cause a phase shift between the two interfering beams and this phase shift is measured by the sensor. From the measurement of the phase shift of the two beams, one can measure the variation of the distance between the mirrors. (Courtesy of the Virgo Collaboration)

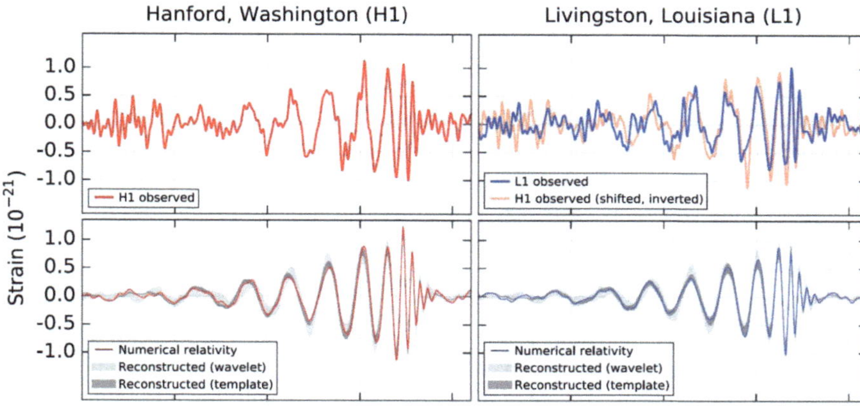

Fig. 4 Signals captured by the two LIGOs and analyzed by the LIGO-Virgo collaboration. The ordinate axis shows the ratio between the measured gravitational signal *h* (i.e., the distance variation between the interferometer mirrors) and the arm length of the interferometer. (l = 4 km for both LIGO detectors). The order of magnitude of the measured distance variations is: $10^{-21} \times 10^3$ m $= 10^{-18}$ m. To compare the two amplitudes, the Hanford signal was shifted in time by 6.9 ms, corresponding to the time interval between the detection of the signal at Hanford and the detection at Livingston (due to the finiteness of the wave propagation velocity). The sign of the signal has also been inverted, due to the opposite orientation of the arms of the Hanford interferometer with respect to that of Livingston (Courtesy of the LIGO-Virgo Collaboration)

8 The First Detection

On February 11, 2016—about one hundred years after the first prediction made by Einstein—the physicists of the LIGO-Virgo international collaboration announced to the world the first direct detection of a gravitational wave. Of the three large third-generation detectors existing in the world, only the two American antennae were in operation on September 14, 2015. On that date, a particularly intense gravitational signal hit the Earth, causing a measurable vibration in the LIGO interferometers.

To analyze the collected data and identify beyond any doubt the nature of the signal, it took several months of feverish work by the approximately 1000 researchers of the LIGO-Virgo collaboration.

The measured signal has a characteristic trend, as can be seen from Fig. 4 (Abbott et al. 2016).

In the very first fractions of a second, the frequency and amplitude grow rapidly over time. Making an analogy with sound waves, the shape of this gravitational signal resembles the *chirp* made by some birds. Birds' chirp is an acoustic signal in which frequency and amplitude increase rapidly in time. The *chirping pattern* characterizes the gravitational signal emitted by a binary system in the final phase of its life, just before the merger.[10] The two celestial bodies spiral around each other faster and

[10] It is worthwhile noticing that the sound waves that our ears perceive have frequencies in the range of 20 Hz–20 kHz. Detectable gravitational waves have frequencies from 10 kHz dow.: There

faster, emitting a gravitational wave that has an instantaneous frequency equal to twice the orbital instantaneous frequency. In the observed signal, the frequency of the radiation increases from 35 to 250 Hz in less than two-tenths of a second.

The signal captured by LIGO was emitted 1.3 billion years ago by a system of two black holes, which were rotating around each other at half the speed of light, in their last two-tenths of a second before they merged. Once the final black hole is formed and a spherical symmetry is reached, the signal rapidly goes to zero: the gravitational emission ceases.

Based on the shape of the detected signal and the theoretical models calculated within the theory of General Relativity, it was possible to establish that the two original black holes had masses equal to 29 and 36 solar masses, concentrated in two spheres with a smaller diameter at 200 km. Moving at an astonishing speed, they merged to form a final black hole of 62 solar masses. The missing three solar masses were emitted in the form of gravitational radiation energy.

9 Cosmic Messengers and a New Era of Astronomy

Gravitational waves can pass almost undisturbed through huge amounts of matter, giving up an infinitesimal part of their energy along the way, as if they were traveling through empty spaces. This feature that hid them from observation for more than 50 years is also the reason for their enormous interest in astrophysics and cosmology. By interacting so feebly with matter, they can travel enormous distances without losing information about the sources that generated them. Ultimately, they are very precious messengers which may come from extremely remote astrophysical objects.

Gravitational waves are providing completely new and complementary information with respect to electromagnetic waves. They are particularly precious for studying those phenomena, which have no electromagnetic counterpart, such as, for example, binary systems of black holes. Suffices to say that the first gravitational signal observed was also the first direct proof of the existence of black holes and the very first observation of a binary system made up of these monsters of the cosmos.[11]

In August 2017, a new exciting and unprecedented event has seen LIGO and Virgo as protagonists. The three interferometric antennae detected the first gravitational signal radiated from a collision between two neutron stars (Abbott et al. 2017). About 2 s later, a gamma-ray signal was observed by the space telescopes Integral—the ESA's International Gamma-Ray Astrophysics Laboratory—and the Fermi Gamma-ray Space Telescope, run by NASA. Alerted by LIGO and Virgo, the astrophysical community started an epical search. In the following hours, tens of ground-based and

is therefore an overlap of frequency ranges, which means that sound simulations of gravitational radiation provide a natural way for physicists to "sense" gravitational waves (analogously, a false-color photo highlights the X-ray emitted by an astrophysical source).

[11] For more details about black holes and their detection, see in this volume the chapter: Gravitational Waves and Black Holes by MariaFelicia DeLaurentis and Paolo Pani.

space-based telescopes began to scan a special patch in the sky identified by Virgo and LIGO, to single out the electromagnetic source of such a special gravitational wave signal. The source was identified about ten hours later: a luminous spot in the Hydra constellation, which was not there before. In the galaxy NGC4939, at about 130 million light-years from the Earth, two neutron stars weighing about one solar mass each had merged, radiating not only gravitational waves, but also all kinds of visible and invisible electromagnetic waves. The *kilonova* was the first astrophysical event to be observed both through its electromagnetic and gravitational emission. A new era of multi-messenger astronomy was thus inaugurated, which now includes, alongside electromagnetic waves and neutrinos, also the precious messengers of gravity.

References

B.P. Abbott et al., (LIGO Scientific Collaboration and Virgo Collaboration), Observation of Gravitational Waves from a Binary Black Hole Merger. Phys. Rev. Lett. **116**(6), 061102 (2016)

B. P. Abbott et al. (LIGO Scientific Collaboration and Virgo Collaboration), GW170817: Observation of Gravitational Waves from a Binary Neutron Star Inspiral. Phys. Rev. Lett. **119**(16), 161101 (2017)

A.S. Blum, R. Lalli, J. Renn, The reinvention of general relativity: a historiographical framework for assessing one hundred years of curved space-time. Isis **106**(3), 598–620 (2015)

A.S. Blum, R. Lalli, J. Renn, *The Renaissance of General Relativity in context. Einstein Studies,* Vol 16. (Birkhäuser Verlag GmbH, Springer Nature, Cham, 2020)

L. Bonolis, A. La Rana, R. Lalli, The renaissance of general relativity in Rome: main actors, research programs and institutional structures, in *Proceedings of the Fourteenth Marcel Grossmann Meeting on General Relativity* eds. by M. Bianchi, R.T. Jantzen, R. Ruffini (World Scientific, Singapore, 2018)

H. Collins, *Gravity's shadow* (The Search for Gravitational Waves (University of Chicago Press, Chicago and London, 2004)

A. Einstein, Näherungsweise Integration der Feldgleichungen der Gravitation (Approximate solution of the gravitational field equations), in *Sitzungsberichte der Königlich Preussischen Akademie der Wissenschaften, Berlin* (1916), p. 688–696

A. Einstein, *Über Gravitationswelle, Sitzungsberichte der Königlich Preußischen Akademie der Wissenschaften,* (Berlin) 1918, pp. 154–167

A. Einstein, N. Rosen, On gravitational waves. J. Franklin Inst. **223**, 43–54 (1937)

R.A. Hulse, *Nobel Lecture* (1993)

R.A. Hulse, J.H. Taylor, Discovery of a pulsar in a binary system. Astrophys. J. **195**, L51–L53 (1975)

D. Kennefick, Controversies in the history of the radiation reaction problem in general relativity, in *Humanities Working Paper, N° 164* (California Institute of Technology, Pasadena, 1996)

D. Kennefick, Einstein versus the physical review. Phys. Today **58**(9), 43 (2005). https://doi.org/10.1063/1.2117822

D. Kennefick, *Traveling at the Speed of Thought: Einstein and the Quest for Gravitational Waves* (Princeton University Press, Princeton & Oxford, 2007)

A. La Rana, Virgo and the emergence of the international gravitational wave community, in *The Renaissance of General Relativity in context. Einstein Studies*, Vol. 16, eds. by A. Blum, R. Lalli, J. Renn, (Birkhäuser Verlag, Springer Nature, Cham, 2020), pp. 363–406

A. La Rana, L. Bonolis, The beginning of Edoardo Amaldi's interest in gravitation experiments and in gravitational wave detection, in *Proceedings of the Fourteenth Marcel Grossmann Meeting on General Relativity*, eds. by M. Bianchi, R.T. Jantzen, R. Ruffini (World Scientific, Singapore, 2018)

A. La Rana, EUROGRAV 1986–1989: the first attempts for a European Interferometric Gravitational Wave Observatory. Europ. Phys. J. H. **47**(3), (2022)

A. La Rana, L. Milano, The early history of gravitational wave detection in Italy: from the first resonant bars to the beginning of the Virgo collaboration, in *Atti del XXXVI Convegno annuale della SISFA (Napoli, October 4–7, 2016)* ed. by S. Esposito (Pavia University Press, Pavia, 2017)

P.-S. Laplace, Sur le Principe de la Gravitation Universelle, in *Oeuvres complètes de Laplace VIII* (Gauthiers-Villars et fils, Paris, 1776), pp. 201–275

H. Poincaré, Sur la dynamique de l'électron. Compte Rendu De l' Académie Des Sciences **140**, 1504–1508 (1905)

H. Poincaré, La dynamique de l'électron, in *Œuvres de Henri Poincaré vol. IX, Gauthiers-Villars et fils, Parigi* (1908), pp. 551–586

B. Schutz, *Gravity from the Ground Up. An Introductory Guide to Gravity and General Relativity* (Cambridge University Press, 2012), pp. 309, 319

J.H. Taylor, *Nobel Lecture* (1993)

J.A. Tyson, Gravitational radiation search, in *Testimony of Dr. J. Anthony Tyson Before the Subcommittee on Science of the Committee on Science, Space, and Technology* (United States House of Representatives, 1991), p. 6

J. Weber, Evidence for the discovery of gravitational radiation. Phys. Rev. Lett. **22**(24), 1320–1324 (1969)

J. Weisberg, J. Taylor, Relativistic binary pulsar B1913+16: thirty years of observations and analysis, in *Binary Radio Pulsars*, eds. by F.A. Rasio, I.H. Stairs (ASP Conference Series, San Francisco, 2005), p. 25

Notes on Revolutions in Physics in the Twentieth Century

Francesco Vissani

Abstract Often, the presentation in teaching paths of what is known in physics resembles a description of a series of incremental steps, giving the false impression of sequential progress in which historical and philosophical contexts play no role. In this essay, we collect a series of notes on the revolutions in physics in the twentieth century to show that things are quite different. From these remarks and observations, the role of a complex and lively cultural intermingling with philosophy and mathematics, as well as some specific features of the various developments, emerge clearly; moreover, the essential role of some specific individuals becomes evident. It is argued that it would be interesting to make a systematic comparison between the way the revolutions in fundamental physics took place in the last century and the way this discipline is organized today.

Keywords Revolutions in physics · Twentieth century physics · General relativity · Quantum mechanics · Particle physics

1 Introduction

The early 1900s saw radical changes in physics. Einstein's theory revolutionized our ideas, unifying space and time, energy and mass, eventually blossoming into a new view of gravity. Quantum mechanics made it possible to penetrate the atom and, combined with relativity, allowed us to understand the world of elementary particles. In this contribution, we gather some notes on the nature of these cultural processes.

F. Vissani (✉)
Laboratori Nazionali del Gran Sasso, INFN, Via Giovanni Acitelli, 22, 67100 L'Aquila, AQ, Italy
e-mail: vissani@lngs.infn.it

M. Streit-Bianchi et al. (eds.), *New Challenges and Opportunities in Physics Education*, Challenges in Physics Education, https://doi.org/10.1007/978-3-031-37387-9_9

The rationale for such a discussion is not so much to expound or to make syntheses of a series events—which would take a lot of space to do with any degree of completeness, and for which we refer to Pais (1986), but rather:

- to give an idea of how complex the path of knowledge is;
- to entice the reader to compare conditions at the beginning of the last century with those of today.

This chapter aims at offering an introduction to a broad topic, through selected notes, which could be the subject of further developments.

We will focus on so-called fundamental physics, in particular relativity and quantum mechanics. However, let us remark that some fundamental problems came to be discussed through the study of apparently much more mundane problems. To illustrate the point, just consider a few examples:

- the observations of the surfaces of stars apparently were only related to problems of astrophysics but led us to talk about new elements, atomic spectra, etc.;
- the discussion of the nature of the heat emitted by our planet and the sun was premature before realizing the existence of nuclear transmutations;
- the diagnostics of the interior of the sun by neutrinos, which was supposed to be an almost obvious verification of the ideas about nuclear energy, led us to the only experimental proof that the standard model of elementary particles and interactions is incomplete.

See Appendix A for a story on the attribution of "fundamental" character to certain research and not to other ones. But now let us turn to the discussion.

2 On the Nobel-Winning Paper of Einstein

It is very stimulating to begin attempting a historical evaluation of Einstein's work on the quanta of light, which earned him the *Nobel Prize* in Physics. His idea that there is a "particle of light", which can behave like a wave (consistently with Maxwell's views), reopened in a very original way the debate between Newton and Huygens on the nature of light—who, as everyone remembers, had defended the corpuscular and wave interpretation, respectively.

Even more, this work began in more than one sense to give substance to the quantum theory, initiated by Planck: on the one hand recognizing the value of a so-called atomistic viewpoint on the light, and the other hand avoiding oppositions between the two viewpoints, that are at first sight antithetical (Pais 1949).

This position would later stimulate de Broglie to proclaim a similar form of duality for matter particles. This, along with Heisenberg's principle, will offer Bohr the basis for proclaiming the principle of complementarity. From this point of view, it is not at all surprising that there was a strenuous resistance by the scientific community to accept Einstein's *Lichtquanten* hypothesis, and in particular, from the German world. This persisted until Compton's discoveries (1922) and, in some cases, even after: his position was truly revolutionary.

3 A Specific Aspect of Connections Between Physics and Math

One of the revolutions in the physics of the twentieth century, which developed fully after quantum mechanics and relativity, occurred thanks to the initial impulse of Hermann Weyl. I am referring to the proposal to extend general relativity also to a "gauge" invariance (Weyl 1918).

This will subsequently stimulate Yang and Mills to propose the first non-Abelian theories (Yang 2013), to non-gravitational forces: I mean the so-called standard model, which summarizes most of the assessed facts concerning particles and interactions, apart from the evidence for neutrino masses, recalled just above.

It is well known that Weyl's proposal was stimulated by Einstein's works, but it should also be recognized the inspiring role for him of certain mathematical considerations. I refer to the Erlangen program, proposed by the mathematicians Klein and Lie, which anticipated a more important role of the groups of transformations than it was previously recognized.

Note that this kind of approach to the construction of physical theories, where mathematical considerations play a central role, is perhaps the closest to the more recent ones.

4 The Role of Society in Relation to Scientific Discoveries

When we consider the role of external stimuli, it seems that the revolutions of the last century differ greatly from one another. For example, is it true that Max Planck was asked by the German Bureau of Standards on how to make light bulbs more efficient, to achieve maximum efficiency with minimum power consumption. Similarly, Poincaré was a permanent member of the French Bureau of Longitude.

It has been occasionally argued that some of these considerations helped shaping Einstein's views on special relativity. However, the effect of practical or empirical considerations on the foundation of the theory of general relativity seems to be minimal; the intellectual debts are much more related to the mathematical developments of the time. Furthermore, on closer inspection, even the studies on the electrodynamics of moving bodies (that Einstein puts at the center of his earlier works) seem to belong to more a chapter of mathematical physics than to a textbook of electrical engineering.

Finally, it seems useful to recall how many scientists, protagonists of the major revolutions in physics, were actively engaged in communicating their results to the public, in full continuity with what was done in the nineteenth century. Planck, Heisenberg, Fermi, Pauli, Schrödinger, Gamow, Landau, and many more come to mind, in addition to somewhat obvious cases such as that of Einstein.

Let us remark that the last consideration is to be considered quite distinct from the myth-making processes some of them underwent in the mass media.

5 The Intricate Path to the Acceptance of Atomism

Part of the nineteenth-century scientific community (Boltzmann in the lead) believed that the atomic hypothesis was essential for understanding the structure of matter, but it is worth remembering that the *Nobel Prize* which concludes the discussion is the one given to Jean Perrin and dates to 1926, less than a century ago. A well-known opponent of the concept of the *atom* in science was Mach. Also, leading chemists like Ostwald did not believe in atoms. Again, it is remarkable that the theoretical basis of Perrin's study is the one of Einstein's famous papers from 1905.

In this respect, an interesting observation about the thinkers who were the architects of the scientific revolutions of the last century can be made. Several of them had solid philosophical backgrounds. I am thinking about Planck, Bohr, Einstein, Schrödinger, and Poincaré and Boltzmann before them. In this respect, they were more like Galileo or Newton than the typical modern scientists.

One revolutionary aspect is the idea that elementary particles exist, the natural evolution of atomism, and how the transformations to which they are subjected are put together. The first striking case concerns the quanta of light, which can disappear completely in a reaction, e.g., in the photoelectric effect, or on the contrary, are created (i.e., appear) in a de-excitation of an atom. This type of conceptualization is still puzzling today, as Feynman testified in a famous speech (Feynman 1968), see Appendix B, but very similar reservations have been repeatedly expressed by Einstein himself.

It is not difficult to imagine what a cultural shock it was to recognize that something similar also occurred to matter particles, as was first argued by Enrico Fermi, with his theory of weak decays. For a very lucid recent testimony of one of the protagonists of the time, see (Yang 2012).

However, it is curious to note how rare it is today that similar considerations are brought to the fore or even marginally noted in popular reports on particle physics. For researchers, the point is (believed to be) self-evident, for the layperson it remains obscure and difficult if not impossible to accept.

6 On the Importance of Visualization and Analogies

Talking of atoms, a very interesting issue concerns the possibility to visualize the theory that we have constructed to describe the observed facts. In general, it is recognized that this feature—visualization—is of limited use, if not entirely excluded, when dealing with the microworld, even if, under some circumstances, certain visualizable patterns may retain some heuristic value.

The first example concerns the planetary model of the atom, proposed by Jean Perrin in 1901. The reason why Thomson contrasted it with a completely different model (1904) is to avoid having moving electrons, which, according to Larmor's results (1897), should radiate energy. Bohr's atom (1913) does not solve these problems even for the hydrogen atom, except by authority, forbidding the existence of orbits without angular momentum. Heisenberg's model (1925) does not seem entirely different in this respect, although the mathematical consistency of his matrix description allows us to conceive an orbiting electron without angular momentum, which is consistent with observations. In the case of Schrodinger's wave mechanics (1926), the situation is similar, with the only difference that it made in part possible to imagine a sort of stationary wave, arranged around an (almost) point nucleus.

A second example concerns spin (1925). The first idea was put forward by a young German physicist, Kronig, using the metaphor of a spinning ball. Pauli severely criticized it, based on knowledge at the time, and Kronig decided not to proceed to expose it to the public. However, a few months later the same idea was published by two dutchmen, Uhlenbeck and Goudsmit. In the meantime, ways were found to overcome the problems pointed out by Pauli and others, and in 1927, Pauli himself outlined the mathematical model of the electron spin that we use today.

Today, and despite their inaccuracy, we continue to use the metaphors of the planetary model and spinning ball in popular presentations for educational purposes, also because in several situations it is difficult to do much better. Interestingly, for what concerns the possibilities of visualization, the case of the spin is the hardest, but on the other hand, the mathematical description of the electron spin using a two-dimensional wave function allows connection with other aspects, such as the statistical properties.

We are willing to accept non-visualisable descriptions whenever we gain significant advantages, but history shows that some less refined or rigorous presentations have been (and often still are) fruitful for teaching or doing science.

7 On the Difficulty to Foresee the Future of a Scientific Revolution

Sometimes we tend to take for granted the value of the things we have learned, completely forgetting how difficult and tiring it was to acquire them. To understand this, I propose a few examples.

7.1 Three Subtle Implications of General Relativity

It is amusing and perhaps instructive to remember that Einstein doubted

1. the expansion of the universe,
2. the existence of black holes, and even
3. that gravitational waves are physical and observable.

It is to some extent ironic that these are among the main predictions of his theory and are often presented as such. On the one hand, his attitude gives credit to an honorable tradition of critical thinking; but on the other hand, it underlines how difficult it is to make progress in science, even for the founders themselves.

For the record, the first scientists who obtained significant results in these topics in the context of general relativity were

1. Friedman (1922)
2. Schwartzschild (1916) and then Oppenheimer and Snyder (1939)
3. Brinkmann (1925) but this was completely messed up by the subsequent discussion, to which Einstein also contributed, till 1957 (Hill and Nurowski 2017).

7.2 Some Superseded Opinions on Atomic Energy

Let us conclude with three opinions on atomic energy by some of the recognized protagonists of these scientific advances. What stands out is that their ideas, formulated less than a century ago, are hopelessly dated and by now superseded by scientific results, with which even children are familiar today.

- *There is no likelihood man can ever tap the power of the atom. The glib supposition of utilizing atomic energy when our coal has run out is a completely unscientific Utopian dream, a childish bug-a-boo.*

  —Robert Millikan, 1928
- *There is not the slightest indication that [nuclear energy] will ever be obtainable. It would mean that the atom would have to be shattered at will.*

  —Albert Einstein, 1932
- *The energy produced by breaking down the atom is a very poor kind of thing. Anyone who expects a source of power from the transformations of these atoms is talking moonshine.*

  —Ernest Rutherford, 1933

Note that these opinions assume implicitly that the nuclei of atoms will repel each other due to electrostatic forces. They were abandoned only after full acceptance of the discovery of neutrons (1932), which can penetrate the nucleus much more easily.

8 What Are the Drivers of Physics Discoveries?

The above observations, however fragmentary, highlight a number of aspects of the cultural processes that affected twentieth-century physics. It is not easy to indicate precisely what were their main features, nor to give a satisfactory overview of these processes. However, the different visions of what science is allow to usefully highlight one aspect or another. Let us sketch some of them, without claiming completeness, purely by way of example:

- A heroic view of science recognizes the fact that there have been a few obvious protagonists (the geniuses).
- If instead, we take more interest in social facts, we will notice how these processes are linked to the destiny (realization, failure, positioning, etc.) of certain national identities.
- If we place more emphasis on culture and its dynamics, we may notice the many links and interrelationships between physics and other disciplines—of great importance those with mathematics, chemistry, astronomy, and natural philosophy.
- If, on the other hand, one prefers to focus on laboratory observations and experiments, one will be inclined, for example, to highlight the Michelson–Morley experiment, in discussing relativity, or the studies of spectroscopy in discussing atomic theory.
- When looking at the role of tradition, one cannot miss the lively but respectful confrontation that took place with specific previous elaborations (e.g. between Faraday and Maxwell), and also, for the whole science since Galileo, the application of the hypothetical-deductive method developed in Hellenism.
- It is important to notice the enormous role played, in achieving radical changes of point of view, by some very young scientists, occasionally led by other less young scientists, such as Sommerfeld, Bohr, and to some extent Born, Fermi, etc.

Each of these interpretations is to some extent useful if considered critically (i.e., bearing in mind its setups and aims, and realizing its limitations and even dangers). Each interpretation would lead to some course of action, to replicate an undeniably successful period, that we sometimes like to call the '20th-century revolutions in physics.

9 Countries, Individuals, and Their Impact on Scientific Progress

The role of certain scientific personalities is striking when considering the scientific revolutions in physics that took place in the twentieth century. Einstein, Bohr, Heisenberg, Schrödinger, Dirac, Fermi, Pauli, Majorana, and many others come immediately to my mind.

A curious case regards Italy. This country did not participate in the major revolutions in physics of the twentieth century before Enrico Fermi entered the arena. The most important contributions (before him) came from scientists of other countries, such as Austria-Hungary, Prussia, the UK, France, etc. The role of individual protagonists—also from other countries—should be underlined. See (Hund 1974) for an excellent overview of the aspects connected to quantum mechanics.

Note the interesting fact, that the Italian mathematicians instead—such as Ricci, Levi–Civita, Bianchi, etc.—had leading roles in the formalism needed for general relativity.

10 Remarks

It is difficult to make a precise comparison between how these "revolutions" in fundamental physics took place and how physics is organized today. However, this is important and interesting.

For example purposes, I limit myself to briefly expanding on the aspects set out in the last previous section, focusing on three specific considerations.

- It is unclear to me whether physics today places the highest importance on individual thinkers. The Nobel Prize in Physics, for example, is still awarded to up to three people, indicating some recognition of the role of individuals. However, the more recent Fundamental Physics Breakthrough Prize seems more inclined toward recognizing the role of collaborations. For instance, in 2015, 2016, and 2018, large collaborations were explicitly recognized and awarded. These changes are still relatively mild and can even be considered fair, but there is little doubt that they are indicative of the direction in which physics is transforming.
- Panels of scientists or large collaborations that oversee experiments ensure the achievement of research objectives in the best possible way. However, there is no doubt that scientists now work in vastly different conditions compared to those of the early twentieth century.
- During the revolutions in physics of the twentieth century and before, leading physicists spoke directly about their scientific plans or achievements; just think of the interviews or public conferences of Boltzmann, Planck, Rutherford, Einstein or Fermi, or the writings, usable by the public, of Bohr, Heisenberg, Schrodinger, Gamow, or Feynman, see e.g. (Boltzmann 1974). The public seems still to appreciate the active scientists who are engaged seriously in communicating science, but this seems less common today.

We will not proceed further in the discussion: It does not seem particularly sensible to rush into such an important discourse. But perhaps the elements gathered in the text and these last remarks are sufficient to argue that there are significant differences between our time and that of the scientific revolutions of the last century, and perhaps it would be interesting or even useful to make a systematic comparison.

Appendix A: On Stances Toward Research in Fundamental Physics

As mentioned at the outset, in this short essay we focus mainly on fundamental physics topics.

However, although there is some consensus on what is meant by this term, the concept of *"fundamental"* is not self-defined and can cause some confusion. For instance, it may obscure the role of the link between the various disciplines for progress, it may lead to overlooking the more complex passages in the history of science, and it may present a distorted picture of knowledge, exaggerating the importance of the state achieved in a certain moment, etc.

An amusing recent episode, which sheds light on these considerations, offering food for thought on modern procedures and ways of attaching importance to certain scientific topics and not to others, is provided in an article by White (2007), where the words *"fundamentalist"* occurred in the title shows its polemical nature.

This work was intended to oppose the strong pressure from other scientific communities to convey mainly or exclusively the resources of astronomers on a single research topic, Dark Energy, allegedly more fundamental than others.

Appendix B: The Photon, an Elusive Concept

It is interesting to report the opinion of Feynman on the meaning of the same point, expounded to a wide public (Feynman 1968). To fully appreciate his words, let us briefly recall the context. These remarks are from the above-mentioned speech, where Feynman repeatedly states that he owes a great and welcome debt to his father, who guided him in the development of a rich and original vision of science. As soon as he grew and concluded his scientific education, his father turns to him, begging for clarification on a question he has never been able to answer. Here is the exchange between father and son, as reported by Feynman (Feynman 1968).

> (...) It was a wonderful world my father told me about.
>
> You might wonder what he got out of it all. I went to MIT. I went to Princeton. I came home, and he said, "Now you've got a science education. I have always wanted to know something that I have never understood, and so, my son, I want you to explain it to me." I said yes.
>
> He said, "I understand that they say that light is emitted from an atom when it goes from one state to another, from an excited state to a state of lower energy." I said, "That's right."
>
> "And light is a kind of particle, a photon, I think they call it." "Yes."
>
> "So if the photon comes out of the atom when it goes from the excited to the lower state, the photon must have been in the atom in the excited state." I said, "Well, no."

> He said, "Well, how do you look at it so you can think of a particle photon coming out without it having been in there in the excited state?" I thought a few minutes, and I said, "I'm sorry; I don't know. I can't explain it to you."
>
> He was very disappointed after all these years and years of trying to teach me something, that it came out with such poor results (...)

Let us forget for a moment his eccentric attitudes and let us put aside the role *homo novus* that he—an American—happened to cover, among the many theoretical physicists from the old continent. Behind the brilliant prose to which we are accustomed, it is apparent that Feynman is very interested in philosophical questions, even though he professes the opposite in various circumstances; what he stigmatizes, apparently, is a certain type of thinking, which instead of addressing the problems and dilemmas posed by our attempt to reconcile our experience and our image of the world, is content with itself. We are therefore led to the conclusion that, also in this respect, Feynman is in full and perfect continuity with many of his predecessors who accomplished the scientific revolutions of '900. In fact, we can find no valid reason not to consider him among the main protagonists of physics in the latter part of the twentieth century.

We conclude with a sentence from a famous letter of Einstein to Michele Besso (12 Dec. 1951):

> All those 50 years of careful pondering have not brought me closer to the answer to the question: 'What are light quanta?' Today any old scamp believes he knows, but he's deluding himself.

References

L. Boltzmann, *Theoretical Physics and Philosophical Problems* (D. Reidel Publishing, 1974). ISBN: 978-9027702494; M. Planck, *L'immagine della scienza. Saggi sulla fisica moderna* (Castelvecchi, Collana Le Navi, 2018). EAN: 9788832822892 2018; E. Rutherford, *The Newer Alchemy* (Cambridge, 1937). ISBN 13: 9781107440425; A. Einstein, *The World as I See It* (General Press, Paperback, 2018). ISBN-10 938811812X, ISBN 13: 978-9388118125; E. Fermi, *Atomi, Nuclei E Particelle* (Collana Universale Bollati-Boringhieri, 2009). ISBN: 9788833920108; Bohr, N. *Atomic Theory and the Description of Nature* (Cambridge University Press, 2011). ISBN 13: 978-1107628052; W. Heisenberg, *Physics and Philosophy: The Revolution in Modern Science* (Harper, 2007). ISBN 13: 978-0061209192; E. Schrodinger, *Mein Leben, Meine Weltansichte* (Verlag Zsolnay, 1985). ISBN 3-552-03712-8; G. Gamow, *My World Line: An Informal Autobiography*, 1st edn. (Publisher Viking, 1970). ISBN-10 0670503762, ISBN 13: 978-0670503766; R. Feynman, *QED. The Strange Theory of Light and Matter*, 1st edn. (Princeton University Press, 1989). ISBN: 978-0691024172

R.F. Feynman, What is science? Phys. Teach. **7**(6), 313 (1968)

A. Friedman, Über die Krümmung des Raumes. Z. Phys. **10**(1), 377 (1922). Translated in Gen. Rel. Grav. **31**, 1991–1999

C.D. Hill, P. Nurowski, How the green light was given for gravitational wave search. Not. Am. Math. Soc. **64**(7), 686–692 (2017)

F.H. Hund, *The History of Quantum Mechanics* (George G. Harrap & Co. Ltd., London, 1974). ISBN 10: 024550902X, ISBN 13: 9780245509025

J.R. Oppenheimer, H.S. Snyder, On continued gravitational contraction. Phys. Rev. **56**, 455 (1939)
A. Pais, Einstein and the quantum theory. Rev. Mod. Phys. **21**(3) (1949)
A. Pais, *Inward Bound: Of Matter and Forces in the Physical World* (Clarendon Press, 1986). Reprint 1988. ISBN 10: 0198519974, ISBN 13: 978-0198519973
K. Schwarzschild, On the gravitational field of a mass point according to Einstein's theory. Sitz. Kön. Preuss. Akad. Wiss. Berlin Phys. Math. Klasse 189 (1916). arXiv 9905030
H. Weyl, Gravitation und Elektrizität. Sitz. Kön. Preuss. Akad. Wiss. 465 (1918) Translated in http://www.neo-classical-physics.info/uploads/3/4/3/6/34363841/weyl_-_grav._and_electr.pdf
S. White, Fundamentalist physics: why dark energy is bad for astronomy. Rep. Prog. Phys. **70**, 883 (2007)
C.N. Yang, Fermi's beta-decay theory. Int. J. Mod. Phys. A **27**, 1230005 (2012)
C.N. Yang, *Hermann Weyl's Contribution to Physics. in Selected Papers of Chen Ning Yang II* (2013), p. 78

Physics and Cultural Industry

Media, Imagery, and Formats for Dissemination in Italy: The Case of "The Shifters"

Maria Chiara Di Guardo, Emiliano Ilardi, and Laura Poletti

Abstract The present chapter examines the potential of the "scientification" of the cultural industry as an opportunity for science communication, particularly in the context of school and university education. In an age where the boundaries between science and fiction are increasingly blurred, it becomes crucial for science communicators and science journalists to play a role in mediating and facilitating a proper dialogue between these two realms. In this contest characterized by a multimedia and multichannel era, science communicators should not restrict themselves to traditional formats designed for television broadcasts, even if these formats are adapted for digital platforms. This chapter aims to describe the model that inspired "The Shifters: The Third Mission" (www.theshifters.it), a format developed by CREA—CenteR for Entrepreneurship and innovation Activities at the University of Cagliari (Italy). The format employs an innovative approach to transmitting scientific content, combining fiction and research, cinema and blogs, digital tools, and various forms of media. The case of the University of Cagliari demonstrates how the need to engage a wider and more diverse audience through clear and diversified messages can provide a more comprehensive and engaging experience for the public. The format proposed by The Shifters highlights how a detailed understanding of the target audience, comprehensive analysis of content, and careful design of connections between different media can be key elements of an effective dissemination strategy for research results.

Keywords Dissemination · Science communication · Storytelling · Transmedia

---

M. C. Di Guardo (✉)
Department of Economics and Business, Via Sant'Ignazio, 74, 09123 Cagliari, Italy
e-mail: diguardo@unica.it

E. Ilardi
Department of Social and Political Sciences, Via is mirrionis, 1, loc. "sa duchessa", 09123 Cagliari, Italy
e-mail: ilardi@unica.it

L. Poletti
Research and Territory Management, Via Ospedale 121, 09124 Cagliari, Italy
e-mail: laura.poletti@amm.unica.it

M. Streit-Bianchi et al. (eds.), *New Challenges and Opportunities in Physics Education*, Challenges in Physics Education, https://doi.org/10.1007/978-3-031-37387-9_10

1 Introduction

In the first half of the nineteenth century, between France and England, the embryonic nucleus of modern cultural industry developed, strongly linked to the birth of serial and consumer publishing that began to be structured into "genres": science fiction, historical novel, the romance, the thriller, the adventure novel, the gothic, and the horror (Ragone and Ilardi 2023). This new way of producing series has as an effect the need of the cultural industry to have continuously new raw materials to be able to model narrative. And where does he find all this new content he needs if not in the world of academic research? The historical novel plunders historiography; detective story and horror chemistry, biology, and medicine; the adventure novel geography, biology, archaeology, and anthropology. And of course, the tougher sciences like physics will be the reservoir of science fiction ideas.[1]

2 Physics and the Cultural Industry

The progressive transfer of scientific knowledge from the academy to the cultural industry takes place in a different way depending on the literary genre of reference. It is much stronger, more detailed, and more realistic for chemistry, biology, and medicine (think, for example, the wide use that makes these disciplines the detective story from Sherlock Holmes to Hercule Poirot), than physics. Science fiction will take many decades to notice that the real revolution is represented by the Theory of Relativity and Quantum Mechanics. For at least a century, from Jules Verne until the 1960s, science fiction *mainstreamed essentially* a Newtonian physics in which time and space were considered as separate, separable, or interchangeable entities, totally reducible to each other (especially treating time as if it were space). And this even though new theories on space–time had already been enunciated in the first twenty years of the twentieth century. For a century, therefore, the physical laws on which the narrative that produced a mass sci-fi imagination was based have substantially ignored the great discoveries of the twentieth century giving rise to two generally independent subgenres: the *cosmic* journey (*man* exploring space *or aliens* reaching Earth) *and time travel* (from the present to the past and future or from the future to the present). The only great novelty of twentieth-century science fiction, especially since the Second World War, is the centrality of particle physics (radioactivity) due to the development and use of atomic energy in the war and

[1] There is a famous anecdote to which the encounter-clash between the cultural industry and the world of university research can be traced. In 1819 is published what is conventionally considered the first best seller in the history of publishing: Ivanohe Walter Scott is guilty of spreading a stereotypical image of the Middle Ages that still survives today with figures like Robin Hood, Richard the Lionheart, the Black Knight, medieval jousts, etc. After publication, British academic historians accused Scott of inventing a non-existent Middle Ages by plundering their medieval research. And Scott answered them, paraphrasing the Humphrey Bogart of the famous film Deadline—U.S.A.: "That's the fiction, baby. The fiction. And there's nothing you can do about it. Nothing".

civil: a subgenre that, however, immediately moves into the biological environment with the popular theme of mutation. And there is always more biology than physics behind the alien colonization of Earth. There is engineering and computer science more than physics behind the robot and computer strand from Isaac Asimov to 2001: A Space Odyssey (1968). And above all physics is practically absent in dystopian science fiction—from Huxley's New World (1932) to Philip K. Dick and James G. Ballard—which has essentially political-social criticism purposes.

It is only perhaps with the great success of Planet *of the* Apes (1968) *and Star* Trek (1966) *and Doctor* Who (1963) that Einstein finally peeks into the mass imagination and space and time begin to be connected in the same narrative. But it is just hinted. The separation between essentially Newtonian science fiction and, on the other hand, contemporary physics, which in the twentieth century goes totally beyond Newton, has always made it difficult an effective science dissemination. For decades science communicators have never had the simplified metaphors offered by the cultural industry to explain to the public the theory of relativity or quantum mechanics. He was himself, think of Piero Angela in Italy, who often had to build examples, simulations, or animations to make more understandable the increasingly complex and counterintuitive concepts of contemporary physics. The real problem is in fact to invent new metaphorical constructs to describe advances in science often far from common sense (Stancampiano 2022).

3 The "Scientification" of the Cultural Industry

But things have changed especially in the last twenty years with the consolidation of a digital media sphere and an increasingly pervasive and globalized cultural industry based on entertainment platforms (van Dijck et al. 2018). The new digital cultural industry takes to the extreme its traditional voracity of new content, and this could only produce a dizzying increase in its "vampirization" against academic scientific knowledge. Platforms such as Netflix, Amazon Prime, Sky, Disney Plus, or (for those who play video games) Steam, Epic Games are in fierce competition with each other, and the winner is the one who manages to dispense at a continuous pace as many products as possible in quantity and quality. This leads them to monitor constantly and in detail any discovery or advancement that takes place in the science field. It is therefore no coincidence that one of the perhaps most interesting phenomena of the last twenty years is the increase in the rate of realism within science fiction stories. And this "scientification" of the cultural industry can be a great opportunity for dissemination and especially for school and university teaching.

Let us take as an example one of the greatest "topos" of science fiction: time travel (Farina 2016). From Herbert G. Welles' *The Time Machine* (1895) to R. Zemeckis' *Back to the Future* (1985), it was not important to know exactly on what physical basis the time travel was based; essentially, it was an engineering issue: There was a scientist, generally eccentric and bizarre, who invented the time machine, ingeniously combining chemical elements and energy sources. What mattered was the narrative effects, and the temporal paradoxes it caused.

If, on the other hand, we analyze the most recent narratives regarding time travel, we realize that it needs to be told "scientifically"; "scientifically" obviously does not mean that fiction offers scientific truth, but that the narrators must explain to the public in a credible way how they have managed to violate or exploit the laws of physics or to make feasible some hypotheses that physics considers possible but not feasible by man, or at least not in this historical era.

The most emblematic example is the work of the British director Christopher Nolan whose objective is to bend the plot to the laws of physics and not vice versa as happened in previous science fiction. In *Tenet* (2020), it is the second law of thermodynamics and the possibility of reversing the arrow of time to unhinge the linearity of the film story. For *Interstellar* (2014), he even used the advice of Nobel Prize winner Kip Thorne to verify the narrative correct use of Einstein's Theories of General and Restricted Relativity. It is a fact that the increase in scientific realism in the stories of the cultural industry has had the effect of a proliferation of requests for advice from scholars and researchers, which shows the progressive rapprochement in recent years of the two areas of fiction and academy.

Signs of a "realistic" representation of the laws of physics are found today in many successful TV series such as, *Fringe* (2008–2013), *Travellers* (2016–2018), *Inverse* (2022) even a sitcom like *The Big Bang Theory* (2007–2019). Not to mention video games, starting with that pioneering title and watershed that was Half *Life (1998)* in which the protagonist is just a theoretical physicist. And then there is the journey into the memories of *Assassin's Creed* (2007), the parallel universes of *Bioshock Infinite* (2016), the space–time distortions of *Quantum Break* (2016), *and Steins*; Gate (2009), the quantum and gravitational theories of *Portal (2007).* The more general increase of realism in videogame narratives has led designers to create game mechanics and simulated realities more and more adherent to the laws of physics. Think, for example, of the bullet drop phenomenon in so-called shooter video games that requires players, especially professionals, to know the laws of ballistics and therefore of gravity (*Battlefield 3, Max Payne, Sniper Elitema* also simple games like *Angry Birds*); or to the thermodynamic component of some games in which the gamer must take into account the temperature that can expand or reduce the size of the object that must use (*Frostpunk*s).

Those who want to dedicate themselves to the dissemination and/or teaching of physics must, therefore, consider two interconnected phenomena: The first one, long-lasting, has always seen the cultural industry plunder the hard sciences to the spasmodic search for new contents. The second one of short duration shows a significant increase in the rate of scientific realism (and therefore of physics) within the stories. The reasons may be multiple: a sharp increase in the voracity of content by digital entertainment platforms but also the demand, by today's users, of ever more complex and intricate plots capable of including even the tiniest details of scientific discoveries.

This closer connection between science and fiction has two opposite consequences. On the one hand, it offers today's science communicators, compared to the past, an audience much more literate about the progress of physics and at the same time a large pool of content and metaphors on which to build its specific narrative popularization or teaching. On the other hand, however, it leads to a dangerous proliferation of increasingly refined and credible pseudoscientific theories if not a conspiracy because they are often modeled on the complex narratives of the cultural industry.

4 The Science Communicator

It is precisely in the age, in which science and fiction risk blending excessively until the boundaries disappear, that the figure of the science communicator is necessary, and the science journalist, able to mediate and properly dialogue fiction and science.[2]

This new figure is being born above all from below, and often it is the same academics to embody it by opening dedicated channels on social media. In Italy, for physics for instance, it is worth mentioning the popular channels: "Curiuss" (www.youtube.com/@Curiuss/featured), "link4universe" (www.youtube.com/@link4universe), "La fisica che ci piace" by Vincenzo Schettini (www.youtube.com/@LaFisicaCheCiPiace/videos) which is especially popular on TikTok, or that of Amedeo Balbi (https://www.amedeobalbi.it/). Especially the latter, a researcher at the University of Rome Tor Vergata, has no problem in using film and television products for dissemination (Interstellar, *Tenet*, etc.). He understood that the greatest difficulty in understanding many of the theories of physics is their extreme abstraction or distance from common sense. The cultural industry produces new metaphors and similarities in a context in which theories are seen at work concretely, and therefore carries out an embryonic educational activity of which school, university, and social contest should consider. Young people today who consume science fiction or fantasy, who saw *Star Trek Discovery* (2017–), *Black Mirror* (2011–), a film by Christopher Nolan, read the popular manga *Dr.* Stone (2017–2022) or have played video games know the laws of physics that underlie space/time better than previous generations who read Frank Herbert, watched Back to the Futuroo played Pac Mane Mario Bros. That is, they possess a much more realistic scientific imagination than previous audiences, which they could rarely meet in their Einstein readings or quantum mechanics. Of course, it is pure imagination, not organized, often fallacious because it is subordinated to storytelling or gameplay; but that cannot be ignored by the science communicators as a fundamental tool of connection with its audience and communicative effectiveness; instead, it should be a privileged starting point for any didactic and informative action. If the cultural industry takes scientific knowledge and takes it into the domain of fiction, the communicator should accompany users in the opposite direction showing them the boundaries, albeit porous, between the two areas.

---

[2] https://www.ilpost.it/2022/01/08/giornalismo-scienza/?fbclid=IwAR00dXHE40IYP73LEzacdM2UfoolTrdEaJfRMVBSbkDvVHgWuxsDfCwvn3w.

5 The Shifters

Based on these observations, it is reasonable to argue that in a multimedia and multichannel era, science communicators should not limit themselves to using only traditional formats that are broadcast on television. In this context, the recent experience of "The Shifters: The Third Mission"[3] is particularly noteworthy. It is a multimedia format developed by Center for Entrepreneurship and Innovation Activities (CREA)[4] at the University of Cagliari, which combines fiction and research, cinema and blogs, digital tools, and various forms of media to transmit scientific content developed within the university.

The main question that *The Shifters* aims to answer is whether it is possible to explain something as difficult and complex as physics, or more in general academic research, in a simple yet not simplistic way, without compromising the importance of the underlying message of discovery and research. The search for an answer to this question involves a shift away from using science solely for entertainment purposes and toward a more evolved approach that recognizes the growing interest in the evolution of science, the impact it has on everyday life, and the curiosity and enthusiasm demonstrated by an increasingly engaged audience in recent years. The project explores the potential of digital platforms and various forms of media to achieve this goal, using metaphors, stories, and digital technologies that are familiar and associated with relaxation and leisure. Like web series, books, and films, science and physics have a plot and specific characters that are capable of arousing interest, wonder, and active participation.

The Shifters is, therefore, a pilot project of transmedia communication of a scientific nature that is proposed as a new model of science popularization, starting from a declared mixture of fiction and science. The project involves activities of "collective contamination" between university research, science, and fantasy (including various types of actors, with different levels and modes of involvement and interaction) making the most of research through a transmedia narration that gives its users the opportunity to navigate and deepen the information and knowledge validated by university researchers related to a specific topic based on their information needs, without imposition and according to their level of curiosity and interest.

The transmedia communication model describes the dissemination of a message across multiple platforms and media. Each platform and medium contributes to creating a unique version of the message, but together they form a coherent experience for the audience. This model emphasizes the importance of coherence and continuity of the message through the different media.

Transmediality can be used to describe science in several ways:

(1) Integration of different platforms: use different platforms, such as videos, animations, infographics, and text, to explain scientific concepts in a more complete and accessible way.

[3] https://www.theshifters.it.

[4] https://www.crea.unica.it.

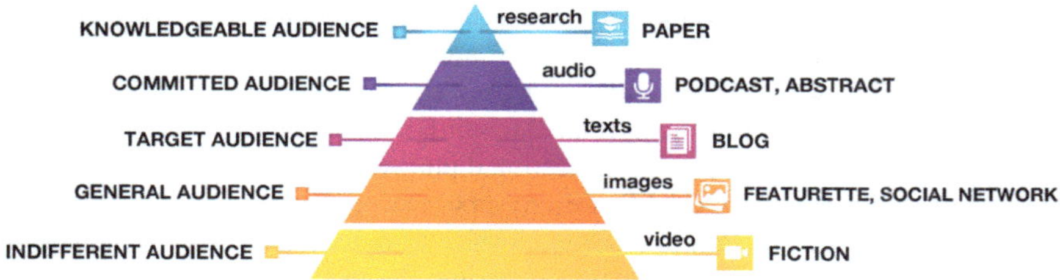

Fig. 1 Our elaboration I model "The Shifters"

(2) Immersive Narrative: Create an immersive experience that incorporates visual, audio, and interactive elements to engage audiences in scientific discovery.
(3) Use of social media: use social media to spread scientific information and involve the public in discussions and debates on scientific issues.
(4) Interdisciplinary collaboration: Working with professionals from different fields to create transmedia experiences that describe science in a more engaging and accessible way.

In line with this objective, *The Shifters* is at the same time a web portal that contains different media for different levels of information. A transmedia portal built according to an informative pyramid (Fig. 1) that sees at the base level of information an anthological web series that contains the research but does not reveal it explicitly. Today, in fact, it is possible to find on the platform the web series composed of three episodes, each of which autonomous and self-contained, to be enjoyed independently, with a different scientific focus, air pollution, dependence, smart cities (inspired by the 2030 UN Agenda[5]). The web series investigate the developments and interactions that scientific research, including physics, can have on society, through a story that represents a fiction, a story created by human intelligence to activate curiosity and interest. Each episode hides inside information and elements born from the scientific research developed by the University, has a duration of 10–12 min, and can be easily enjoyed on the web as a short film. Inside the platform the web series is accompanied by a series of structured documentaries, featurettes, which tell the background on how research and narrative find a point of contact within the individual episodes. The featurette is where the researchers unveil the scientific research elements concealed within the narrative, which then give rise to levels of information that become increasingly more in-depth: from podcasts and blogs to a higher level of information consisting of scientific papers that have inspired the fiction and can be downloaded and consulted for free by the most curious and interested users.

Different messages for different audiences and especially different media, with the aim of involving the public and open to real science dissemination. To engage the public and achieve true science dissemination, it is important to tailor messages to different audiences and media. However, when communicating complex concepts in

[5] https://unric.org/it/agenda-2030/.

a way that is accessible to non-experts, it is crucial to remember that every communicative step involves selecting, reordering, and transforming information. From a structural point of view, the model is organized according to different levels of involvement: At the base of the pyramid, scientific papers aim to provide in-depth information, podcasts have a more educational purpose, the blog tries to inspire the public and social networks establish a conversation, up to the pure entertainment obtained through the creation of a fiction, a short film, telling a "story" engages the most distracted audience and urges them to begin the journey of deepening. In this way the contents are not produced randomly but according to a clear process aimed at obtaining value.

The format is influenced by the platform on which it is presented, the objective of science communication and the audience to which it is intended. Through *The Shifters*, users can take part in the experience with different levels of participation[6]: those who are content to follow the story and those who read the blog because they want to deepen the topic or who download the scientific papers to reach the highest level of depth. This model of disclosure can be represented through a pyramid (Fig. 1).

Broadly speaking, an attempt is made to classify several users from their initial interest in science topics and to associate the optimal communication tool with each user. The lower the interest, the more difficult it is to attract attention to pass the content, and therefore the more creativity and impact of the tools deployed to convey the message must be increased.

6 Conclusions

Starting from the observation that in the last twenty years there has been an increase in demand for "scientific" realism in science fiction stories, a greater attention to "likelihood" and the impact that the speed of technological change can have on modern society. Observing how the development of physics, and of quantum physics, is opening to new "worldviews" related to uncertainty and uncertainty, which also extends to space and time, with the scientific development of some suggestive theories that are inspiring the cultural industry suggesting the existence of many parallel or multiverse realities, where each reality could have its own quantum laws. It is possible to highlight a need linked to transmediality: which can be described as the combination of a model in which narrative lines develop horizontally, within the same medium, but also vertically, passing from one medium to another and leaving the task to the public to deepen and create the right connection to have a global vision of the story narrated. Along this line of reflection are experiments such as *The Shifters*, which tend to respond to a narrow audience (like that of experts), characterized by a high level of interest in science and research, providing content with a very high degree of depth and the consequent use of tools such as science publications that have a low degree of involvement. At the same time, the format proposed by The Shifters

[6] https://www.theshifters.it/come-navigare-in-the-shifters/.

is structured to involve a large audience with a low level of scientific competence, that through the media tool like the web series puts more emphasis on the fiction content than the scientific one to involve the public more to create an access door for this type of audience and then let it explore independently the different levels proposed by the transmedia narrative, moving at its discretion from one medium to another according to its information needs.

In a complex context, science popularization must respond to the challenges creatively brought by innovation. The web series is one of the possible choices for communication, but it is not the only one. The model illustrates that, when targeting a skeptical or secular audience, it is not necessary to create complex content with technical and articulate language. Instead, simple messages exploiting science and research's narrative power through video formats can lead to greater public involvement. However, it is crucial for science communicators to clarify and not blur the fine line between entertainment and research.

References

F. Farina (ed), *Viaggi nel tempo* (Einaudi, Torino, 2016), p. 229. ISBN-10 8806229508, ISBN-13 978-8806229504

G. Ragone, E. Ilardi, *L'editoria in Italia. La quarta generazione* (Liguori, Napoli, 2023) (in press)

D. Stancampiano, "Girandole, cannibali e Big Bang. La metafora nella divulgazione della fisica", in *Confini e sconfinamenti*, Eds I. Candelieri, C. Daffonchio (EUT Edizioni Università di Trieste, Trieste, 2022), pp. 205–220

J. van Dijck, T. Poell, M. de Waal, *The Platform Society: Public Values in a Connective World* (Oxford University Press, 2018), p. 240. ISBN-10 0190889772, ISBN-13 978-0190889777

Teaching, Communication, and Dissemination for Society

Matteo Tuveri, Elisabetta Gola, and Matteo Serra

Abstract The dissemination of scientific culture in society is assuming new forms. On the one hand, digitalization and the affirmation of social networks imply the implementation of a new communication strategy to fulfill people's need of knowledge. On the other hand, science is not free of interconnections and is moving toward an interdisciplinary approach, mixing disciplines and languages. As a result, the language of science is changing and so is the language of science communication. What are the new challenges of science communication? How can science be communicated and taught in the new millennium to promote the learning of science in society and to strengthen cultural awareness of scientific issues? In this chapter, the authors discuss the features and the evolution of teaching, communication, and dissemination of science, offering new (technical and digital) strategies to build an effective way to use the potentiality of natural language to spread and teach science in our society. Based on the scientific literature, the arguments are shown in a form of dialogue, inspired by the famous Galilei's "Dialogue Concerning the Two Chief World Systems".

Keywords Teaching · Science communication · Dissemination · Physics education · Outreach · Scientific culture · Learning of science · Language · Scientific awareness · Scientific literacy

---

M. Tuveri (✉)
Dipartimento di Fisica, Università degli Studi di Cagliari, Cittadella Universitaria di Monserrato, 09042 Monserrato, Italy
e-mail: matteo.tuveri@unica.it

Istituto Nazionale di Fisica Nucleare, Sezione di Cagliari, Complesso Universitario di Monserrato, 09042 Monserrato, Italy

E. Gola
Dipartimento di Pedagogia, Psicologia, Filosofia, Università degli Studi di Cagliari, via is Mirrionis 1, 09123 Cagliari, Italy

M. Serra
Istituto Nazionale di Fisica Nucleare Sezione di Cagliari, Complesso Universitario di Monserrato, 09042 Monserrato (Cagliari), Italy

M. Streit-Bianchi et al. (eds.), *New Challenges and Opportunities in Physics Education*, Challenges in Physics Education, https://doi.org/10.1007/978-3-031-37387-9_11

1 Introduction

The history of Western culture has been an evolution toward the conquest of scientific knowledge at least since the birth of philosophy. In this process, the explanations and reasoning progressively had been moving back from the sphere of religion, myth, and magic to become more and more founded in a scientific method. Yet the two spheres continue to coexist, and we still find processes connected to both spheres in our societies, media, and educational institutions. We can think, for example, of the hybrid arguments about issues that are fundamental to our existence: vaccines, nutrition, even meteorology. Many people are willing to hold on them beliefs that are not based on scientific knowledge, but on acts of faith (not necessarily toward a deity, but toward a leader, an opinion leader, or an influencer). One only must review the abundant literature on fake news to realize the extent of the problem. To fight against this problem, we are assisting at the advent of schools, media (especially newspapers, but also many social profiles) and professionals specifically concerned with popularizing science working hard to counteract information that is false, unfounded, or based on approaches outside science.

For several decades now, scientific research has changed its face in the relationship with society. Today science is no longer a world unto itself, where researchers live in their ivory towers disconnected from the real world, but is a reality now closely connected to all major players in civil society, from politics to industry and to the communication world. People ask science to satisfy their needs, and, in this process, science is perceived as synonymous with technology (Gouthier and Ioli 2016). The link between the researchers and the public, the recipient of the many initiatives for the dissemination of scientific research, is particularly delicate and important. Here, too, conditions are very different from the past: if once the communication approach was based mainly on the so-called "deficit model" (see Simis et al. 2016 and refs therein), which aimed at a simple transfer of information from experts to the public (considered a completely passive subject), today the keyword in science communication is "interactivity" (Trench 2008). As a result of the explosion of social networks and the quick evolution of the media, today the public can dialogue and directly interact with experts, thus finally becoming an active part in the process of dissemination of the research results and their implications on society. This interaction is not always easy, peaceful, and effective, as the communication related to the recent COVID-19 pandemic also demonstrates. However, it is impossible to turn back: the relationship between researchers and non-expert citizens is traveling in a direction where dialogue and direct interaction will be increasingly dominant. Partly because of this, it is now almost a requirement for scientists to complement their research work with a serious and effective commitment to public science communication. An effort that must aim not only to disseminate the results of one's research but also to make people understand more generally the mechanisms of research and the scientific method.

A key point in generating a virtuous and profitable relationship between scientists and citizens is to analyze which way the two perceive science. It is generally diffused the idea that only the science that can make use of mathematics is the real science. This is a common belief today as it was in the past, especially in academia. Aristotle had already realized this when formulating the principles of logic, dialectics, and rhetoric (Piazza 2004). Human sciences are usually not even considered scientific: either they have hard-won hybrid status, such as sociology, which distinguishes between quantitative and qualitative methods (Corbetta 2014), or they are considered just disciplines that tend toward science, such as psychology (Engel 2000).

If, on the one hand, the use of mathematics as the language of science dates back to the works of Galilei and Newton, on the other hand, many studies have underlined the role of imagination in building scientific knowledge. The use of metaphors to imagine the world of the invisible and, therefore, to build an (even detailed) description of phenomena plays a central role in science. Einstein himself, in a letter to Jacques Hadamard, mentioned the importance of visualizing the world through images: "The words or the language, as they are written or spoken, do not seem to play any role in my mechanism of thought. The psychical entities which seem to serve as elements in thought are certain signs and more or less clear images which can be 'voluntarily' reproduced and combined". (Hadamard 1954). Lakoff and Núñez (2000) maintained, for example, the centrality of conceptual metaphor to a full understanding of many ideas in mathematics, as the concept of actual infinity, specifically.

Referring to physics, one of the most famous cases is how knowledge of the atomic world was conveyed through the image of the solar system (Rutherford–Bohr model, 1913). Soon, Bohr himself and other scientists realized that the mathematics used to represent the motion of the planets could not explain the motion of electrons in atomic orbits, nor the phenomenology of atoms bigger than Hydrogen (Rabatzis and Ioannidou 2015). However, the mathematics used today in Quantum Mechanics was developed based on this "false" metaphor, which nevertheless forms the basis of this branch of physics. The use of metaphors as a guide to exploring nature is still evident in other areas of physics. This is the example of Einstein's General Relativity, where new words such as "black holes," "Big Bang," or "primordial soup" have been invented to talk about the final state of matter after the explosion of a massive star, the origin of the universe or the initial state of matter and energy in the universe, respectively.

Analogies, metaphors, and other linguistic resources are quite used when scientists and professionals communicate science to people. The role of Stephen Hawking or Einstein itself in outreach through their books (Hawking 1988; Einstein and Infeld 1938) has been fundamental to bringing high-energy physics themes into society. The possibility to imagine the invisible mediating through language tricks inspired many generations of high school students, leading them to be passionate about physics and, in some cases, to work in this field. Many other examples could be given in this direction (Root-Bernstein and Root-Bernstein 1999; Postiglione and Angelis 2021).

Appropriate use of the various forms of language (mathematical and natural, formal, and informal) is important also in education, especially in didactics. The interplay between formal and non-formal language becomes evident when we pass from formal traditional teaching and learning of science (typically in schools, with final evaluation and certification) to non-formal (such as in summer school, master-class and similar for the former, with a limited evaluation of learning) and informal environments (such as in museums, science shows, social networks and similar with no evaluation at all) (Oriji and Uzoagu 2019; Ucko 2010). In the context of informal learning of science, the experience of learning a specific subject is tacit and implicit, placed in the background of the experience of playing with and freely enjoying science (Hall 2009; Ward et al. 2013; Michelini 2005). In all these approaches, a student-centered environment for learning (Land et al. 2012), regardless of its shape and context, is needed. In this way, the language of science assumes new features, adapting and transforming according to the audience.

Simplifying the language without losing the rigor of the discipline is the most difficult challenge for every communicator, being him/her a scientist, a teacher, or a professional in communication. The more learning is devoted to non-experts, the more images should be produced to visualize and think about the phenomena, fostering the audience's motivation toward science and its creativity (Petrescu 2016; Gobet 2015). Such a process should be well defined by also communicating the limits of such analogies, thus preventing possible misunderstanding and false images of the world (see Teixid et al. 2019; Watkins 2014 and refs therein as an example). The use of a hybrid forms of languages, where science contents are communicated through visual thinking (Coll 2005), goes in the direction of using new digital environments for teaching and learning. This is the case with social networks, where new opportunities are now appearing to share science with a large audience (Otte and Rousseau 2002; Zaidieh 2012). Nowadays, offering a contemporary vision of science to high school students and teachers should be a common goal of outreach activities and informal learning of physics mediated by researchers. Indeed, science is evolving, and its forms and languages are changing. To give new instruments to learn science and physics in an enlarged context, mixing knowledge, techniques, and methods from different disciplines is becoming a priority for scientists and society (Pan et al. 2012; Pluchino et al. 2019; D'Este and Robinson-García 2023; Spelt et al. 2009; Davies and Devlin 2007).

For all these reasons, we present here an overview of the state of the art of the evolution of science teaching, dissemination, and communication for society. Inspired by the "Dialogue Concerning the Two Chief World Systems" of Galilei (1953), in the following, the authors of this chapter discuss the world and the vision of science, offering a scientific argumentation on how the language of science is changing. The discussion is divided into four sections referring to formal and non-formal languages, that is, how science is modifying its habits to find new answers to old and present questions, as well as, how the communication of science is adapting to these rapid changes. The dialogue ends with a discussion on the role of interdisciplinary in science communication and in the teaching of science at school, giving insights into recent methodologies now explored in the field of education. All along the dialogue, they offer some reflections and analysis based on the literature cited in the bibliography. MT stems for Matteo Tuveri, EG for Elisabetta Gola, and MS for Matteo Serra. The chapter ends with a summary of the contents dealt with by the authors and reflections on the future of science communication and education.

2 The World and the Vision of Science

MT: "What is science? Physics, mathematics, and chemistry are good examples of what, at least in the collective imagination, is thought to be a good representation of what science is or should be. Is it something based on the language of mathematics, something very rational, capable of offering solutions to problems, and leading our minds out of our common sense to discover what nature is made of, or something else? So it is, for scientists as well as for everyone else. But science is also perceived as capable of solving people's problems, and it does with the help of technology. In some sense, in the common imagination, science is technology, built for satisfying everyone's needs. For someone else science is also perceived as a difficult subject, often involved in studies about the invisible, not so urgent nor so important. So, what can we say about science?".

EG: "The world of science is a vast and heterogeneous field, not least because we should first ask ourselves what we mean when we speak of science. Certainly, if we must consider a prototypical case, we can point to disciplines such as physics, mathematics, and chemistry, on which there are no controversial positions on their nature: they are science and they deal with scientific subjects".

MS: "I would also add that science does not produce any irrefutable truths, but rather a set of "provisional" truths that can always be challenged by discoveries and insights: correctly explaining this and other key features of research is essential to improve the public perception of science, and scientists themselves must do this, in addition to professional popularizers and communicators".

MT: "So, if science is not infallible, it does not produce eternal truths but rather local and temporary models of nature, corroborated by experiments and rigorous studies on specific subjects, why there is not a social agreement on that? What is missing in our society? Maybe the communication itself?".

EG: "Indeed, the skill profile and how science is disseminated, taught, or communicated differ from discipline to discipline. "Physics is a science": this is a statement on which everyone agrees. The existence of a clear method such as the Galilean "scientific method" based on experiments, verification or falsification of someone's theory or hypotheses about a phenomenon ensures the status of science. On the contrary, life sciences, such as biology and even more medicine, are fields in which the actual scientific method coexists with 'artisanal' or 'artistic' acts that are an integral part of the knowledge for those who engage in these disciplines. Human sciences are usually not even considered scientific: either they have hard-won hybrid status, such as sociology, or they are considered just disciplines that tend toward science, such as psychology".

MT: "Maybe, we should communicate not only the discipline and its results but also the world of science. Shouldn't we?".

MS: "Indeed, we still too often refer to the research as something impersonal, forgetting that scientific research is made by people (researchers), with all their baggage of human stories, talents, and weaknesses. The "humanization" of science in the eyes of the public can be one of the missing ingredients, a key step in fostering and improving the dialogue between the research world and citizens, helping to decrease the level of distrust in science that still exists".

3 The Language of Science

EG: "There is a common belief now as in the past that the sciences that can make use of mathematics are the real sciences. Thinking about it, Aristotle had already realized this when formulating the principles of logic, dialectics, and rhetoric, and it emerged in all its epistemological force in the authors who explored the possibility to build a language specifically for science (a kind of meta-mathematics) and capable of maintaining the 'truth' in the transition from one statement to another. These languages, universal and perfect, have been sought since at least Ramon Llull in the Middle Ages, and have been explored with greater frequency and commitment in the age of the scientific revolution by Bacon, Descartes, and Leibniz".

MT: "However, mathematics and formal descriptions alone are not sufficient to build science, nor the language of science. What is scientific progress if not a texture of past knowledge mixed with some new idea to build the new theories? There have been moments in history in which no discoveries at all were possible. Someone invoked revolutionary people to change the status quo, bringing new revolutionary ideas into the game.[1] There is also a smooth way to see it by the introduction of accumulation points of knowledge well interpreted by people who, studying the past and invoking the transversal thinking, made a synthesis of contents to build the future.[2] But knowledge passes through any form of language, to build it and to transmit it. Language, in turn, transforms over the years and, thus, passes through knowledge. Looking at scientific problems in a different way led to the emergence of new mathematics, and new terms in the natural language to refer to it. Therefore, new concepts to explain and communicate a new vision of the world manifest in time".

EG: "If we look at the history of science, we can notice that it is imagination, in most cases, and not a formal automatism, that makes knowledge progress. We were under the illusion for a long time that language was a compositional and deductive system, and that we could almost automatically move from axioms to theorems. But today we know that our possibilities of expression are anchored and bound to the way we are structured, that is, they are 'embodied'. That is why it is no coincidence that we

[1] See Kuhn (1962) for details.

[2] See Morganti (2016) for details.

speak of progress as something that moves forward, for example, because our motor patterns translate into symbolic and metaphorical patterns. Conceptual metaphors are central to a full understanding of many ideas in mathematics, as the concept of actual infinity".

MS: "The communication of physics, together with other scientific disciplines, is emblematic in this case, providing an interesting reference case study. Indeed, those who communicate physics and its discoveries have always had a major problem to overcome. Unlike other disciplines, often the 'objects' of physics cannot be perceived with our senses. No one has ever 'seen' with their own eyes an electron, 'touched' a gravitational wave, or 'felt' a neutrino. This is a big problem because all forms of communication play primarily on the solicitation of the audience's senses".

MT: "This is, for example, the case of Quantum Mechanics (with new terms in correspondence of a new phenomenology such as 'quantum', 'wavefunction', 'entanglement', 'wave-particle duality', ...), or even of General Relativity ('spacetime', 'warping of space and time', 'gravitational waves', 'black holes', 'primordial soup' and so on), in physics, where new terms had to be invented to understand, describe, and communicate the manifestation of nature to scientists and people, too. Quantum mechanics looks in detail at the smallest scales of the universe, whereas General Relativity looks at the largest ones. Natural language, as well as the mathematics, are involved in both cases to describe two invisible sides of the same coin. The former manifests itself with new images to represent and describe new scientific images of the world. The latter glues the experiment with logic and a rational description of phenomena, building models and falsifiable predictions of peculiar features of nature. In both cases, a language is needed, and language (natural or mathematics) and learning are deeply interconnected. They are intertwined subjects and knowledge is a result of their interplay. Isn't it?".

EG: "Yes, they are. One of the most important results of this interplay is one common expedient used by scientists to visualize the invisible: using images, more specifically, metaphors. Imagination and, metaphors, have always played a central role in science. Physics offers a good example in this direction. Let us think of the most famous images of physics, that is the atomic world as the solar system, black holes, or the primordial soup. None of them exactly represents the actual manifestation of nature in specific conditions (electrons and atoms, the gravitational collapse and the formation of a black hole, or the initial instants of the universe, respectively), but they help in imaging the invisible world. Metaphor is a cognitive process that we constantly use to try to understand the things we are exploring, to describe them, and to find explanations and laws. A theory is not formulated from scratch, but we conceive new hypotheses using something we already know, including expectations and biases. In the case of metaphors, we refer to an area of inspiration and project the structure, the relationships among the elements, onto the realm we want to know. In this process, we find confirmations and rejections, and in this process, new knowledge emerges. The sciences that can make use of mathematics are those that have come closest to the goal. However, if we look at the history of science, we can notice that it is imagination, in most cases, and not a formal automatism, that makes knowledge progress".

4 The Communication of Science

MT: "What is emerging from our discussion is that knowledge passes through languages. Science has its languages: the one based on mathematics and the one based on the natural language upon which the scientific understanding and learning of phenomena are built. The former is often thought of as exclusively for experts, due to its technicality. The latter is the most fluid one: it can easily assume the shape of a formal language when it is used in a specific and technical way (e.g., scientific English, the explanation of a mathematical demonstration, or the exact description of a phenomenon according to a general model). It is also the instrument to communicate science to non-experts, in outreach events, magazines, tv-shows, social media, newspapers, etc. In all these cases, the speaker is asked to talk simply, emptying the communication from all the technicalities and the specific formalism of the discipline. Thus, we are left with two ways to explain the concept which depends on the public to whom we are referring".

EG: "And the principles of communication are the backbone of many professionals: journalists, teachers, writers of manuals, social media managers, physicians, and even those who work as clerks in universities and public sector administration in general. However, they rarely receive specific training in scientific communication (and in some cases, e.g., teachers and clerks, in communication tout-court). But this is necessary to be able to design and implement an effective communication: it is not just a matter of translating concepts in words, knowledge and language are two sides of the same coin".

MT: "Speaking metaphorically, who will be the final user of this coin? On the one hand, we said that there are people involved in communicating science, such as journalists, teachers, or science communicators. On the other hand, the recipient of the communication, the one who will make use of all this information, is the public, made by scientists or people in general. Thus, in the process of disseminate, explaining, and telling science, the public is at the center of the communication and the language plays the role of the theater where communication is on stage. So, how to communicate science, especially to the public, to society?".

MS: "Well, we can make some specific examples of good practices of communicating science to the public. In 2016, the LIGO and Virgo collaborations first announced the discovery of gravitational waves,[3] ripples of space–time produced by violent astrophysical events, predicted by Albert Einstein's general theory of relativity. To

[3] See Abbott et al. (2016).

bring the public closer to this great achievement, during the press conference where the discovery was announced, LIGO and Virgo researchers played an audio transduction of the gravitational wave signal, which was very helpful to understand more precisely the 'shape' of the signal. In 2019, researchers from the Event Horizon Telescope collaboration instead published a 'picture' that quickly went around the world: it was the image of the shadow of the supermassive black hole at the center of the M87 galaxy.[4] It was the first time one could literally 'see' a black hole (or at least its edge): needless to say, the effect on the audience was extremely powerful. The same effect was then replicated (if not enhanced) in 2022,[5] with the publication of the first image of the black hole at the center of our galaxy, the Milky Way. In formal or non-formal communication, solicitation of the audience's senses, whether it is an image, a video, an audio track, or even an object to touch (think for example about hands-on installations in science museums) is fundamental. This could be a good practice in communicating science, especially to non-experts".

MT: "I completely agree with you. Let us imagine what happens in the case of formal communication, even from experts to experts. In this case, the communication plan is based on involving the capacity of giving a sense of reality to abstract symbols, or languages, as in the case of formal and technical scientific communication, for example, when mathematics is used. The solicitation of imagination using metaphors as well as analogy is even stronger in the case of informal learning or dissemination of scientific knowledge (outreach events) to people. Playing with objects and participating in hands-on and mind-on activities is important to help people in learning, even in informal contexts such as science shows or museums. An immersive experience which involves all the senses is more effective than others and, in some sense, is welcome when communicating science to people".

EG: "When we refer to physics as science, those who 'disseminate' are the teachers (in all schools at all levels, but especially in those schools where 'physics' is a specific subject of study) and the disseminators are necessarily experts in the subject. As soon as we move away from this territory and step in a slightly less mathematical topic, this coherence is immediately broken. To give an example, it is not only doctors who argue, speak, and give references to vaccines, therapies, and foods. And the arguments are no longer deductive demonstrations, but conclusions, inferences, and metaphors. However, these types of argumentations may or may not be considered part of a scientific method".

---

[4] https://eventhorizontelescope.org/press-release-april-10-2019-astronomers-capture-first-image-black-hole.

[5] https://eventhorizontelescope.org/blog/astronomers-reveal-first-image-black-hole-heart-our-galaxy.

5 Toward an Interdisciplinary Approach in Communicating Science

MT: "A conceptual change is happening in the world of science. Scientific disciplines are moving to interdisciplinarity, where new solutions to solve old and still open problems can be found. Physicists collaborate with biologists to tackle the problem of pandemics, or with pharmacologists to find new drugs against bacterial infections. They collaborate with engineers and informatics to build new super and quantum computers which are opening the doors to new technological applications. Physics has always had an interdisciplinary approach, even in the past, where it was strictly connected with philosophy. That was the time when revolutions in our understanding of nature happened, as in the case of General Relativity and, again, of Quantum Mechanics. After some years, the two disciplines took different paths, both kidnapped from their intrinsic technicalities. But, I guess, this interdisciplinarity could be restored even today. As already mentioned, the history of science taught us that from the mixing of different perspectives it emerges a new understanding of the world around us. Why do not try to combine different visions of what we know to offer a complete reading of the world of science? A similar experiment could be done starting with schools, where a universal form of learning is offered[6] in students' curricula. It could start a new education era when new methods to explain phenomena around us and new strategies to do it could emerge in school and academia. Therefore, a new language will be needed".

EG: "Which language? Language itself is changing. We were under the illusion for a long time that language was a compositional and deductive system, and that we could almost automatically move from axioms to theorems. But today we know that our possibilities of expression are anchored and bound to the way we are structured, that is, they are 'embodied'. That is why it is no coincidence that we speak of progress as something that moves forward, for example, because our motor patterns translate into symbolic and metaphorical patterns".

MS: "In this regard, the examples cited above, and the resonance these discoveries had in the media show that one can succeed in effectively communicating a major physics discovery, despite the inherent difficulties that we have mentioned above. With one crucial sticking point: it is always essential to explain to the audience how that specific representation of discovery was conceived. So, relative to the examples cited above, to say clearly that the 'sound' is not produced by the gravitational wave, but it is an audio transduction of the signal, or that the picture taken with the EHT is not exactly a picture of the black hole, but a reconstruction of the image of its shadow".

---

[6] Such kind of experiments have been made during the "Gravitas" project, an outreach and educational initiative led by the Cagliari Division of the National Institute for Nuclear Physics (INFN), in Italy. You can find more information here: https://dark.infn.it/eventi-pre-festival/. You can also have access to all the webinars at the dedicate YouTube playlist: https://www.youtube.com/playlist?list=PL94cdNBLY9XqD3V_YqEjVyQPXmspf2-k8.

MT: "The limit of representation, mental or real, is very important, both in communicating and teaching science. We are full of metaphors in our language which help us in imagining the world, even when the world is invisible to our sight. But without the right rational and methodological instruments, those images can create many misunderstandings. Let us think of the famous exhibit of the elastic sheet to represent and describe gravity. This is a wonderful analogy based on how we think space-time behaves when matter (or energy density) is present. But it is an analogy, not a real representation. It is a wonderful didactic instrument (and a play, too), but it brings many limits that the final users should know. Nowadays, the use of metaphors and symbols is a usual procedure in scientific communication. This is partly due to the transition we are now observing in our society, which is moving towards a purely visual and digital environment. The smartphone is always in our hands and from there stay informed about what happened around us. Science is not immune from this process and science communicators have understood it. Therefore, the use of images, both pictorial and linguistic, e.g., metaphors, is increasing. This, in turn, accommodates users' needs of being informed but, at the same time, is leading communication to model itself to assume new (effective) forms. The audience is always at the center of the communication, as we already said".

MS: "And that is how it is. We are seeing a multiplication of ways and channels by which science can be communicated: in addition to traditional tools - such as journal articles, radio, and TV programs—science is now being disseminated (among other things) through podcasts, video platforms such as Twitch, and of course on social media (with YouTube, Instagram, and TikTok leading the way). This is an ever-changing landscape where it can often seem difficult to find one's way around, but it offers an undoubted advantage: there is something for everyone, for all targets. The multiplication of the offer allows anyone, from the youngest to adults, from those who are unfamiliar with science to experts, to easily find their favorite communication 'channel', whether it is a short video on Tik Tok or a prime-time TV documentary on the generalist television. And it is certainly positive and interesting to see that, in addition to professional popularizers and communicators, several researchers and scientists have decided to put themselves on the line in the new media, in some cases becoming real influencers of science communicators. In Italy, an interesting case is the chemist Dario Bressanini, whose videos, and popular content on the topic of food science have led to the creation of a community of hundreds of thousands of followers on major social networks.[7] This is a very interesting trend that can contribute to the hoped-for humanization of science, which we mentioned earlier, and makes inroads with very young targets".

MT: "In the blowing of this wind of change of research and science, the school cannot be forgotten. Teaching is changing and we are now observing that the boundary between science communication and teaching is thinning more and more. Social media are becoming a place where people can be informed about everything, and researchers and teachers are using them to teach science. This is the case with

[7] https://www.instagram.com/dario.bressanini/.

YouTube, where a lot of instructional video and channels exists to explain how nature works.[8] The same is happening on Instagram[9] or TikTok, where the main goal is still to inform and not to teach. Smartphones are also becoming a pocket laboratory thanks to dedicated apps which consent to perform real experiments. This is the case of PhyPhox,[10] just to make an example. Visual resources for teaching and learning science are not only on social media but on the web in general. There are many examples of virtual laboratories which allow teachers to perform online experiments or websites which give educational supplementary material for learning science at school.[11] They offer immersive scenarios or a virtual (but quite real, we could say) reproduction of experiments easily performable in the classroom. A great innovation for the teaching and learning of science which, in turn, ensures a certain level of inclusivity in education and schools. This is not bad at all!".

6 Conclusions

The language of science is evolving, adapting to the discoveries and the audience's needs, both of experts and non-experts. The teaching and communication of science are also interesting in this phenomenon, being strictly related to the research. People are becoming more social, and the teaching of science cannot forget it. Social networks propose a visual world and science communicators should be aware of this to better spread scientific knowledge, as well as scientific cultural awareness. Images, imagination, and creativity play a fundamental role in science and research, too, and the dissemination of the latter's content is influenced by this fact. Nevertheless, if on the one hand science uses both formal (mathematics) and informal (natural language) language to communicate between experts, effective communication with the public is more oriented to the use of natural languages, where analogies, metaphors, and visual contents are predominant to simplify and explain contents. In this way, the problem of accuracy in the dissemination of science arises, and this is an issue for the science communicators.

Moreover, the science communicator is becoming more and more influential in fostering interest and motivation in the public toward scientific truth. In this process of mixing languages and cultures along the lines of social interconnection, assuming an interdisciplinary approach is fundamental. Going beyond the division of the "two cultures," the arts, or humanities on one hand and science on the other hand (Snow 2001), bringing the two worlds together instead, is very important for the common

[8] See https://www.youtube.com/@SteveMould or https://www.youtube.com/@LaFisicaCheCiPiace (in Italian) as an example.

[9] See https://www.instagram.com/emilia.science/ as an example.

[10] https://phyphox.org/.

[11] See https://www.frontiers-project.eu/gravitational-wave-astronomy/ and https://phet.colorado.edu/it/ as an example. The former offers educational resources for gravitational waves astronomy. The latter spans over all scientific topics covered in school, from physics to biology and chemistry and so forth.

goal of disseminating the scientific content, using as many tools as possible to explain complex and uncertain topics and to give reasoned, motivated, and argued interpretations (Ervas 2021). The clarity for the audience, coupled with the accuracy of the matter presented, must always be an essential starting point, valid for whatever medium one decides to use (whether "classic" or new): this has always been the main challenge in communicating science, and it must continue to be so in the future.

References

B.P. Abbott et al., Observation of gravitational waves from a binary black hole merger. Phys. Rev Lett. **116**, 061102 (2016)

R.K. Coll, The role of models/and analogies in science education: implications from research. Int. J. Sci. Educ. **27**(2), 183–198 (2005)

P. Corbetta, *Metodologia e tecniche della ricerca sociale* (Bologna, Il Mulino, 2014)

M. Davies, M.T. Devlin, *Interdisciplinary Higher Education: Implications for Teaching and Learning* (Centre for the Study of Higher Education, Melbourne, 2007)

P. D'Este, N. Robinson-García, Interdisciplinary research and the societal visibility of science: the advantages of spanning multiple and distant scientific fields. Res. Pol. **52**(2), 104609 (2023). https://doi.org/10.1016/j.respol.2022.104609

A. Einstein, L. Infeld, *The evolution of Physics. The growth of Ideas from Early Concepts of Relativity and Quanta*, (Editor C.P. Snow. Cambridge University Press) 1938

P. Engel, *Philosophie et psychologie* (Gallimard, Paris, 2000)

F. Ervas et al., *Scripta manent. Dieci lezioni sulla scrittura argomentativa* (Milano, Mimesis, 2021)

G. Galilei, *Dialogue Concerning the Two Chief World Systems* (trans. S. Drake) (The University of California Press, Berkeley, Los Angeles, CA, 1632) (1953)

F. Gobet, Cognitive aspects of learning in formal and non-formal contexts: lessons from expertise research, in *British Journal of Educational Psychology Monograph Series II: Number 11, Learning Beyond the Classroom* (2015), pp. 23–37

D. Gouthier, E. Ioli, *Le parole di Einstein* (Edizioni Dedalo, 2016)

J. Hadamard, *The Psychology of Invention in the Mathematical Field* (Dover, New York, 1954)

R. Hall, Towards a fusion of formal and informal learning environments: the impact of the read/write web. Electron. J. e-Learn. **7**(1), 29–40 (2009)

S. Hawking, *A Brief History of Time* (Bantam Books, 1988)

T.S. Kuhn, *The Structure of Scientific Revolutions* (University of Chicago Press, Chicago, 1962)

G. Lakoff, R. Núñez, *Where Mathematics Comes From* (Basic Books, New York, 2000)

S.M. Land et al., Student-centered learning environments: foundations, assumptions and design, in *Theoretical Foundations of Learning Environments* (Routledge, 2012), pp. 3–25

M. Michelini, The learning challenge: a bridge between everyday experience and scientific knowledge, in *Proceedings of the Third Int GIREP Seminar "Informal Learning and Public Understanding of Physics"*. Ljubljana, Slovenia, 5–9 Sept 2005, pp. 18–38

M. Morganti, *Filosofia della fisica. Un'introduzione* (Carocci Editore, 2016)

A. Oriji, I.F. Uzoagu, Lifelong learning in a technology-driven society: the needs, the benefits, and the challenges. Eur. J. Educ. Stud. **6**(9) (2019). https://doi.org/10.5281/zenodo.3595250

E. Otte, R. Rousseau, Social network analysis: a powerful strategy, also for the information sciences. J. Inf. Sci. **28**(6), 441–453 (2002). https://doi.org/10.1177/016555150202800601

R. Pan et al., The evolution of interdisciplinarity in physics research. Sci. Rep. **2**, 551 (2012). https://doi.org/10.1038/srep00551

A.-M.A. Petrescu et al., *The Role of Non-formal Activities on Familiarizing Students with Cutting-Edge Science Topics. ERD 2016: Education, Reflection, Development*, 4th edn. (2016). https://doi.org/10.15405/epsbs.2016.12.56

F. Piazza, *Linguaggio, verità, persuasione, La retorica del Novecento* (Carocci, Roma, 2004)

A. Pluchino et al., Exploring the role of interdisciplinarity in physics: success, talent and luck. PLoS ONE **14**(6), e0218793 (2019). https://doi.org/10.1371/journal.pone.0218793

A. Postiglione, I. De Angelis, Students' understanding of gravity using the rubber sheet analogy: an Italian experience. Phys. Educ. **56**, 025019 (2021)

T. Rabatzis, D. Ioannidou, The role of models and analogies in the Bohr atom, in *One Hundred Years of the Bohr Atom*, ed. by F. Aaserud, H. Kragh (The Royal Danish Academy of Sciences and Letters, 2015), pp. 360–376

M. Root-Bernstein, R. Root-Bernstein, *Sparks of Genius: The Thirteen Thinking Tools of the World's Most Creative People* (Houghton Mifflin, New York, 1999)

M.J. Simis et al., The lure of rationality: why does the deficit model persist in science communication? Public Underst. Sci. **25**(4), 400–414 (2016). https://doi.org/10.1177/0963662516629749

C.P. Snow, *The Two Cultures* (Cambridge University Press, London, 1959) (2001)

E.J.H. Spelt et al., Teaching and learning in interdisciplinary higher education: a systematic review. Educ. Psychol. Rev. **21**, 365 (2009). https://doi.org/10.1007/s10648-009-9113-z

P.B.A. Teixid et al., Evaluating the rubber sheet spacetime analogy by studying ball movement in a bent trampoline. Eur. J. Phys. **40**, 045005 (2019). https://doi.org/10.1088/1361-6404/ab1a5c

B. Trench, Towards an analytical framework of science communication models, in *Communicating Science in Social Contexts*, ed. by D. Cheng, M. Claessens, T. Gascoigne, J. Metcalfe, B. Schiele, S. Shi (Springer, Dordrecht, 2008). https://doi.org/10.1007/978-1-4020-8598-7_7

D.A. Ucko, The learning science in informal environments study in context. Curator Mus. J. **53**, 129–136 (2010). https://doi.org/10.1111/j.2151-6952.2010.00014.x

T.W. Ward et al., Effective learning: lessons to be learned from schooling, in *Effective Learning in Non-formal Education*, Org. by T.W. Ward, W.A. Herzog Jr. (Michigan State University, East Lansing, 1974), p. 38 (2013)

T.R. Watkins, Gravity & Einstein: assessing the rubber sheet analogy in undergraduate conceptual physics. Master thesis (2014). https://scholarworks.boisestate.edu/cgi/viewcontent.cgi?article=1889&context=td

A.J.Y. Zaidieh, The use of social networking in education: challenges and opportunities. World Comput. Sci. Inf. Technol. J. (WCSIT) **2**(1), 18–21 (2012)

Use of Art in Scientific Education

Margarita Cimadevila and Wolfgang Trettnak

Creativity and imagination are basic qualities for a scientist, since science, in reality, is an art.
Ángeles Alvariño. Oceanographer

Abstract A magnificent channel of scientific dissemination that excites the curiosity of the public, especially the young, is art. Art and science are similar in that both involve inspiration, creativity, research, meditation, and experimentation. Their combination can lead to novel results in both fields and reach the public in a subtle way, as beauty, poetry, and creativity are mixed with scientific facts or investigative methods. An example of this is the work of artists M. Cimadevila and W. Trettnak through which they will make a brief tour. In this chapter, we will see how the two artists make use of the art to approach topics from the world of physics as diverse as luminescence or particle physics, as well as incorporating other topics of social interest such as the role of women in science or ecological topics such as marine pollution by plastic. In their work, they mix and intertwine these themes, showing their beauty or even the problems related to them, disseminating them in many ways in artistic proposals in which they combine art, science, physics, ecology, sustainability, equality, didactics, and dissemination.

Keywords Physics · Art and science · Education · Didactics · Dissemination · Gender equality · Sustainability · Ecology · Young audience · Scientific culture

M. Cimadevila (✉)
ARSCIENCIA, Sada, A Coruña, Spain
e-mail: cimadevila@arsciencia.org

W. Trettnak
ARSCIENCIA, Werndorf, Austria
e-mail: wolfgang@trettnak.com

M. Streit-Bianchi et al. (eds.), *New Challenges and Opportunities in Physics Education*, Challenges in Physics Education, https://doi.org/10.1007/978-3-031-37387-9_12

1 Science and Art: An Introduction

Art and science are similar in that they involve inspiration, creativity, research, meditation, and experimentation. Their combination can lead to novel results in both fields and reach the public in a subtle way, as beauty, poetry, and creativity can be mixed with scientific information or even uncomfortable facts. Throughout the ages, art and science have been intimately linked. However, today's societies have perceived them as different areas with little in common. The digital revolution and the new technological means at our disposal open fields that are explored by art and science together within a new universe of scientific-artistic creation, in such a way that it is becoming difficult to conceive art without the world of science and vice versa.

Art, with its chameleon-like character, and its ability to amaze and reach all kinds of audiences, can be used as a powerful tool for communicating science and scientific culture, not only directly in schools and classrooms but subliminally in all kinds of media and environments (Miller 2014). According to our personal experience, art can show science in a direct, attractive, visual, and immediate way, causing an impact difficult to forget in the viewer, which can last in his imagination for a long time since "an image says more than a thousand words". It is an excellent way to make problems known (in a simple, clear, and impacting way), a valuable tool to fight against them and a magnificent channel of scientific dissemination that awakens the curiosity of the public, especially that of young people (see Chap. 4) (Cimadevila and Trettnak 2020).

The artistic work can take different forms: sculpture, painting, video, installation, performance… and can deal with scientific topics as diverse as particle physics, luminescence, or wave equations; the limit is really the artist's inspiration (Fig. 1) (Miller 2014). Through making scientific dissemination, it can become an exciting and highly rewarding work.

The combination "science and art" has an attractive *innovative and universal* character, breaking barriers, and crossing cultures, ideas, and borders. Its *transversality* allows it to address scientific issues combined with other topics such as gender equality, eco-art, and research, being also *interdisciplinary* since its materials can be used at all levels of other areas, such as environmental education, citizenship education, humanities, social sciences, or languages (see Chap. 4).

2 Goals and Target Audience

The main goals of the science and art combination that we have been pursuing are:

- Disseminate the world of science and art, showing their relationship.
- To promote scientific and artistic vocations.
- To promote creativity and the scientific method in the students.
- To create innovative materials in the field of science and art.

(a) (b)

Fig. 1 **a** Cimadevila exhibiting at a physics conference at the University of Santiago de Compostela, Spain. **b** Trettnak at the exhibition *Licht und Schatten* at Graz University, Austria. *Photographs* © Cimadevila and Trettnak 2017, 2015.

Young people are the most important target audience since they are the future of our society, and their training in science is necessary and essential. For this, the artists depend on the teachers who are the second most important target audience, not in vain their work is the one that guides and orients young people. The activities that combine science and art are very well received by students and teachers, who actively participate in them through workshops, competitions, science fairs, and conferences (Fig. 2).

(a) (b)

Fig. 2 Cimadevila conducted a particle physics workshop in the international course *The World Classroom* for high school students from Bologna, Italy, and Dallas, Texas: **a** The sketch for a mural and **b** the final product. *Photographs* © Cimadevila 2007.

The approach to classrooms at all levels is essential in the dissemination of the binomial art and science, from universities, institutes, and secondary education to preschool classrooms of the smallest schools. It is very important to sow the seed of curiosity in the small viewers.

One must not forget the policymakers, politicians, and disseminators since they are the actors who translate into rules and laws the needs of science and the public. Ultimately, the public is the target audience, who, with its taxes, finances a large part of the world of science and art.

3 Dissemination

Popularizing science can be done also through artistic work and using very different paths. The first and most obvious is the exhibition of the work in art galleries and museums; however, its exhibition in places outside the artistic circuit, such as a science contest, a school institution, a shopping mall, social networks, or a leisure place, often reaches a wider and more varied audience. But we must never lose sight of the fact that it is not only necessary to reach the public with an artistic proposal, but also to attract its attention to the scientific theme on which the work is based.

A very important point of dissemination is the creation of teaching materials, which should be oriented to both students and teachers. On the basis of photos or videos of the artistic work countless "traditional" materials such as books, brochures, and posters can be created. Special mention should be made of the new interactive digital materials that combine science and art, which are fantastic and an inexhaustible source of fun and learning for students. Their possibilities are endless: games, puzzles, brainteasers, riddles, and didactic units. However, not all are positive points, since their ephemeral nature, due to the rapid evolution of technologies and software, makes very interesting materials of recent creation disappear or stop working. This fact is worthy of careful reflection, not only on the ephemeral nature of the materials but also on the fragility of our digital culture and its possible difficulty of transmission in the distant future.

A Digital Paradox

In 2013, I found my grandfather's doctoral thesis from the early twentieth century on the Internet, which was a great joy for me. However, incredibly, it is no longer possible to access the digital materials *Ciencia y Arte 1 and* 2, created by me in 2014/15 for secondary school students to enjoy science and art, and subsidized by the government of Galicia (Fig. 3).

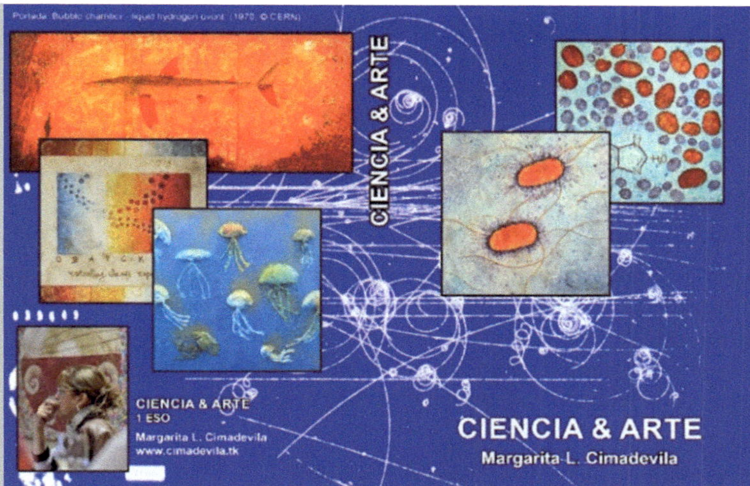

Fig. 3 Cimadevila, *Ciencia y Arte 1*, the cover of CD casing. © M. Cimadevila 2014. All right reserved

What is the reason? A program used in the creation of the materials is no longer operational and no longer supported. It has certainly upset me to see a work, to which I have devoted a lot of time and effort, disappear without a trace or to see that recovering it might be costly, time-consuming, and difficult.

Maybe stone would have been a better support material!

Margarita Cimadevila

4 Linking Science and Art

Let us now move on to the practical, showing how linking science and art has been realized by us, Margarita Cimadevila[1] (Cimadevila 2023), and Wolfgang Trettnak[2] (Trettnak 2023), artists with scientific training whose artwork has among its objectives to raise awareness of the world of science.

The source of inspiration for our individual work is science, contemplated from such different points of view as teaching (Cimadevila) and research (Trettnak). The artwork has been created on subjects as diverse as particle physics, discoveries of *CERN* (*European Organization for Nuclear Research*), luminescence, bionics, or

[1] The artist, Margarita Cimadevila has a degree in Chemistry from the University of Santiago de Compostela and worked as a Physics/Chemistry teacher and director in different institutes of Galicia.

[2] Wolfgang Trettnak received a Ph.D. in Chemistry from the University of Graz. He undertook applied research on sensors and biosensors for several years and published a number of scientific articles. In 2002 he became a freelance artist.

electric fish, and has a strong informative and didactic charge. Never losing the scientific point of view, the work of both of us also evolved incorporating other themes, both social and environmental.

4.1 Margarita Cimadevila

The artwork, deeply influenced by the roots of her land Galicia and her experience as a science teacher, pursues since many years now a triple objective: *didactic, informative, and artistic*, as well as fighting for *gender equality* and the *defense of the environment*. Science and art are mixed and intertwined in her artistic proposal, in which, from a feminist and always didactic perspective, she

- disseminates the world of science,
- gives visibility to the role of women in science,
- denounces environmental problems.

Through the artworks, she shows the problems of these issues and disseminates them by combining *science, didactics, equality, ecology, sustainability, recycling, dissemination, and art.*

4.1.1 Science and Art, CERN

As a result of her participation in the *Teachers' School (HST 2003)* at the *CERN* in Geneva, her artistic work underwent a radical change and focused on mixing her two worlds: science and art. In 2004, she showed at *CERN* her first painting exhibition with this double theme (*Particle physics I*), being her sources of inspiration the *HST* conferences and the *CERN* facilities (Fig. 4) (High School Teachers at CERN 2004). Her artwork has been shown in different events related to *CERN* and particle physics, such as the *International Particle Accelerator Conference 2011* and she has organized multiple events and programs to introduce the world of *CERN* to students and teachers.

4.1.2 Equality, Science, and Art

Making visible the role of women in science and fighting for gender equality in science and in life have been the fundamental objectives of two of her exhibitions:

- *CIENCIA EX AEQUO* (Cimadevila 2009; López Díaz and Cimadevila 2013) shows and values the work of women in science throughout time. It pays tribute to the unjustly forgotten women scientists, who, by doing magnificent work and

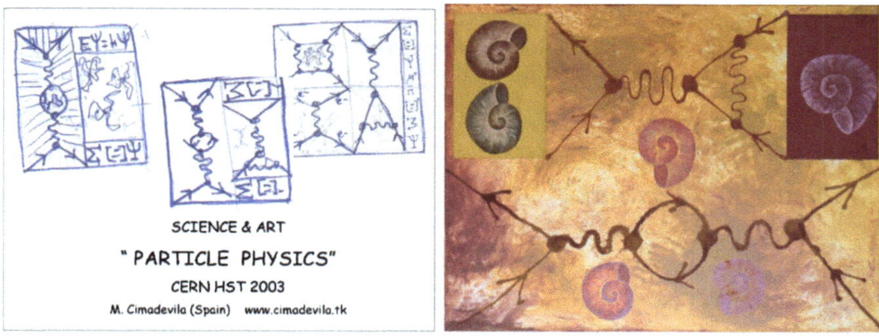

(a) (b)

Fig. 4 Cimadevila, *Feynman diagrams*, **a** sketches and **b** final painting. *Photographs* © Cimadevila 2003, 2006. All rights reserved

deserving recognition for it, were ignored, forgotten, or relegated to the background. The exhibition is based on the scientific work of twelve pioneering women scientists of the last century (Fig. 5).

- *Avant-garde in Science* is about men and women who were at the forefront of science and life, working in equality in the early twentieth century. It gives recognition to those men who rose above pettiness, recognized the intelligence of women, and favored their incorporation into science, working with them on an equal basis.

The source of inspiration has been the scientific field on which these scientists worked, which makes that, although all the paintings deal with science, they touch on topics as different as pulsars, *Escherichia coli*, or tryptophan or nummulites, accompanied by brief scientific and biographical explanations (Fig. 5).

Fig. 5 Exhibition *CIENCIA EX AEQUO* of Cimadevila at the *Innovation Convention 2014* organized by the European Commission at Brussels, Belgium. *Photographs* © Cimadevila 2014. All rights reserved

Recycling and reuse of materials are always present in her works, and many of the paintings have been made with leftover textile materials and worn-out clothes, which give them an imprint difficult to achieve with new materials.

4.2 Wolfgang Trettnak

Trettnak draws much of his inspiration from the work done as a research scientist. For many years, he worked on the use of fluorescent and phosphorescent dyes in sensor applications (Trettnak and Wolfbeis 1993; Trettnak et al. 1991), which also deeply influenced his artistic work and resulted in the creation of luminescent objects and installations. Luminescent dyes are not only fascinating, because of their uncommon color behavior, but also allow to play with our visual perception. This makes them very attractive for catching the attention of the spectator (Fig. 6). Similar fascination results from living bioluminescent organisms, which he tries to imitate not only by using luminescent materials but also by light-emitting diodes as light sources.

Sensors play an important role in clinical applications, implants, and artificial limps. The developments in these fields and in robotics were the inspiration for Trettnak to introduce electronic components in his artwork, thus resulting in series of paintings showing "bionic" men and animals. Animals, especially fish, which use bioelectricity as a means of communication, for hunting or defense, were the basis for the creation of a whole series of paintings on "*Electronic fishes*" (Fig. 7).

Many of these works have been presented with information aside, which give hints on the background and short scientific explanations. In addition, Trettnak has given lectures with a popular science character on all these subjects combined with his artwork. For example, he lectured in schools and at universities (Trettnak 2015),

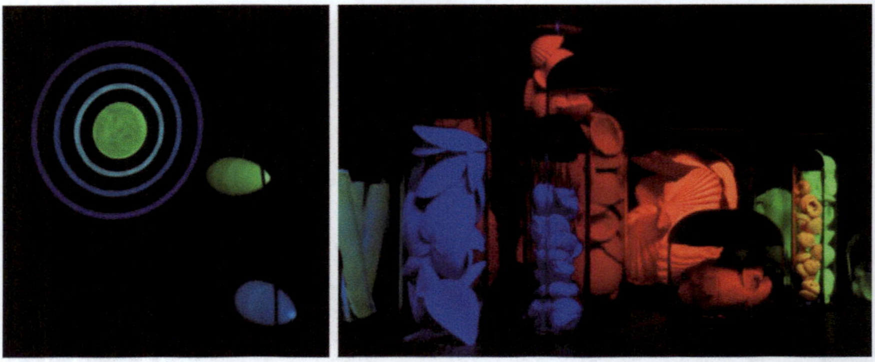

(a) (b)

Fig. 6 Trettnak, luminescent objects **a** *Particles hitting the target* (hanging mobile) and **b** In vitro *III*. Details. © Trettnak. 2014, 2015. All rights reserved

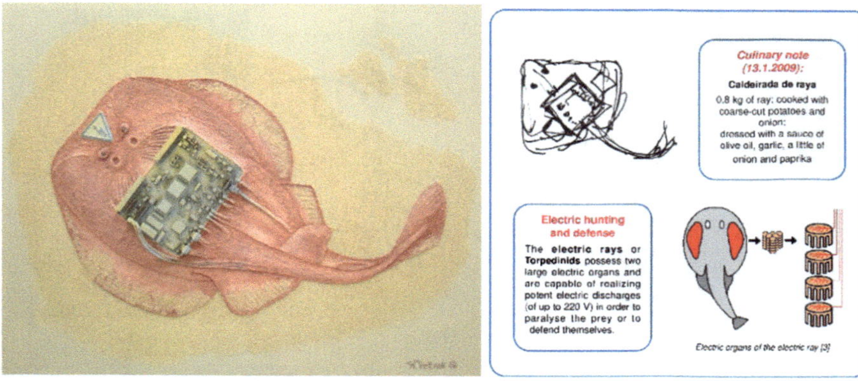

Fig. 7 Trettnak, **a** painting *Electra* and **b** short scientific explanation and cooking recipe (Graetz 2006). © Trettnak 2008, 2009. All rights reserved

at science and art courses for teachers (Dialogue between Science and Art 2023), or at scientific conferences.

Another issue, which is addressed by him, is linking with environmental topics. Recycling and reuse of materials play an important role in his work, not only reuse of electronic components, but also, for example, of textile waste. The problems of electronic or textile wastes are very serious and still need to be resolved. A special problem of concern has been marine plastic pollution and its consequences, which stands in the focus of the joint work of Cimadevila and Trettnak with their eco-art exhibition *MARE PLASTICUM* (Cimadevila and Trettnak 2020; ARSCIENCIA 2023).

But how to win the attention of the spectator or visitor of an exhibition, which may have a scientific background or contain even worrying information? Here humor can help, an aspect which is often present in Trettnak's work.

4.3 *Cimadevila/Trettnak—Eco-Art: MARE PLASTICUM*

Since 2011, Cimadevila and Trettnak worked, with four hands and two heads, on the exhibition *MARE PLASTICUM* (Cimadevila and Trettnak 2020; ARSCIENCIA 2023) about the pollution of the oceans by plastic and its environmental consequences: entanglement, plastic ingestion, species mortality, and toxicity. Its aim is to inform about the problem and raise public awareness (Streit-Bianchi et al. 2020; Verany 2015; Del Rosso 2015). The exhibition with their artwork and scientific

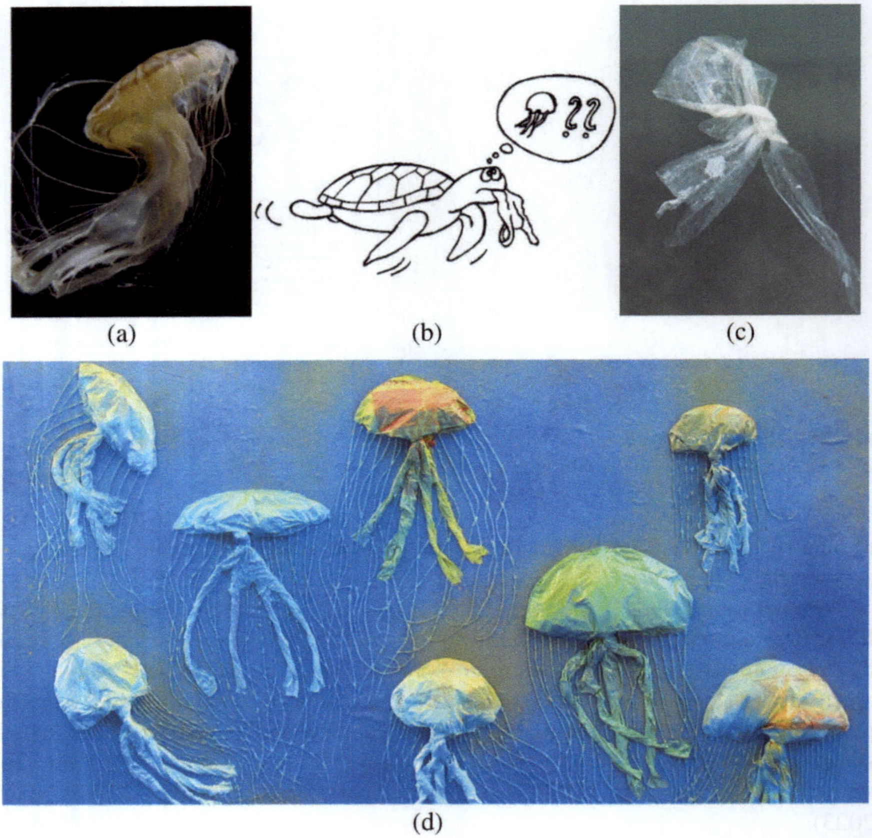

Fig. 8 *MARE PLASTICUM*. Sea turtles eat plastic bags because they mistake them for jellyfish. **a** Jellyfish, **b** turtle, **c** plastic bag, **d** *Jellyfish*, painting made with plastic bags (Detail). *Photographs* **a** public domain (Thurner 2023); **b–d** © Cimadevila and Trettnak 2017. All rights reserved

explanations offers a vision that is both real and imaginary and raises a cry of alarm: *It's time to act!*

The artworks made with plastics collected by them on the beaches of Galicia (Fig. 8) and the scientific information, comics, posters, mobiles, and videos that complete the exhibition, have been shown, for example, at: World Congress *Our Ocean 2017 Malta*, *MUNCYT* (*Museo Nacional de Ciencia y Tecnología de España*) A Coruña, *CONAMA 2016* (National Environmental Congress) Madrid, *CERN* Geneva, and *National Museum China* in Beijing (Cimadevila and Trettnak 2020).

For the originality in showing a pressing problem in a creative way, but without losing sight of the scientific rigor, the 1st prize for sustainability is awarded to the work:

MARE PLASTICUM—PLASTIC FISHES by Margarita Cimadevila and Wolfgang Trettnak from ARSCIENCIA of a Coruña.

Translated from: *Acta de la comisión de CIENCIA EN ACCIÓN* (Ciencia en accion 2023)

28 de mayo al 15 de julio 2015

4.4 Exhibitions

Exhibitions have been organized in all types of museums, centers, schools, universities, events, and national and international forums. It is intended that the public is not limited to static contemplation, but actively participates through guided tours, lectures, and workshops, or even by using on occasion "augmented reality" to include virtual objects through the mobile phone, in order to make the exhibition more interactive.

They have had wide dissemination and reach, thanks to the relevance and high public attendance of centers such as the *Oceanographic Museum of Monaco*. This has made it possible for their message to reach a wide and diverse public. But they have also been shown in modest halls, local meetings, and educational centers, whenever the authors have been invited to spread their message.

4.5 Conferences, Congresses, and Fairs

Events such as conferences, congresses, or fairs, national such as *CONAMA 2016*, or global ones such as the *Our Ocean 2017* Congress, both face-to-face and virtual, are a good way to reach more specific audiences, in addition to opening new forms of collaboration and dissemination. The authors have participated in a wide range of such events: conferences on physics and particle physics, environmental congresses, congresses related to gender equality in science and technology, teacher´s congresses, science fairs, and many more.

4.6 Courses

Another important role is played by courses for teachers and students. For example, both authors were involved in the organization of the course *Science and Art: so similar, so different!* in Galicia (2014) and participated in the course *Dialogue between Science and Art* organized by Michal Giboda in the Czech Republic (Dialogue between Science and Art 2023).

International course for teachers
Science and Art: So Similar, so Different!
(*Comenius-Grundtvig Course* of the Europ. Union; 2011–2014)

The organization of courses for teachers on science and art are a magnificent and enriching experience, where the exchange of ideas and points of view of teachers of different nationalities, levels, subjects, ideologies, and countries give rise to new projects that contribute to promote the popularization of science through art, creating links between different cultures difficult to forget and that remain through time (Fig. 9).

Fig. 9 Teachers in the course *Science and Art: so similar, so different! Photographs* © Niki Alexakou 2012. All rights reserved

(a) (b)

Fig. 10 **a** Workshop for students in Sada, Spain and **b** conference at a primary school in Betanzos, Spain. *Photographs* © Cimadevila and Trettnak 2011, 2020.

4.7 Educational Activities

Educational activities intended for primary schools, secondary schools, and university are lectures, guided tours, workshops, participation in competitions, and traveling exhibitions. They are very gratifying for their good reception by teachers and students. Good examples are the participation in competitions such as *Ciencia en acción (Science on Stage)* (Ciencia en accion 2023; Science on Stage Europe 2023), or in educational awareness raising events, such as the *European Maritime Day* (2023). The creation of artistic work with students and teachers also belongs to these kinds of activities (Figs. 2 and 10).

4.8 Didactic Materials

Some didactic materials have been created for students, teachers, and the public such as videos, games, posters, comics, bookmarks, and brochures without forgetting the incorporation of new technologies through interactive digital materials (Fig. 3), all of them in several languages: English, Galician, Spanish, French, German, and Chinese. Books, catalogs, and other publications were edited, such as *CIENCIA EX AEQUO* in collaboration with the *Office for Gender Equality of the University of A Coruña* (Cimadevila 2009) or the book *Mare Plasticum—The Plastic Sea* with Springer publishing house (Streit-Bianchi et al. 2020).

4.9 Media Presence

The exhibitions of the authors have been echoed by various media such as: television and radio programs dedicated to art, equality, science, and environment, reports on websites, scientific journals, and non-specialized press. Their work has been present also in various social networks, websites, blogs, etc., and, of course, in their own websites and social media (Cimadevila 2023; Trettnak 2023; ARSCIENCIA 2023). Although this might seem a trivial issue in today's multimedia culture, it is still of importance and should not be ignored, since, consequently, resulted contacts to schools, museums, and other institutions. On the other hand, events such as the *European Maritime Day* were completely organized with schools via the own websites (ARSCIENCIA 2023; European Maritime Day 2023).

5 Final Remarks

It should be noted that all activities presented here have been done so far with very limited funds without special financial support, sponsors, or subsidies from the public sector. In general, more important than money is the enthusiasm and personal engagement of the people involved in organizing art and science events. Art and science are efficient means to spread science in society as demonstrated by some of our examples and even more by the interest now important scientific institutions such as *CERN*, *ESO* (*European Southern Observatory*, Chile), *INFN* (*Istituto Nazionale di Fisica Nucleare*, Italy), or *Ars Electronica* (2023) (Austria), and many other institutions around the globe have demonstrated in setting up science and art events for the general public or collaborations between scientists and artists. Furthermore, also universities are moving forward, to give just one example, in Linz (Austria), the *Johannes Kepler University* and the *University of Applied Arts Vienna* have created an inter-university called *Art x Science School for Transformation* (Transformation Studies 2023).

Many of the activities and events presented in Chap. 4 have been channeled through the association *ARSCIENCIA*. The international association ARSCIENCIA (ARSCIENCIA (2023) founded in 2013, is a nonprofit socio-cultural entity, whose purpose is to promote and organize all kinds of activities on science and art, showing the relationship between these two worlds and paying special attention to the presence of women in science.

Their long trajectory in the world of science and art makes Cimadevila and Trettnak firmly convinced that art is a powerful tool for promoting scientific culture and to raise awareness in scientific topics. They both consider as magic to allow students to see science through art and to create art because of science.

The authors hope that their journey is a convincing example to encourage in the classroom the collaboration between artists, art teachers, science teachers, and students to make up projects on topics such as, e.g., the climate change or environmental pollution, where the students can profitably and leisurely deepen their scientific knowledge on physics, chemistry, ecology, biology, mathematics, etc., by using creative methods based not only on traditional methods, but also on the most appealing available digital tools.

References

ARS Electronica. https://ars.electronica.art/. Accessed 24 Feb 2023

ARSCIENCIA (2023). https://www.arsciencia.org/. Accessed 9 Feb 2023

Ciencia en accion. https://cienciaenaccion.org/. Accessed 11 Feb 2023

M. Cimadevila, *Ciencia Ex AEQUO* (Concellería de Igualdade, Ayuntamiento de A Coruña, Galicia, 2009)

M. Cimadevila (2023). http://www.cimadevila.tk/. Accessed 9 February 2023

M. Cimadevila, W. Trettnak, The exhibition MARE PLASTICUM: art and science for the environment, in *Mare Plasticum—The Plastic Sea, Combatting Plastic Pollution Through Science and Art*, ed. by M. Streit-Bianchi, M. Cimadevila, W. Trettnak (Springer International Publishing, 2020)

A. Del Rosso, De poissons en plastique, Le Bulletin du CERN, Numero 24–25/2015

Dialogue Between Science and Art. http://www.sciart-cz.eu/. Accessed 11 Feb 2023

European Maritime Day. https://maritime-day.ec.europa.eu/index_en. Accessed 11 Feb 2023

Galicia (2014). https://www.arsciencia.org/the-science-art-course-2014/

A. Graetz, Figure 6b based on image *Elektroplax_Rochen.png* (Creative Commons Attribution-Share Alike 3.0 Unported) (2006). https://commons.wikimedia.org/wiki/File:Elektroplax_Rochen.png. Accessed 10 Feb 2023

High School Teachers at CERN. http://hst-archive.web.cern.ch/hst/2004/work.htm. Accessed 10 Feb 2023

A.J. López Díaz, M. Cimadevila (eds.), *Ciencia/Science Ex Aequo. Oficina para a Igualdade de Xénero* (Universidade da Coruña, 2013)

A.I. Miller, *Colliding Worlds: How Cutting-Edge Science is Redefining Contemporary Art* (W. W. Norton & Company, 2014). https://www.arthurimiller.com/books/colliding-worlds/. Accessed 24 Feb 2023

Science on Stage Europe. https://www.science-on-stage.eu. Accessed 11 Feb 2023

M. Streit-Bianchi, M. Cimadevila, W. Trettnak (eds.), *Mare Plasticum—The Plastic Sea, Combatting Plastic Pollution Through Science and Art* (Springer International Publishing, 2020)

B.S. Thurner, Chrysaora melanaster. https://commons.wikimedia.org/wiki/File:Qualle_Kompa%C3%9Fqualle_2006-01-02_2.jpg. Public domain. Accessed 11 Feb 2023

Transformation Studies. Art x Science. https://www.jku.at/art-science-transformation/. Accessed 24 Feb 2023

W. Trettnak, Optical sensors based on fluorescence quenching, in *Fluorescence Spectroscopy: New Methods and Applications*. ed. by O.S. Wolfbeis (Springer, New York, 1993), pp.79–89

W. Trettnak, *Lumineszenz—Eine leuchtende Welt, Lecture at the UniGraz@Museum* (University of Graz, Austria, 2015)

W. Trettnak (2023). https://www.trettnak.com/. Accessed 9 Feb 2023

W. Trettnak, M. Hofer, O.S. Wolfbeis, Applications of optochemical sensors for measuring environmental and biochemical quantities, in *Sensors: A Comprehensive Survey*, ed. by W. Göpel, J. Hesse, J.N. Zemel, vol. 3/II (VCH Publ., Weinheim, 1991), pp. 931–967 (Chapter 18)

C. Verany, L'invasion du plastique dans les océans en expo, Monaco Matin, 15 Feb 2015

Theatre: The Other Side of Physics

Marco Giliberti

Abstract In the current century, the relationship between physics and theatre is becoming more and more intense and fruitful, so much so that stimulating connections—previously not even suspected—are being discovered between the ways of unfolding reality that are common to both theatrical research and basic physics research. Nevertheless, in most important theatres, it is still difficult to attend scientific performances that are on the bill; on the other hand, there is little theatre in university physics courses, yet. We will analyse the reasons for this situation and highlight the developments of physics theatre from an educational point of view, but above all, we will underline the dual structure of physics and theatre when looked upon from a deep cultural perspective.

Keywords Physics · Theatre · Education · Research

1 The Unreasonable Non-existence of a Physics Theatre as Such

Physics studies eternal and universal laws, which the entire cosmos is expected to obey. Theatre, on the contrary, deals with people and society. It is an art that is immediately consumed, lasting the time of a glance, and soon it is over. It is not like a movie which is always the same, unchanged over the years and the different places. Nor like a book, or a sculpture that lasts for millennia. Theatrical scripts can indeed be re-read identically, and that shows are repeated, but representations are different every time. Theatre is not like physics, whose laws sculpt the structure of the entire universe since the Big Bang. So, what do physics and theatre have in common (Giliberti 2014, 2019, p. 77)?

Although so different, during their history, physics (and more generally science) and theatre have occasionally had some points of contact. In fact, the theatre has

M. Giliberti (✉)
Physics Department "Aldo Pontremoli", University of Milan, Via Giovanni Celoria 16, 20133 Milan, Italy
e-mail: marco.giliberti@unimi.it

M. Streit-Bianchi et al. (eds.), *New Challenges and Opportunities in Physics Education*, Challenges in Physics Education, https://doi.org/10.1007/978-3-031-37387-9_13

sometimes been dealing with the social implications of science, even if it hardly pushed science into the foreground of the play. Even today, when putting the so-called science shows on the billboard is not such a risky operation as it was just a couple of decades ago, this action is still hardly carried out by theatre operators. Attending science-theatre shows at science festivals or science museums is, indeed, much easier than seeing titles of scientific plays in the seasons of big theatres next to those written by famous playwrights.

It is useless blaming this on theatre operators; indeed, few people think science can touch, besides people's minds, also people's hearts. *The Conics* are not the title of a Greek tragedy, and *Broken Symmetries* is not a show about the relationship between men and women. Who among the *habitué* would go out on a winter night to a rerun of *The Magnetic Vector Potential*? Just thinking about such a title makes us smile.

Yet the market is full of so-called scientific shows; and on the Internet, there are commercial proposals for shows addressed to schools with spectacular demonstrations, or of performances promoted by large scientific institutions and research bodies concerning the wonders of astronomy or elementary particles; but they are almost always off the bill. In fact, even if there is a growing number of people working on it, science theatre struggles to emerge from an artistic point of view. The great talk that is going on about scientific theatre is more concerning the way to promote scientific initiatives, or the market positioning of shows which would never find a place in important theatre seasons—but which carve out their place for schools or as events in science festivals, or which are presented as children's theatre for the whole family—than questions concerning theatre criticism or poetics, which would indicate a cultural living genre with important works. Whatever science theatre is—in fact, we have not even attempted a definition of it—from a commercial point of view, scientific theatre is often a good operation, and there are many scientific shows that attract large numbers of spectators and fill theatre halls. Nonetheless, important dramas that could, at least broadly speaking, be called scientific theatre can be counted on the fingers of one hand.

Yet, on a slightly deeper look, the absence of science in the great plays appears strange, unreasonable. Indeed, science is a cultural product of man and society, with its several facets, constitutes an important part of the social framework, and contributes significantly to changing our vision of the world. Why, to paraphrase Richard Feynman[1] (Feynman 2000), does the theatre of the present not speak of it?

At the turn of the century, European society witnessed a crisis in vocations toward scientific studies, also due to a perceived cultural superiority of the humanities over scientific disciplines (Brandi et al. 2005). The great efforts of the political, scientific, and academic world have made it possible to change the situation, and now (2023) sciences are, in general, considered by most people to be extremely important for society. But the human, profound aspect, the one that allows to touch—in a way like art—the strings of the human soul, is still generally misunderstood. It is mainly the utilitarian aspect of sciences that is put forward; sciences are, in fact, considered more

[1] "For far more marvellous is the truth than any artists of the past imagined it. Why do the poets of the present not speak of it?"

useful than the humanities, and, therefore, in this historical moment, perhaps even more appreciated or—given the increasing attention to ethical aspects and dangers related to the use of scientific knowledge—even sometimes watched with fear. Of these aspects, only the last one can, in general, provide inspiration to theatre.

My experience concerning physics shows with "Lo spettacolo della fisica"[2] (Fig. 1), however, leads me to be optimistic and believe that in the near future the way will be found for the great theatre to really deal with science; not only for its human or social, and ethical questions, but, above all, for the specifically scientific, and spiritual dimensions that still wait to be disclosed by a theatrical gaze. At least, this is a challenge to what "Lo spettacolo della fisica" is working on. Scientific theatre has extraordinary potentiality that are still to be explored, and we are convinced that a work both physical and theatrical, addressed to public, that makes physics the protagonist of the text is possible, with plays that will be considered as great as the those admired by all of us for centuries, like the tragedies of ancient Greece or the works of Shakespeare.

In the next sections, we first see a very brief history of scientific theatre (with a focus on the theatre of the last decades), and briefly discuss some of the main uses of science theatre in teaching. We will then try to understand the main cultural difficulties that hinder the artistic development of physics theatre and propose a cure for this situation.

2 A Very Short Account of Science-Theatre History

2.1 Ancient Science Theatre

Rites and theatrical performances are very ancient and have always had a didactic role, transmitting the mythological and cultural heritage, as well as the rules of moral behaviour. The beginning of Western theatre can be identified with the dawn of Greek tragedy, and therefore, at its origin, there is no written text to be staged, but only the celebration. Participants were at the same time actors and spectators, so these two functions were not yet separated.

Ancient Greek authors rarely used scientific concepts. No theatrical text concerning Pythagoras or his discoveries, or Archimedes has come down to us; no text concerning science. For the ancient tragedians, nature belonged to the divine and the myth. In this regard, Aeschylus' *Prometheus Bound* is fundamental in the history of Western thought. For Aeschylus, Prometheus, "the prescient", out of love for men, from Hephaestus stole the fire and gave it to humans. In fact:

---

[2] In 2004, Marina Carpineti, Nicola Ludwig and the author of this chapter (all physicists from the University of Milan) founded the group "Lo spettacolo della fisica" (The show of physics) that, so far (January 2023), has written and performed 8 physics-theatre shows and 3 lesson-shows, reaching more than 150,000 spectators and making more than 400 performances throughout Italy and Europe (see spettacolo.fisica.unimi.it/).

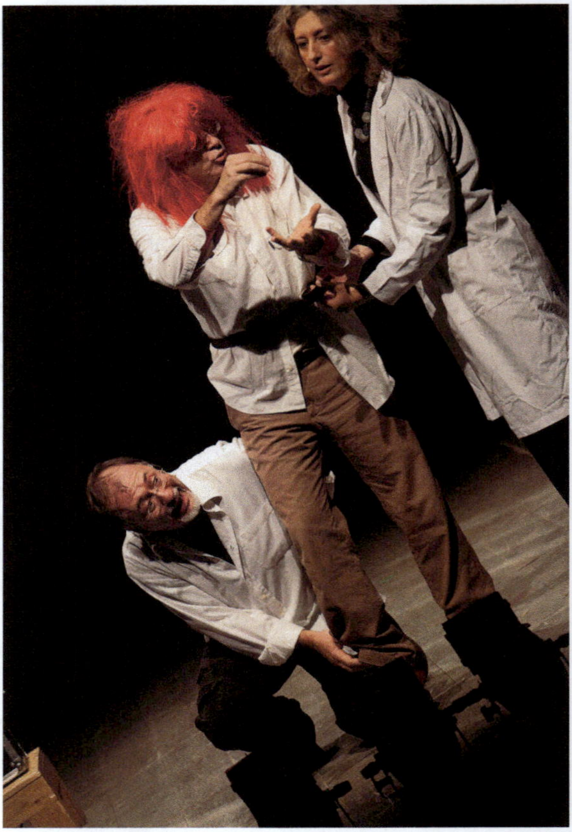

Fig. 1 A picture taken from *Luce dalle stelle* made by "Lo spettacolo della fisica" and performed in 2012 at Teatro Franco Parenti (Milan). *Credits* Fabrizio Favale, Marina Carpineti, Nicola Ludwig, Marco Giliberti

> though they had eyes to see, they saw to no avail; they had ears, but they did not understand; [...] They had no sign either of winter or of flowery spring or fruitful summer, on which they could depend but managed everything without judgment until I taught them to discern the risings of the stars and their settings, which are difficult to distinguish. [...] Yes, and numbers, too, chiefest of sciences, I invented for them, and the combining of letters, creative mother of the Muses' arts, with which to hold all things in memory. (Aeschylus 1926, lines 457–459)

Prometheus gave men thought, conscience, and knowledge; even the first knowledge, that of Mathematics, and is punished exactly for this: science and technology distance the soul from nature and therefore distance mankind from the divine. This theme has been accompanying the evolution of Western culture for two and a half millennia, and many people still experience the rift between spirituality (in a very broad sense) and science, which reveals, discovers, and makes things cold and unpleasant (Carpineti et al. 2011).

Comedies also reflected this view, and playwrights mocked scholars who disrespected gods with their theories. *Birds* by Aristophanes, for example, (we are in the fourth century BC) is a fantastic comedy in which two Athenians decide to build a world of pleasures and delights alternative to Athens. Helped by two birds, they build the city Nubicuculia halfway between the earth and the sky, in which the moral norms are opposite to those of Athens. Some men, including a mathematician, try to reach Nubicuculia. Upon his arrival, Pisthetaerus, one of the two founders of the city, interrogates him. As soon as Pisthetaerus discovered he is a mathematician; he explodes saying "He's a Thales!" and does not accept him in the city. Thus, science is expelled by Aristophanes from the ideal city of all pleasures.

2.2 *Modern Science Theatre*

Taking a big leap forward, it was in the sixteenth century that science, and more particularly mathematics, began to put itself at the service of the theatre. With the birth of perspective, scientific foundations were found for the construction of realistic sets. An example among all is given by the *Teatro Olimpico* in Vicenza, one of the first permanent theatres of the modern era. Built by Andrea Palladio in 1585, it is a sort of small reconstruction of a Roman theatre, but with a scenography of narrow streets in perspective and with the sky painted on the ceiling.

It is interesting to observe that in the sixteenth century and then, even more so, in the seventeenth century, the so-called "cabinets of curiosities" were born. They were ancestors of our natural history museums which showed the visitor strange and fascinating findings from the natural world or artefacts from the past.

The first form of didactic theatre was the work of the Jesuits and dates from this period.

> Jesuit drama [...] had the triple purpose of proposing examples of piety, of teaching the language of the Church, and of educating the offspring of the aristocratic classes in good personal behavior. [The] last two purposes, above all the last one, were essentially realized in the moment of staging: the actor-student learned, by acting, the speaking pose and the decorous control of the gesture. (Molinari 2011)

The didactic aspect of the theatre, therefore, consists of the fact that it was acted out by the student, something whose importance is perhaps then understood for the first time, and which we still find today, and sometimes with purposes of the same type, in our school.

As far as science is concerned—even if they are not exactly theatre—the dialogues of Giordano Bruno's *La cena delle ceneri* are memorable in which physics and mathematics are explicitly discussed and the Copernican system is proposed in a more general form than the original of Copernicus, with the Sun no longer in the centre of the world, the universe being infinite. We also recall the only comedy by Giordano Bruno, *Il candelaio*, set in sixteenth century Naples, in which science (understood above all in the sense of knowledge) proves to be a tool against astrology.

In the seventeenth century, a rather particular type of conference show was born: that of the anatomical theatre. It consisted of dissecting a corpse in front of an audience of students. In the anatomical theatres, lessons were held in which an anatomy text was read and commented, on while the surgeons showed the organs described to the public. In addition to the anatomy lessons, we recall many other lessons held by scientists, starting from the beginning of the seventeenth century, that were also structured as conference shows. Among these, Alessandro Volta's lessons in Pavia were full of spectacular visual effects, and those of Louis Pasteur at the *Sorbonne* which were real theatrical episodes with plays of light and catches. Very famous is the 1864 conference on spontaneous generation, which had the dual purpose of teaching the public and having the consent of colleagues. Spectacular lessons have been present everywhere ever since: the tradition continues. Last, but not least, we cannot forget the Christmas Lectures, initiated in the 1820s by Michael Faraday and aimed primarily at children (particularly famous is *The chemical history of a candle*), they continue each year at *Royal Institution.*

At the end of the eighteenth century, circus shows were born in London, above all thanks to Philip Astley, who performed on horseback with great success inside a circle drawn on the ground. Several circus acts are largely based on the laws of mechanics, with the appropriate use of levers, seesaws, etc., and, in some cases, juggling tricks provide the basis for a certain type of performance which, properly commented, are used today as elements of physics educational shows.

Between the eighteenth and nineteenth centuries, a new fracture was created between humanistic and scientific cultures. The various disciplines specialized, and science moved further and further away from ordinary people. For this reason, there was considerable interest in new means of communication, possibly able of bridging this gap. Among the new languages, there was also that of scientific theatre, in the modern sense of the term. In the beginning, a scientific street theatre was born; real shows staged in squares were organized to intrigue and amaze the public with new applications of science (for instance, spectacular electrical and optical phenomena) (Magni 2011).

Elements of science spectacularization can still be seen today in major science museums, such as the *Deutsches Museum* in Munich, the Museum of Science in Boston, the *Palais de la Decouvert* in Paris, or the *Museo Nazionale della Scienza e della Tecnologia* in Milan; they can also be found in many outdoor events of the numerous "open days" organized by research centres or scientific festivals.

At the end of the nineteenth century, we cannot forget the *Ballo Excelsior*, an apology for the most important scientific and technical discoveries of the century. The premiere was made at *La Scala* in Milan in 1881 and then revived in many theatres worldwide.

2.3 Science Theatre of the Last Decades

Often leaving aside spectacular and promotional aspects or the apology of the science of the first scientific theatre, one of the typical forms of modern scientific theatre concerns the lives of great scientists. Many plays are inspired by the lives of great scientists: who are scientists? How do they behave? Several plays tell the life of mathematicians, physicists, and inventors. Often, however, the focus is on the scientist and his/her facets, and, although science gives an important background for the story, it has, in general, a somewhat secondary role, while the human affairs of the protagonist are put forward.

The staging of important scientific problems, which are particularly suitable for theatrical dynamics due to their ethical interest and the reflections they allow on men and society, is also undertaken. In fact, as the historian of physics Pasquale Tucci writes: "When a practice such as science becomes a widespread cultural heritage, it is no longer the exclusive property of those who produced it but becomes an object of debate in which everyone has the right to participate" (Tucci 2007).

Bertolt Brecht's *Life of Galileo* is one of the most famous science-theatre works. Brecht's text not only tells the life of the man-scientist, with his fears and passions, but also addresses the relationship between intellectuals and power, and the responsibility of the man of science towards society. Galileo's choice to abjure before the Church, and thus avoid torture, is the fulcrum of the action. The first draft of the work dates to 1939. In such a time of dictatorship, Brecht gave life to a reflection on the meaning of truth, be it scientific or not. Galileo is presented as a man who has chosen to deny his discoveries to have the possibility of continuing his research, even at the cost of compromising his precarious health. In the second draft of 1945, after the catastrophe caused by the atomic bomb, Brecht changes a bit his attitude and proves to be much more severe and critical towards Galileo and those scientists that do not worry about the consequences of their discoveries. Although sometimes it is thought otherwise, "Brecht himself recommends not considering the drama a pamphlet against the Catholic Church" (Chiusano 1976).

In the matter of J. Robert Oppenheimer is a title of piece, dated 1964, by Heinar Kipphardt that recounts the trial in which Oppenheimer was accused of treason for delaying the construction of the hydrogen bomb in the United States. From being a national hero, as he had been deemed during World War II, Oppenheimer became a possible threat to his country, so he suffered a period of persecution, which ended only after the advent of the Kennedy administration. The work does put the problem of the freedom to choose of the scientist, and about the boundaries of his freedom, and the relationship between science and politics.

In Michael Frayn's *Copenhagen* (1998), the souls of Werner Heisenberg, Niels Bohr, and his wife meet after death, essentially to answer the question asked at the beginning of the play by Margrethe (Bohr's wife): "Why did Heisenberg come to

Copenhagen in 1941, when Denmark was occupied by the Nazis?". This work against the background of the history of quantum mechanics reflects on the role of scientists in the very delicate military and political situations.

Still more recent is *A disappearing number*, a show, written and directed by Simon McBurney, that tells the story of Srinivasa Ramanujan considering many psychological aspects of the relationship between mathematics and worldview. Srinivasa Ramanujan was born in India in the late nineteenth century and was called "the Mozart of mathematics" for his genius. From a very poor family, he studied as an autodidact. He moved to London to collaborate with Professor Hardy, at Cambridge University, and died of tuberculosis at the age of 32, due to the climate and diet which were so different from those of his original country. In the play, the life of the protagonist is intertwined with that of a character of our times: a young mathematician who loves her work and who is fascinated by the figure of Ramanujan. Mathematics is always present in dialogues, and in scenography, with science seen as poetry: an art form that conquers the human soul.

A different style of the theatre play is *Infinities*, written by John Barrow in 2001 and staged by Luca Ronconi, and marks a fundamental point in the history of scientific theatre. Ronconi had in mind a show that was a meeting point between theatre and science but was neither popular nor didactic. For this reason, the theme chosen was infinity, with which, as Ronconi himself says, we remain on the terrain of hypotheses, suppositions, and logical form (Gregori 2001). The place chosen for the performance was not that of the traditional theatre, but the 2500 m^2 space of the former laboratories of the *Teatro alla Scala* in the suburbs of Milan. Thirty-two actors, with nineteen researchers, animated the five spaces of the exhibition: "Welcome to the infinite hotel!", "Living forever", "The paradox of infinite replication", "Infinity is not a large number", and "Where does this play come from?".

One of the novelties was the way the audience participated in the show. People had to follow the path that led from one room to another and once got to the end, they could start over as many times as they wanted indefinitely! The spectator got out with the sensation of having been the protagonist of a truly infinite show and, therefore, an essential part of the same show. The theme was thus reflected in the same structure of the play with the mathematics that did not develop in a lesson. The experience of *Infinities* broke the classic schemes of theatre and created a new language of communication between actors and the audience.

It may also be worth mentioning the works of the South African artist William Kentridge, who offers reflections on time starting from the ideas absolute time of Newton, and the relative time of Einstein with poetic images and socio-political considerations about the colonial era and with the personal psychological meaning of the idea of time, in an optimistic and joyful show.

3 Few Words About the Importance of Theatre in Physics Education

Research results show that dramatizing scientific learning through tools that promote improvisation and reflection on historical narratives—and, therefore also peer interaction—not only engages students but motivates them and helps them to grasp ideas, concepts, and scientific procedures more effectively. It also helps teachers better understand what students are thinking (McGregor 2014). But theatre has a more general action, in fact, it also promotes interaction among school, family, and society (Ødegaard 2003).

Learning is enhanced by the didactic use of theatre; emotional involvement and dramatization develop scientific imagination, allow personal learning styles, and favour an effectively mediated approach to physics. In general, dramas and comedies present a conflict that must be resolved through the interaction with other characters; within this dynamic, theatre helps to reduce cultural and gender gaps and promote a deeper and more humane scientific culture (Giliberti 2021; Fazio et al. 2021). If the conflict/game/script generates questions about physics, an emotional involvement arises that may concern physics itself, which generates active interest in students.

In general, we can distinguish three types of science dramas for educational use: those that promote science as a product, those that highlight the process and nature of science, and those that transmit science as an institution of society. So that the representation, if it is conceived and implemented by the students, can be exploratory (students begin and experientially engage with science), semi-structured (such as a role play), or structured (determined by the teacher with fixed activities, e.g., from a script).

Theatrical performances are no longer considered only as a communication or dissemination tool and are increasingly starting to be validated by physics education research. In general, the use of theatrical tools can promote active learning. The theatrical approach to inquiry-based science education (IBSE) has also been promoted by European teacher training projects. For example, the Physics Education Research Group of the University of Milan worked on the EU project "Teaching Inquiry with Mysteries Incorporated" together with eleven other European partners plus Israel (TEMI) to bring teachers closer to IBSE through theatre. A science-theatre play, entitled "Light Mystery", has been specially prepared for this purpose and shown in Italy and Europe. A commented version of the English script has also been prepared for use by teachers (Fig. 2). Naturally, the problem of teacher training arises, so that they are prepared for theatre-based learning (TBL).

Even in the courses titled "Preparation of teaching experiences" 1 and 2, held by the author for third-year undergraduate students in mathematics and physics, theatre is widely used to promote understanding of physics (for e.g., geometric optics or oscillations) through minds-on embodied activities. A similar discourse can be made for the scientific theatre laboratory held by Marina Carpineti as part of orientation activities for high school students of the Physics Department of the University of Milan.

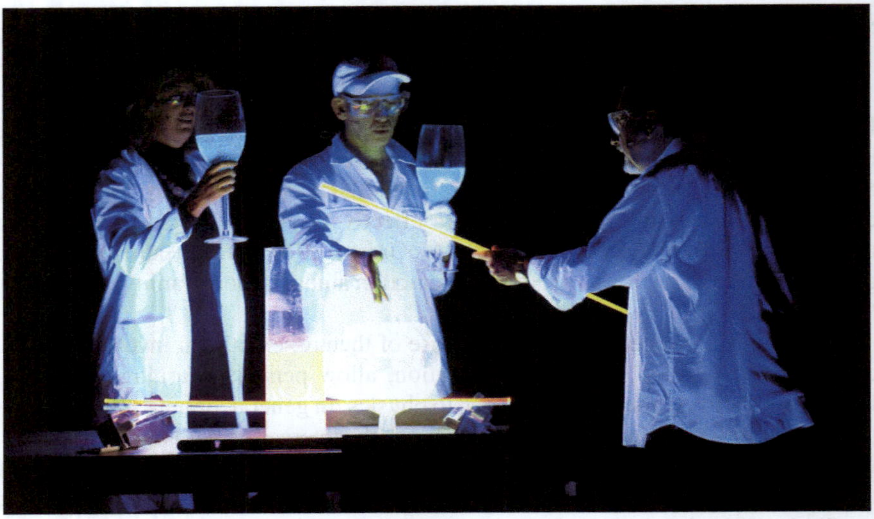

Fig. 2 A picture taken from *Luce dalle stelle* made by "Lo spettacolo della fisica" and performed in 2012 at Teatro Franco Parenti (Milan). *Credits* Fabrizio Favale, Marina Carpineti, Nicola Ludwig, Marco Giliberti

4 Physics and Culture

As we have seen in the very brief previous historical excursus, with very personal choices on scientific theatre, and, perhaps, except for *Infinities*, none of the famous theatrical performances that we could classify as science-theatre put science in the foreground. Science is the background and support, but it is not the protagonist; just as, on the other hand, in Shakespeare's historical dramas, history is not the protagonist. Yes, theatre deals with man, not with science. To better discuss this idea, it is perhaps worth noting how the present situation comes from an erroneous conception of science (particularly of physics) that we have in our culture, and from a too-closed idea of theatre as well. There are, indeed, various facets and meanings of the word "culture"; as far as we are concerned, we will mainly consider three of them: (1) the personal aspect, (2) the social aspect, and (3) the aspect common to a disciplinary group.

In the personal aspect, culture is rich and productive knowledge that gives a person a broad and personal vision of new ideas and reflections, and new interpretative grids of reality. In the social aspect, on the other hand, "culture" represents the set of ways of thinking, social and political structures, customs, idioms, ways of cooking, relating to others, and so on, and in a self-referential way, of considering the very idea of culture, which characterize a society in a certain period. Finally, there is the aspect of culture as it is understood by a particular discipline differentiated group, such as that of physicists or poets or lawyers. Substantially, it consists of the specific social culture and opinions that are, on average, shared by that group. It is from this last specific point of view that scientists often complain that science should be a more consistent part of the citizen's (social) culture. However, they hardly wonder why, and in what sense, should science be part of the main culture of the entire society, and not only of single persons or of specialized groups. To be closer to our theme, in fact, in what sense, should physics manifest itself as intertwined with the vision of the world and life that gives meaning to our lives and opens to the future?

We all know that when topics are meaningful to people—for example because they are perceived as useful or beautiful, or fascinating—interest, attention, and desire to understand are often present. It is completely natural to seek a better and happier life, and it is precisely in this sense that perhaps the most significant aspect of culture manifests itself. In our modern society, we often talk about culture as a citizen's duty, or as a social elevator, but perhaps the most important and characterizing element of culture is, in general, overlooked: in fact, culture is culture if it changes our life, if it speaks to us in-depth, modifying our vision of the world and life, and giving us a chance to experience the pleasure of searching, and, sometimes, even of understanding. Culture is the pleasure of continuous research.

Therefore, if we—physicists like I am—believe that physics should be perceived as a culture in a personal and social sense, and not just for a small group of insiders, we have to wonder if the image and vision of the world that emerges from modern physical science are suitable to enrich our lives; and, therefore, whether science, and physics in particular, can, or cannot, play a role that helps us feel better.

The positive answer that is often given in an automatic and stereotyped way to the question "should we—scientists, educators, politicians, communicators ...—work for a widespread scientific culture?" which is exaggeratedly uncritical. The main motivation that pushes us to respond positively is, indeed, since our society is unquestionably linked to technology, which is mainly the result of science. However strong, and essential this link may be considered, it is not enough.

For many years, we have been experiencing the results of the first quantum revolution which allowed the growth and development of the electronic industry of computer chips, photovoltaic cells, and all laser-based technology. But now, we have already entered the second quantum revolution, which is greatly involving production systems and, hopefully, making better our lives soon. There are now manmade coherent quantum states of radiation or entangled matter particles with new properties and huge possibilities of applications in new types of computer communication systems and sensors (Dowling and Milburn 2003). The European Union is investing more than one billion euros to create the conditions for public awareness of this second revolution and for fostering an educational system able to prepare enough technicians specialized in quantum technologies. Unfortunately, these efforts, together with the general awareness of the importance of physics in technology are not enough for physics to enter our hearts and sprout, thus generating culture for society.

The link between science and technology is already very clearly perceived by many people, even, perhaps, in an exaggeratedly strong way. Science is "increasingly confused, especially by children and adolescents, with the use of high-tech tools [...]. This misleading perception can persist into adulthood and can reduce the interest in science" (Rustichelli and Stefanon 2012). Indeed, research made more than ten years ago on a thousand upper secondary school students near Milan (Giliberti 2010) showed that most of them consider physics important for society, but more linked to technology than to general culture (Carpineti et al. 2011; Tolstrup Holmegaard et al. 2014). Scientists, from this point of view, do not generally help, in fact, they almost aim at showing the numerous social implications of the research they carry out. If this fact is perfectly understandable from the point of view of medical research—as it has become clear in the pandemic—it is, however, less comprehensible from the basic research viewpoint: an example above all is provided by large research infrastructures such as CERN which, in his life has, perhaps, been more advertised for the birth of the Web or the development of magnetic resonance technologies, than for the research activities about particle physics taking place there. In short, it almost seems that physicists would apologize for the funds spent on basic research, rather than show how much living and vital culture is there in their research, and how great the pleasure of understanding which, in this way, is made available to everyone. When they indulge in the social implications of physics, they often try to fascinate young people with the wonders of applied physics and fail to realize that in doing so they generate interest, which, however, rarely turns into a real fascination with science, rather pushes young people towards applications.

Luckily, people who love knowledge for itself do not seem so few; at least observing how many echoes had the discovery, right at CERN, of the Higgs boson, or that of gravitational waves made by LIGO and VIRGO and the recent "images" of black holes taken a few years ago. The first glimmers of a trend reversal towards fundamental science and research can perhaps be already glimpsed.

Society needs to invest in the value of knowing well before it can spend that knowledge. The greatest cultural aspect of physics for society is more regarding its ability to ask questions about nature than about the answers we obtain. Indeed, we have more questions today than we ever had in the far past when we knew much less. Physics, in the modern sense of the term, was born when we began to understand that a stone falling from the mast of a ship has a trajectory with respect to an observer on the ship, and a different one with respect to an observer stationary on the quay, with neither of them that has more right than the other to claim to see the truth (Bruno 1584). Indeed, the general cultural meaning of physics mainly consists in its drive to transcend common thinking, to transgress the usual interpretation of reality, and understand that questions that have seemed well posed, and decidedly sensible for centuries, are, on the contrary, meaningless. Our opinion is that the fundamental cultural root of physics starts from here, from the need to go beyond questions like "what is the trajectory of the stone?"; or "which is its speed?"; or, as it happens when considering the relativity of simultaneity, beyond questions like "what is happening now on Andromeda?"; or, in the quantum mechanical domain, well beyond the question "is it a wave or a particle?". Without, however, giving up the search for the truth; a search that will become significant when it leads us to the construction of a precise theory within which a precise meaning to "true" answers to well-posed questions will be given.

5 Theatre and Physics

Equipped with the considerations of the previous sections, we can now face the relationship between theatre and science with greater awareness. Preliminary to any real scientific theatre is the understanding that science, while trying to describe the laws that govern the universe, is, at the same time, also speaking about humans, and how they see the world. And since "theatre is an art dealing with the way great themes—geography, history, philosophy, love, …—are coagulated in the human soul, to really have a science-theatre play, it must speak about the world of science, for example of physics, in a such a way that makes people spectators of the show, while seeing and listening about science, also seeing and listening about their lives. On the contrary, people are often used to seeing science as something distant from themselves; they think: "I am not a scientist, after all!", feel far from that world, and get out of it".[3]

Whatever the difficulties of creating a theatre show that speaks to man by talking about science, an awareness serves as our guide; it is the observation that theatre is the ideal tool to address themes that are at the roots of scientific research and questions at the basis of its meaning. Not only a tool for proposing a discussion in society, but because the very conceptual structure of science is intrinsically and naturally of a theatrical nature.

"Science is reality re-imagined" (Ogborn 2011, p. 1). "A scientific explanation is a story. It is a story about how some imagined entities, taken as real, would by their nature have acted together to produce the phenomenon to be explained" (Ogborn 2011, p. 14); passing from the provisional—i.e., ad hoc models—to the permanent-but-transient -i.e., established theories. It follows that the same understanding is storytelling. The form of the narrative respectful of the permanent-but-transient is a theatre in a natural way. A scientific paper is a script: there is a story, a problem, and a challenge to tell. Theatre speaks of the world, and, in turn, the world is the representation that we give of our understanding.

Theatre and physics are like two dual aspects of the same research. The former is ephemeral, but immortal, the latter absolute and universal, but transient. The word "theatre" comes from the Greek verb $\vartheta\varepsilon\alpha o\mu\alpha\iota$, which means to look, to contemplate, with the root of the verb that means admiration or wonder. But we also must observe that $\vartheta\varepsilon\alpha o\mu\alpha\iota$ in Greek is a deponent verb, and, therefore, it has in itself a passive *nuance* that can perhaps be better translated with "being spectators, contemplators".

Physics is a metaphor for the world, and theatre is a metaphor for reality; therefore, physics theatre is a metaphor for a metaphor, a *metaphor squared*. The usual analogies explain new things in terms of simpler and more known things; physics, on the contrary, explains obvious things—like that heavy bodies fall—and less obvious things, like gravitational waves, in terms of abstract constructions that are not obvious at all, but that can be described with precision and studied in every detail since they

[3] From a private video communication with Flavio Albanese (16th January 2023).

are constructed with the fundamental aid of mathematics and "invented" by humans. The usual metaphors carry out a symbolic transposition that makes the speech more significant, such as the phrase "the sea of lavender sways with the blow of the wind" which considers the lavender fields as a sea, making a comparison that easily reinforces ideas. In physics "the electron is a wave function that evolves in the Hilbert space using a unitary operator" is a phrase that makes the idea extremely effective only after long and passionate personal training. Physics operates in reverse, so to speak, like an inverse metaphor.

Concerning gravity, Newton said the famous *Hypoteses non fingo*. In the *General Scholium*, at the end of the *Principia*, he wrote that:

> It is sufficient that gravity exists, acts according to the laws we expound and explains all the movements of the celestial bodies and of our sea.

For this, Leibniz accused Newton of returning to occult phenomena and Huygens said that, therefore, it was not even worth the effort to make complicated calculations about gravitational forces, if, after all, one did not even know its cause.

But physics evolved in the direction Newton had shown, moving forever away from common sense and "understandable" explanations. Since then, the usual explanatory similarities no longer apply in physics or physics education. Since then, it is no longer even possible to carry out a scientific theatre based on popular understanding.

But it will be precisely starting from this awareness that a real physics theatre can be born. It will be his main task—at least, this is the work that I personally, and within the group "Lo spettacolo della fisica" intend to do—to try to make the scientific metaphor direct. This time, therefore, it will no longer work by starting from common reality, but from real, actual physics in all its complexities, with all its language and all its ideas; by doing the reverse work of what it is used to do, which starts from reality to allow reflections and interpretations; by doing, therefore, it too, an inverse metaphor. Thus, theatre and physics will discover each other chasing each other along the faces of the Möbius strip of the human universe.

Acknowledgements I am very grateful to my Ph.D. student Luisa Lovisetti for her helpful suggestions, interesting comments, and attention in revising the text.

References

Aeschylus, *Prometheus Bound*, trans. by W.A.H. Loeb. Classical Library, vols. 145 and 146 (Harvard University Press, Cambridge, MA, 1926)

M.C. Brandi, L. Cerbara, M. Misiti, A. Valente, Giovani e scienza in Italia tra attrazione e distacco. JCOM **04**(02), A01 (2005)

G. Bruno, De immenso et innumerabilibus (1584), https://wq-it.wikideck.com/Giordano_Bruno#cite_note-38. Accessed 07 Apr 2022

M. Carpineti, M. Cavinato, M. Giliberti, N. Ludwig, L. Perini, Motivare allo studio della fisica attraverso il teatro. JCOM **10**(1) (2011)

I.A. Chiusano, *Storia del teatro tedesco moderno, dal 1889 ad oggi Collana*, trans. by the author (Torino, Einaudi, 1976), p. 219

J.P. Dowling, G.J. Milburn, Quantum technology: the second quantum revolution. Philos. Trans. R. Soc. **361**(1809), 1655–1674 (2003). https://doi.org/10.1098/rsta.2003.1227. Accessed 30 Jan 2023

C. Fazio, M. Carpineti, S. Faletič, M. Giliberti, G. Jones, E. Mcloughlin, G. Planinšič, O.R. Battaglia, Strategies for active learning to improve student learning and attitudes towards physics, in *Teaching-Learning Contemporary Physics: From Research to Practice*, ed. by B Jarosievitz, C. Sükösd (Springer, Berlin/Heidelberg, Germany, 2021), pp. 213–233. https://doi.org/10.1007/978-3-030-78720-2_15

R. Feynman, *Six Easy Pieces* (Basic Books, New York, 2000), p. 99

M. Giliberti, La Percezione della Fisica negli Studenti di Scuola Secondaria di secondo grado: Indagine Statistica collegata allo Spettacolo Teatrale TRACCE, in *Proceedings of the Frascati Physics Series-Italian Collection, Collana: Scienza Aperta, II—ComunicareFisica2010, Conference "Comunicare Fisica e altre Scienze"*, Frascati, Italy, 12–16 Apr 2010

M. Giliberti, *Fisica a Teatro, collana Lo scrigno di Prometeo* (Aracne, Roma, 2014). ISBN: 8854877069

M. Giliberti, Fisica e teatro: spunti di riflessione per un teatro di fisica. Nuova Second. **8** (2019)

M. Giliberti, Stories and theatre for teaching physics at primary school. Educ. Sci. **11**, 696 (2021). https://doi.org/10.3390/educsci11110696

M.G. Gregori, *Movimento metafora del tempo, conversazione con Luca Ronconi*, trans. by the author (2001). https://lucaronconi.it/storage/filemedia/infinities.pdf

F. Magni, *Il teatro scientifico per la scuola, in Attori del sapere—Un progetto di teatro, scienza e scuola*. Scienza Under 18 eds (Scienza Express Edizioni, Milano, 2011), p. 60. ISBN-10: 889697320 and ISBN-13: 978-8896973202

D. McGregor, Chronicling innovative learning in primary classrooms: conceptualizing a theatrical pedagogy to successfully engage young children learning science. Pedagogies **9**(3), 216–232 (2014)

C. Molinari, *Storia del teatro*, trans. by the author (Editori Laterza, Bari, 2011), p. 46

M. Ødegaard, Dramatic science. A critical review of drama in science education. Stud. Sci. Educ. **39**(1), 75–101 (2003). https://doi.org/10.1080/03057260308560196

J. Ogborn, Science and commonsense. Rev. Bras. Pesqui. Educ. Ciênc. **6**(1) (2011). https://web.phys.ksu.edu/icpe/publications/teach2/Ogborn.pdf

F. Rustichelli, M. Stefanon, The European project "immersion in the science worlds through the arts." Il Nuovo Saggiatore **5–6**, 46–49 (2012)

TEMI (Teaching Enquiry with Mysteries Incorporated), European Community's Seventh Framework Programme (FP7/2007-2013) under grant agreement n° 321403-2012-1. Available online: https://cordis.europa.eu/project/id/321403. Accessed 30 Jan 2023

H.H. Tolstrup, M.L. Møller, L. Ulriksen, To choose or not to choose science: constructions of desirable identities among young people considering a STEM higher education programme. Int. J. Sci. Educ. **36**(2), 186–215 (2014). https://doi.org/10.1080/09500693.2012.749362

P. Tucci, Teatro e Scienza, in *Se una notte, guardando Marte...*, ed. by M. Caldini, trans. by the author (Imperia, Ennepilibri, 2007), p. 5

Digital Challenges for Physics Learning

Digital Challenges
for Physics Learning
Springer

Research-Based Contribution on ICT as Learning Challenges in Physics Education

Marisa Michelini and Alberto Stefanel

Abstract The new technologies and in particular those of information and communication are tools and methods in physics education, to have experience in specific areas and proper physics methods. A computer equipped with sensors and appropriate software constitutes a powerful acquisition system for the didactic lab of good efficiency, reliability, and accuracy, which allows immediate data visualization, and integrates different measuring instruments, software, and hardware. This allows students not only to make good measurements in the laboratory, but also to experiment with methodologies typical of physics and modern research laboratories, such as in-person and remote measurement activities, computerized data analysis, and modeling. These opportunities are amplified nowadays by the possibility of using smartphones as measuring instruments thanks to free apps that make use of the sensors with which they are equipped. New possibilities for active teaching, both in-presence and distance learning, have opened up by new technologies for physics education and, in particular, for student learning. Some of the main ones are discussed in the present contribution for experimental, modeling, simulation, and active student involvement activities.

Keywords Information communication technology · High school students learning · Active teaching strategies · Experimental laboratory

M. Michelini · A. Stefanel (✉)
Physics Education Research Unit (PERU), University of Udine, Via delle Scienze 206, 33100 Udine, Italy
e-mail: alberto.stefanel@uniud.it

M. Michelini
e-mail: marisa.michelini@uniud.it

M. Streit-Bianchi et al. (eds.), *New Challenges and Opportunities in Physics Education*, Challenges in Physics Education, https://doi.org/10.1007/978-3-031-37387-9_14

1 Introduction

The rapid evolution of new technologies of information and communication (ICT) is producing a rapid and continuous change in society, which has induced a profound change in many professional figures (Gatti et al. 1998; Euler and Müller 1999; Bao and Koenig 2019; Grayson 2020). Schools are involved at different levels in this change, and the figure of the teacher is among those expressing the greatest need for innovation. The challenge is to offer young people a richer education in disciplinary content and methodologies, specific, and transversal skills. These can intercept and enhance the new skills of digital natives and are adequate to deal with the growing complexity of society, of the choices to which citizens are called, and of the world of work with which today's students will measure themselves (Grayson 2020; Messeri 2004; European Union 2006; Galliani 2009; Michelini and Stefanel 2019).

ICT is an integral part of working methods in physics (Euler and Müller 1999; Mazur 1991; Teodoro 2005). Physicists at CERN invented the web for scientific collaboration, and all physics research currently relies on and contributes to tools and methods that heavily employ ICT (Bao and Koenig 2019; Final Report Hope Project WG2 2017). At the experimental level, for designing experiments, outcomes are simulated, and equipment is tested with simulators, data are acquired, analyzed, and fitted using a computer. At the theoretical level, models are built to simulate behaviors and identify new phenomenologies. Therefore, ICTs are a tool and a method in physics education, to have experience in specific areas and proper physics methods (Mazur 1991; Ogborn 1986; Michelini 1988, 1999; Titus et al. 1998; Euler 2001, 2004). Moreover, they offer several opportunities for schools, which are more in line with a need for educational innovation, rather than simply a demand for information literacy (Euler 2001, 2004; Riel 1998; Swan and Miltrani 1998; Esquembre 2002; Vendramini and Michelini 2018). Physics education must include moments of personal operational involvement, exploration of ideas and reality, application of hypotheses, inquiry approach to the construction of interpretations (McDermott 1993, 2001; Abd-El Khalick et al. 2004), and individual and collective activities in and out of the classroom (Pontecorvo et al. 1995; Caravita 1995; Caravita and Hallden 1995). ICTs provide contributions on different levels for these objectives (Adams 2010), also giving the tools and methods to implement effective distance learning (Strubbe and McKagan 2021). In the following, we discuss those more aimed at conceptual and active learning, related to experimental laboratory, modeling, and simulations. Examples are taken frequently in our research work. Examples are most often related to the high school level, which is the reference for what concerns the role of ICT in physics learning. They can, however, easily be transferred in their entirety or with little modification to the undergraduate university level as well. Finally, some examples are aimed at showing the role of ICT in learning physics in primary school.

2 ICT and the Computer in the Experimental Teaching Laboratory

Physics education must involve confrontation with experimental fact, as it is a fundamental part of the process by which science constructs knowledge about the world. Educational laboratory activities, while involving methodological elements typical of research, differ from it in terms of their formative and educational roles, according to different modalities: demonstrative experiments for phenomenological recall, hand-on/minds-on phenomenologies exploration; experiments and measurements in educational labs; design to carry out apparatus and experiments; laboratory exercises; interfacing experiences (Michelini 1999; Euler 2001, 2004; Giugliarelli et al. 1994; Hofstein and Lunetta 2004; Johnstone and Al-Shuaili 2001; Millar and Abrahams 2009; Planinsic 2020).

In this framework, the role of the computer and ICT in the laboratory is supported by methodological–disciplinary reasons, practical and social reasons, and didactic reasons for learning (Hofstein and Lunetta 2004; Johnstone and Al-Shuaili 2001; Michelini 1992; Bernhard 2003).

Young people need to experience peculiar methodologies of physics typical of today's research laboratories. Measurements with sensors online with the computer offer students opportunities in this perspective. A computer equipped with sensors and appropriate software constitutes a powerful acquisition system for the didactic lab of good efficiency, reliability, and accuracy, which allows immediate data visualization, and integrates different measuring instruments, software, and hardware. The portability of sensors, interfaces, and PCs allows measurements to be made in laboratories and outside of them, in an ordinary classroom, in cramped or remote environments (natural environments, the car, the plane). Measurements can be made on a wide variety of quantities of interest to both physics and other areas. Software is of increasingly immediate use and aimed at education. Measurements with sensors online with the computer familiarize students with a way of controlling many technological systems in common use and pervasive in the world of work.

On the learning side, ICTs enable new activities, and new disciplinary and methodological goals can be achieved (Giugliarelli et al. 1994; Hofstein and Lunetta 2004; Johnstone and Al-Shuaili 2001; Michelini 1991, 1992; Bernhard 2003; Hofstein 2004). Students are more polarized on concepts and design rather than operational and executive aspects. The laboratory becomes a culturally stimulating environment in which the characteristics of a phenomenon can be analyzed at different levels of depth. Data processing aims to find characteristic relationships, and phenomenological laws and stimulates the search for interpretive models. The type of sensors used, and the data acquisition software are crucial because they determine the possible roles of the user and thus the possible strategies of use. The user must be able to operationally master their functionality and response modes, with room for planning and choices about the experiment to be performed and the quantities to be measured. Mastery over the construction characteristics of sensors or their operating principles is a significant goal only for advanced laboratories. Advantages and new

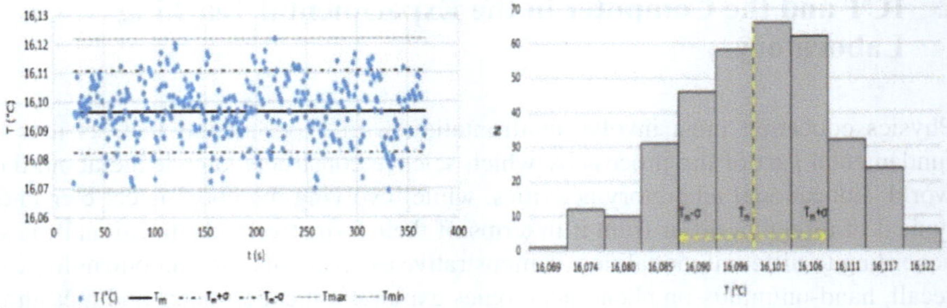

Fig. 1 Graph of the time evolution of the temperature of a sensor immersed in a thermal bath and distribution of the data collected and processed during a university lecture in a few minutes

learning opportunities enabled using sensors interfaced with computers or today also embedded in smartphones are related to their inherent potential, outlined and exemplified in the following.

2.1 *Extending the Scope and Potential of the Laboratory with Online Sensors*

Online sensors make it possible to extend the scope and learning potential of the teaching laboratory in different dimensions. In a simple and time-saving way, a lot of data can be collected for statistical analysis (Fig. 1).

Measurements become possible, which are very laborious or impossible to carry out with traditional methods, such as measurements of the light intensity produced by a source as a function of distance or direction (angle) of measurement, or measurements of the light intensity transmitted by refringent filters and polaroids (Michelini and Stefanel 2006; Freitas et al. 2018; Cescon and Stefanel 2022). Phenomena occurring over very long-time scales (e.g., temperature over a day) or very short ones such as transients can be studied (Fig. 2).

2.2 *Dedicated Systems to Enhance the Laboratory*

Research and Development approaches in physics education (Lijnse 1995; Beichner 2006; Wang and Hannafin 2005) have always contributed to the realization of prototype systems dedicated to specific fields (Ogborn 1986; Michelini 1992; Thornton and Sokoloff 1990; Simpson and Thornton 1995; Sokoloff 2022). The design criteria aim to maximize learning effectiveness: hardware that guarantees robustness and portability of the sensors and interface; high performance in terms of sensitivity, precision, reliability; versatility of use that guarantees the possibility of qualitative explorations

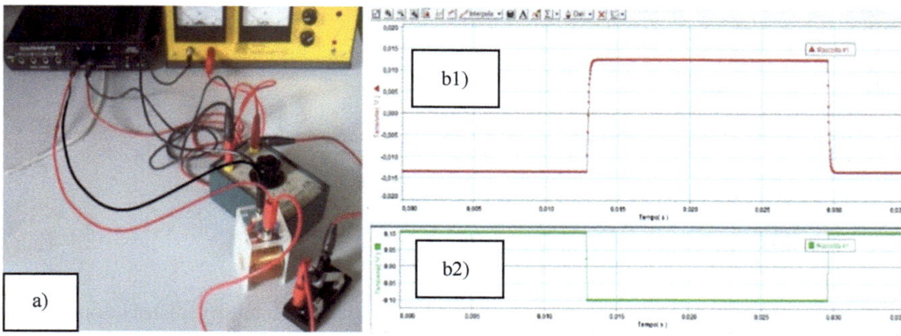

Fig. 2 Apparatus for the study of the transient in the on and off phases of an electric circuit (**a**). Time evolution of the current (**b1** graph) in the on/off transient of an RL circuit supplied with a square-wave generator (**b2** graph)

as well as good measurement performance; software interface of immediate use; clear representation of data; the possibility of varying the detection parameters as required (Hofstein and Lunetta 2004; Johnstone and Al-Shuaili 2001; Michelini 1992).

Examples of this are two systems developed and patented by the University of Udine: TERMOCRONO (Gervasio and Michelini 2006), a USB system that allows the simultaneous acquisition of the temperature of four thermal sensors, displaying their evolution in real time (Fig. 3); LUCEGRAFO (Gervasio and Michelini 2009), connecting via USB to the computer, for the acquisition of the intensity of light on a sensor as a function of its position (Fig. 4) (Santi et al. 1993; Michelini et al. 2014a).

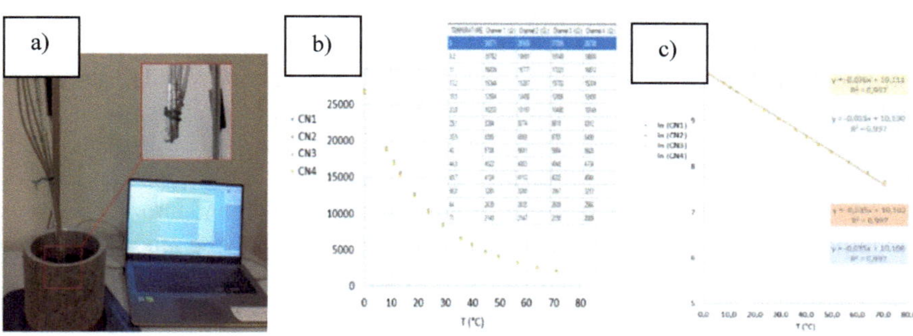

Fig. 3 The calibration procedure of the TERMOCRONO (Gervasio and Michelini 2006) exemplifies an advanced didactic laboratory activity. It is based on the measurement in a thermal bath (**a**) of 15 different temperatures (**b**). The transfer equation $C_{NX} = e^{-kT}$ is obtained by the data fit of the log C_{NX} versus T diagram (**c**)

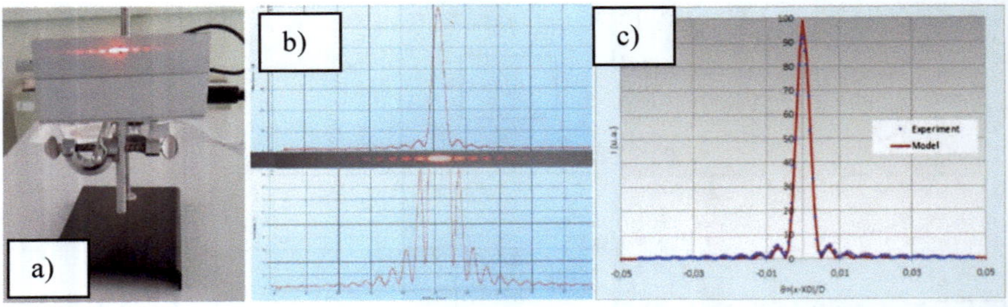

Fig. 4 The LUCEGRAFO (**a**) makes it possible to acquire the intensity distribution of light diffracted by a single (**b**) or multiple slits. The measurement quality makes it possible to fit the experimental graph with the distribution obtained from a model, in which the distribution is the interferential outcome of the spherical secondary waves produced by 51 sources simulating the 0.24 mm slit used experimentally (**c**)

2.3 *Improving the Quality of Measurements*

The availability of reliable and accurate acquisition systems makes it possible to analyze and discriminate nonlinear phenomena. Basic measurements can be deepened with new objectives. For example, the photogate measure of the period of a pendulum can focus on analyzing the dependence of the period of oscillation on amplitude or length (Michelini 1991). The laboratory thus becomes an environment in which students explore hypotheses, experimentally analyze the limits of models, tackle experimental problem-solving, and develop skills (Millar and Abrahams 2009; Michelini 1991; Hofstein 2004). They can use low-cost electronic components and open-source platforms such as Arduino (Freitas et al. 2018; Michelini et al. 2014a; Huang 2015; Oprea 2018; El Hadi et al. 2020), to make homemade devices and instruments with good performance. Objects such as laser pointers and LEDs can be used in measurement systems or apparatuses (such as light sources, or the LED for studying the photoelectric effect) or studied as technological artifacts (LED current–voltage characteristic) (Teodoro 2005; Planinsic 2020). Measurements on the electrical properties of metals, semiconductors, and superconductors (resistivity vs. temperature and Hall effect measurement—Fig. 5) can now be made (Michelini et al. 2008; Kedzierska et al. 2010; Gervasio et al. 2014).

2.4 *Possibility of Integration with Software, Hardware, Web*

Data collection with online sensors is easily integrated with the use of other software for data processing, comparison of experimental results and simulations, data fit with models, and modeling (Heron and Meltzer 2005; Crăciun and Bunoiu 2017). Some software integrates the management of online measurements, video analysis, and modeling. The Coach system was its forerunner (Heck et al. 2009; van den Berg

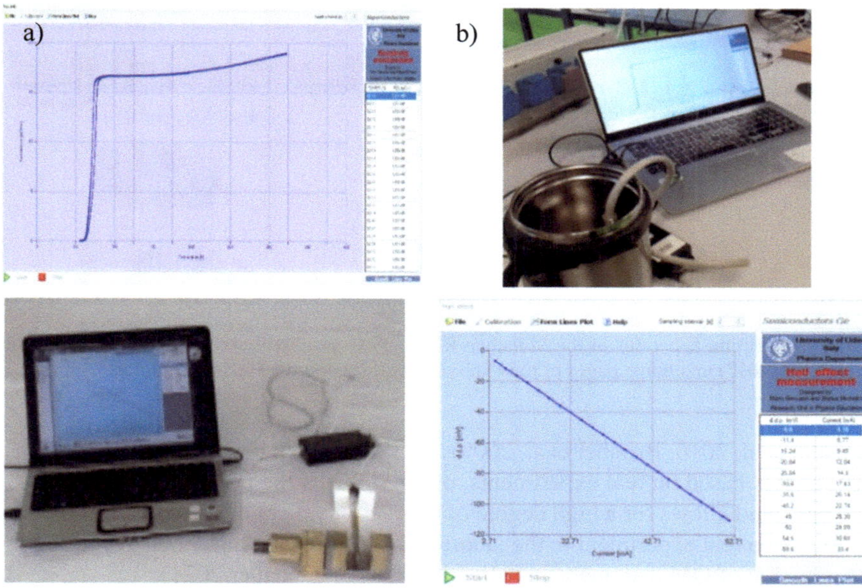

Fig. 5 The USB interface developed for measuring resistivity as a function of temperature (**a** *R* vs. *T* graph for a YBCO sample and the experimental apparatus) and for measuring the Hall coefficient (**b** apparatus and sample graph for Germanium *N*) (Gervasio and Michelini 2009)

et al. 2008). With video analysis, measurements can be made on films of experiments, natural phenomena (i.e., the motion of water in a river), the motion of a ball (van den Berg et al. 2008), the impact of a drop of water on a mosquito (Dickerson et al. 2012), everyday life situations (Beichner 1996), sports situations, such as the run of a sprinter, the flight phase of a high jumper, the trajectory of a ball (Heck et al. 2009; Bradamante et al. 2004). Schools can easily equip themselves with low-cost video cameras and video analysis software (Heck et al. 2009; van den Berg et al. 2012).

The opportunities offered by ICT have made it possible to carry out educational experiments remotely, a mode of work now typical of many physics laboratories (Thoms and Girwidz 2018; Grober et al. 2007, 2013; https://www.i2u2.org/). Expensive or potentially dangerous experiments, such as often those in modern physics, are offered to schools as an expansion of the experimental laboratory (Thoms and Girwidz 2018; Grober et al. 2007, 2013; https://www.i2u2.org/; https://www.ises.info/index.php/en) (Fig. 6). The limitation of this type of laboratory is the need for continuous maintenance and servicing, which requires a commitment of both personnel, financial resources, and physical laboratories. Currently, the Munich site suffers from this situation, remaining a model for the development of RCL laboratories.

A further opportunity for learning is offered by the development and availability of interactive screen experiments (ISE) for schools (Kirstein and Nordmeier 2007;

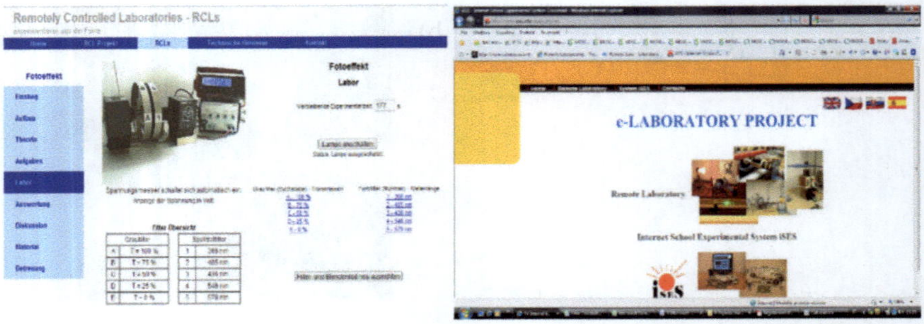

Fig. 6 Diffraction apparatus of the München Remotely Controlled Laboratories—RCLs (Grober et al. 2007, 2013) and home pages of the e-lab project (https://www.ises.info/index.php/en)

Kirstein et al. 2016; Bronner et al. 2009). An ISE is much more than a trivial animation, as it is based on real experiments, offers students the possibility of processing real data, and allows for a high degree of interactivity, very close to that afforded by an RCL or a real experiment in the laboratory. A recent test in schools regarding the introduction of QM documents its effectiveness (Bitzenbauer 2021).

3 Real-Time Lab (RTL): Role of Real-Time Graphs

Alongside the multiple potentialities outlined, the role of the computer as an interface between phenomenal reality and the abstraction of concepts is emphasized here, for the development of theoretical and formal thinking (Michelini 2010).

The visualization of the RTL graph plays a fundamental role in producing that imaginative reduction of the phenomenon itself with its physical description. There is a broad consensus on the positive impact on learning of RTL activities based on POE and PEC strategies (Theodorakakos et al. 2010) as catalysts and mediators of learning both in physics (Bernhard 2001, 2003; Thornton and Sokoloff 1990; Sokoloff 2011, 2016; Sassi 2000; Hake 1998; Sokoloff et al. 2007; Tinker et al. 1999) and mathematics (van den Berg et al. 2012; ESM 2004).

One example is a teaching proposal based on the use of motion sensors for the study of kinematics. Using a PEC strategy, students analyze a sequence of common situations, such as the walk of a person in front of a sensor or the free motion of a toy car on a horizontal (Fig. 7) or inclined plane, to construct the concept of acceleration and correlate the graphs of derived quantities. They acquire mastery of concepts such as those of reference system, position, displacement, and speed, activated by the link between iconic elements of graphs and conceptual meaning (Sperandeo et al. 2002; Corni et al. 2005).

A second example is a course for a basic school that uses temperature sensors as an extension of the senses (Bosio et al. 1996; Michelini et al. 2010). Children explore the thermal properties of systems with their fingers as sensors. They learn to recognize

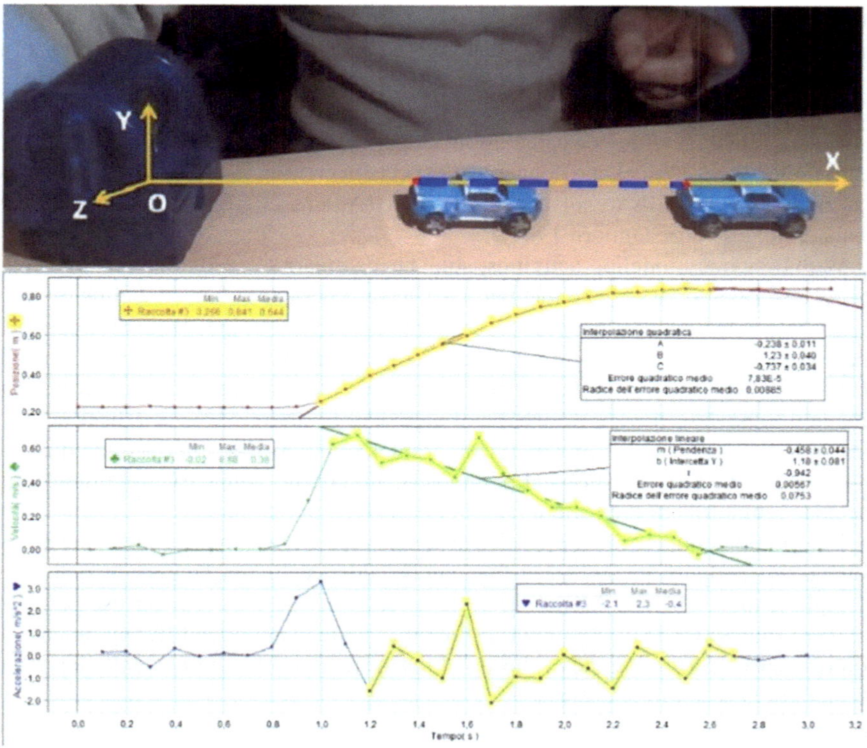

Fig. 7 Kinematical graphs acquired in real time for a toy car moving on a horizontal plane table in front of a motion sensor

the reliability of the information associated with thermal sensation and the parameters on which it depends. They use sensors to carry out similar explorations, learning to recognize thermal states and processes by visualizing the temporal evolution of the temperature of the sensors and thus of the systems with which they interact (Michelini et al. 2010; Stefanel et al. 2002; Benciolini et al. 2002).

4 Computer Modeling Activities for Learning

Modeling is a foundational methodological element in physics as in other sciences (Clement 2000; Hestenes 1987, 1995) and also in the environmental and life areas (Hoskinson et al. 2014; CBE—Life Sciences Education (LSE) 2013). It must therefore also play a role in science education, first of all, so that students have direct experience of how models are constructed and what characteristics and roles they have (Michelini 1991; Clement 2000; Hestenes 1995; Grosslight et al. 1991; Rodríguez et al. 2013). Here, reference is made to dynamic modeling in which the objective is to interpret a phenomenology, describing the evolution of physical quantities of a system, by means of a formal model typically expressed with a differential equation (Ogborn 1984). Dynamic computer modeling allows students to free themselves from most of the technicalities of calculation. It meets the didactic need for students to learn the methodological and conceptual aspects of writing a model, without having the mathematical skills to solve it analytically (Adams 2010; Michelini 1991; van den Berg et al. 2008; Ogborn 1984).

With dynamic modeling activities, students can learn to recognize that the same model describes a class of phenomena. They have experience of the modeling process based on a coherent scheme with descriptive, interpretative, and predictive capabilities with respect to an entire phenomenology. They have experience of how in science the aspects to be looked at of a phenomenon are selected, the quantities that enable it to be described, the conceptual network of relations between these quantities underlying the model, and its formal implementation. They understand the role of system and phenomenological parameters, initial and boundary conditions, and the conceptual role of the formal operators involved in the model.

The implementation of the model in a software environment allows one to go beyond the study of the few cases that can be addressed with the poor formal tools of high school students. Modeling systems such as the mass-spring oscillator or the $1/r^2$ interaction (Fig. 8) allows a student to understand how different quantities contribute to determining the evolution of the system and does not trivially learn the formula of how amplitude varies.

It also provides a working methodology that makes it possible to construct models for the temporal evolution of any quantity (e.g., temperature) from a law or hypothesis on its mode of variation (e.g., a hypothesis of a functional relationship between T and $\mathrm{d}T/\mathrm{d}t$). In the field of natural sciences, for example, one can study how populations of prey and predators vary according to the Lotka–Volterra model. It is possible to extend the method to describe how a quantity varies because of the variation of

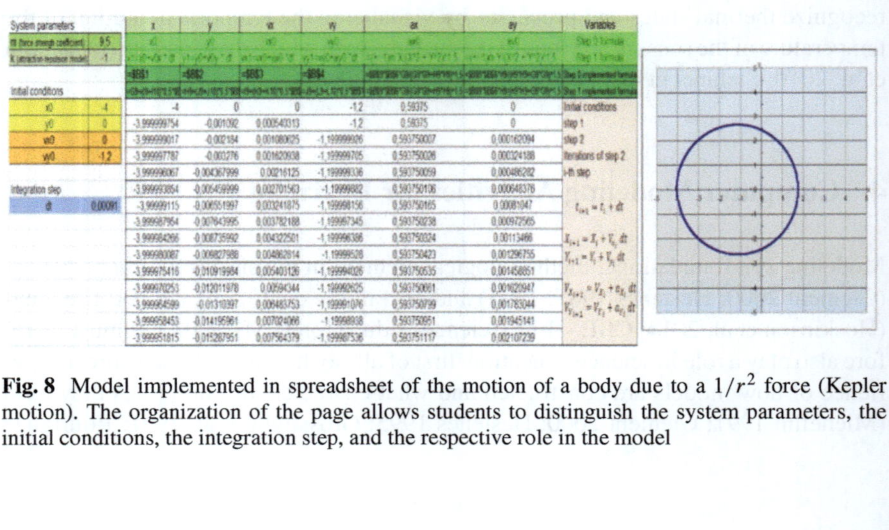

Fig. 8 Model implemented in spreadsheet of the motion of a body due to a $1/r^2$ force (Kepler motion). The organization of the page allows students to distinguish the system parameters, the initial conditions, the integration step, and the respective role in the model

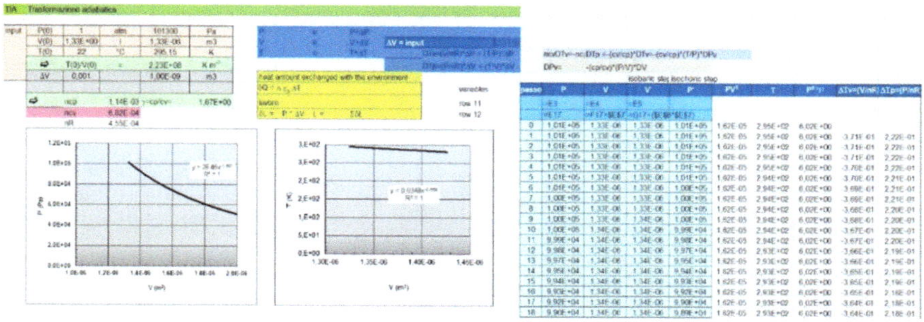

Fig. 9 Modeling an adiabatic transformation of an ideal gas, implemented on a spreadsheet as in previous research (Michelini 1991; Gallo et al. 1989)

another related quantity. Figure 9 shows the implementation in a spreadsheet of the SIGMA modeling environment (Michelini 1991; Gallo et al. 1989), to model the ideal gas polytropic, using isochoric and isobaric steps.

The use of research-based tutorials is necessary to have a significant learning impact (Sperandeo et al. 2002; Linn and His 2000; Cooney 2003; Benacka 2008; Liengme 2014; Kohnle et al. 2015; Singh 2008). Similarly, specific modeling environments have been designed in which the user only has to write the formal elements of the model. Several proposals that were ahead of their time are still valid in their design structure but are obsolete for today's computers (Michelini 1991). Today, environments such as EJS (Brown and Christian 2011; Christian et al. 2010), VPytom (Sherwood and Chabay 2010), and Modellus (Martinez et al. 2010; http://modellus.fct.unl.pt/) are accessible on the Internet, which allows the user to carry out simulations by writing only the model equations (http://www.compadre.org/portal/index.cfm) (Fig. 10).

A different modeling approach is based on the water tank and flux paradigm. An example of this is the Coach modeling environment (van den Berg et al. 2008, 2012).

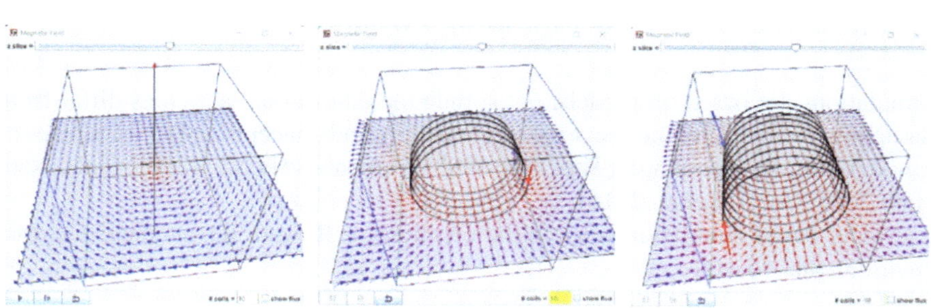

Fig. 10 EJS—simulation (Christian and Hwang 2014) of the field generated by a current-carrying wire twisting back on itself to form a solenoid. The image can be manipulated to visualize the field from different perspectives. The animation shows the dynamics of how the field structure changes, changing the source geometry

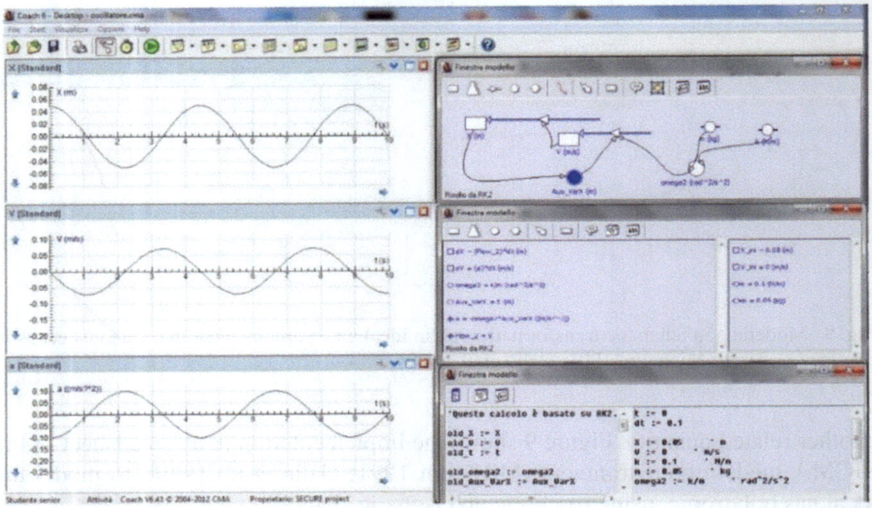

Fig. 11 Model of the spring-mass oscillator in the modeling environment of Coach6. The model can be implemented in iconic form by creating a map like the one above right. The software also allows visualizing the mathematical formal meaning and how it is translated into the programming language

Variables are represented by containers whose content, initially fixed by the user, is varied over time by a "flow," which formally corresponds to the differential operator that increases the variable considered and to which the other quantities of the model contribute according to the functional relationships that characterize it. The model has been constructed through icons as a map (Fig. 11) and automatically translated into analytical form and code.

5 Simulations for Understanding Physics

Simulations are based on models, but simulation educational activities differ from modeling ones because the user explores to understand concepts and the outcomes of a model, without accessing it (Titus et al. 1998; Wieman and Perkins 2005; Finkelstein et al. 2006; Ceberio et al. 2016).

Model-based simulations enable the creation of 2D and 3D interactive virtual environments, to explore:

- a specific situation or experiment (e.g., the simulations in Fig. 12);
- a set of experiments that can be carried out with the same apparatus (e.g., the laboratories virtual laboratories in Fig. 13) (Zollman et al. 2002; https://phet.colorado.edu/it/simulation/legacy/band-structure);

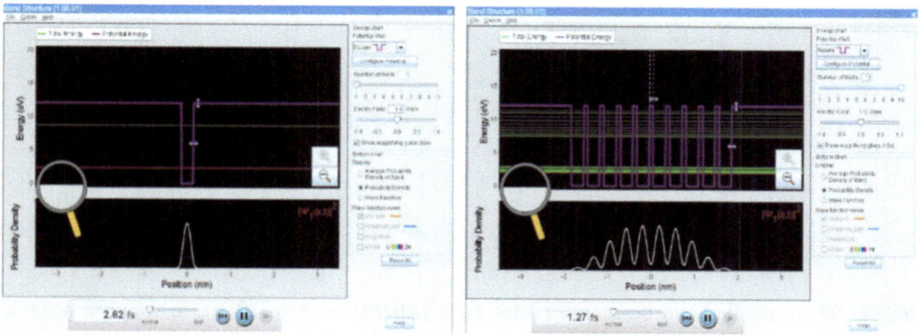

Fig. 12 Applet di PhET of the Colorado Boulder University simulating the band formation in a solid (https://phet.colorado.edu/it/simulation/legacy/band-structure)

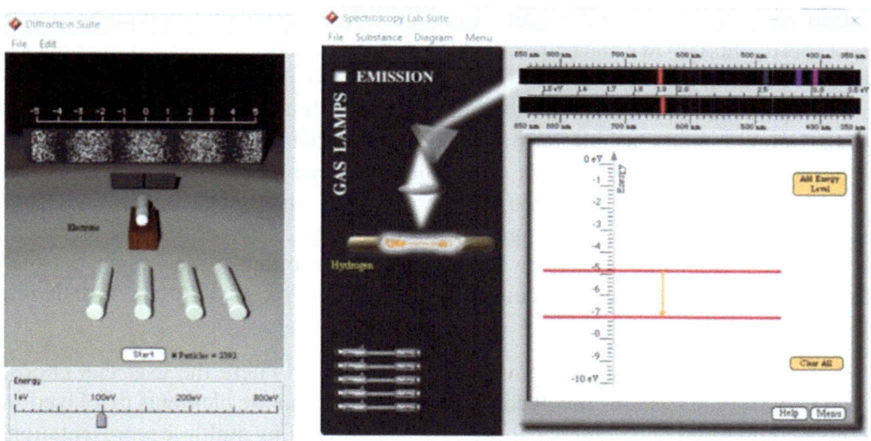

Fig. 13 Applet di Visual Quantum Mechanics VQM (Zollman et al. 2002), for the simulation of a spectroscopy laboratory and a quantum diffraction phenomenon. The applets simultaneously simulate the experimental apparatus, where the user can choose the system to work with (a different spectral lamp, type of source) and the experimental outcome

- a specific phenomenological context (e.g., the JQM software exemplified in Fig. 14, designed as a training ground for experiments ideal with polaroid and birefringent crystals, to study polarization as a quantum property of photons (Michelini et al. 2016));
- a thematic experimental laboratory (e.g., a laboratory where the user disposes of different circuit elements to study arbitrary electrical circuits);
- a simulated world, such as Interactive Physics, in which the user implements objects with different shapes and physical properties, and which evolve according to the physics determined by Newton's principles (Roth 1995; IP 2005).

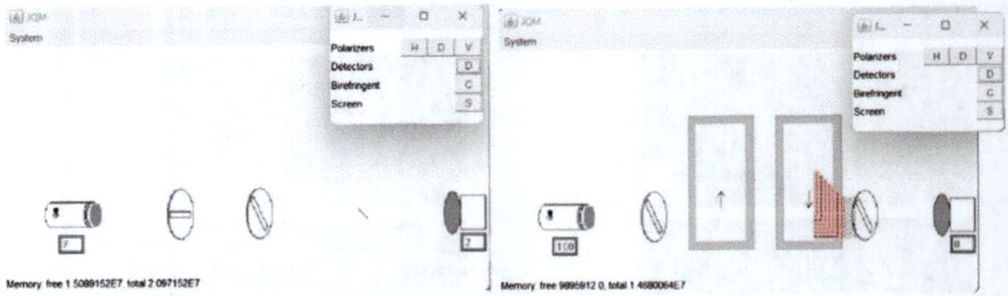

Fig. 14 Situations that can be realized with JQM (Michelini et al. 2016), in which a beam of polarized photons impinges on a polaroid, birefringent crystal. The transmitted photons are detected by detectors arranged to intercept the directions in which they are expected to be transmitted

Several studies have shown that simulation activates the development of models and learning (Adams 2010; Podolefsky et al. 2010), especially when integrated with real experiments (Brown and Christian 2011; Krusberg 2007; Smith and Puntambekar 2010; Klein et al. 2015). They play an irreplaceable role in learning abstract concepts or concepts relating to microscopic phenomena (Zollman et al. 2002; Martin et al. 2010; Richtberg and Girwidz 2018; Michelini et al. 2014b; Mason et al. 2015).

For such outcomes, the design of simulations must be research-based, and the teaching activities to be conducted with inquiry-type strategies (McDermott 1993, 2001) triggered by specially designed tutorials designed (Adams 2010; Christian et al. 2010). Some elements are a particularly important connections between real phenomena and scientific concepts—(model on which they are based; real experiments they can simulate);

- interactivity, through dynamic visual environments, where the user can choose and control the systems involved, their properties, and the elements displayed by good animations and graphical tools capable of providing multiple representations
- the appealing graphic appearance of adequate visual complexity to activate curiosity and attention to aspects of interest, but without overloading or distracting
- graphical interface with intuitive control; ability to select and drag objects that can be grasped virtually; limited control tools easily individualized.

6 ICT Tools for Active Teaching/Learning

The interactive whiteboard has entered all schools around the world. Among its various potentialities, here we emphasize the role that the interactive whiteboard can play in fostering students' involvement (Abu Baker Ilyas and Al-Tabtabaie 2004; Glover et al. 2005; Challapalli et al. 2012), activating shared working memory and interactive and collaborative learning processes (Stoica et al. 2011), and promoting the development of formal thinking (Challapalli et al. 2012; Stoica et al. 2011; Akbaş and Pektaş 2011). Some of its functionalities, such as allowing images and films to be captured and managed, and multilayer representation, enable the linking of phenomena or experiments, and their description in terms of physical quantities (Challapalli et al. 2012; Stoica et al. 2011; Knight 2002). Figure 15 shows the examples of passages from the phenomenon to its interpretative model.

Along with interactive whiteboards, especially in the US, clickers have become widespread (Titus et al. 1998; Sokoloff 2011; Beatty 2004; Corrada-Emmanuel et al. 2007). Clickers are analogous to mobile phones with alphanumeric keypads that students are equipped with and are connected to the teacher's PC via a hub connected to its USB port (Fig. 16). The clicker activities (typically questionnaires) are managed via the interactive whiteboard management software. For each question, the teacher activates the response mode (voting) and defines the duration. At the end, he/she shows the students the distribution of answers. All clickers provide multiple-choice answers and only the more advanced ones also open-ended answers.

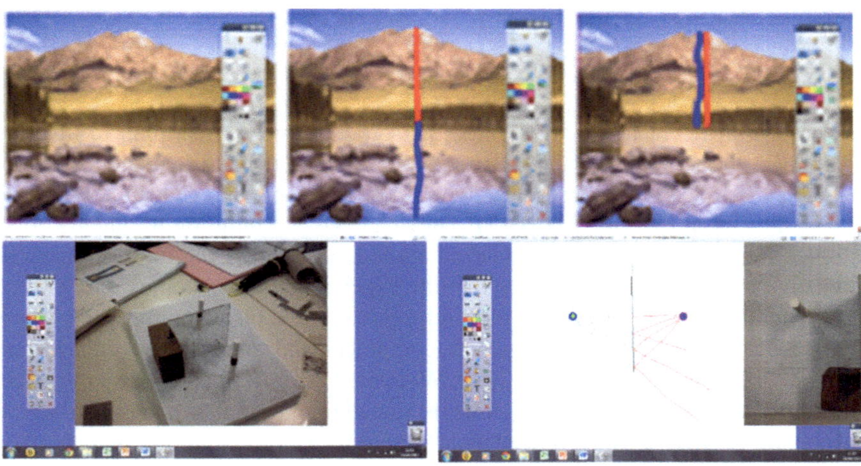

Fig. 15 From the experiment on the image formed by a flat mirror to its model based on the law of reflection (Challapalli et al. 2012)

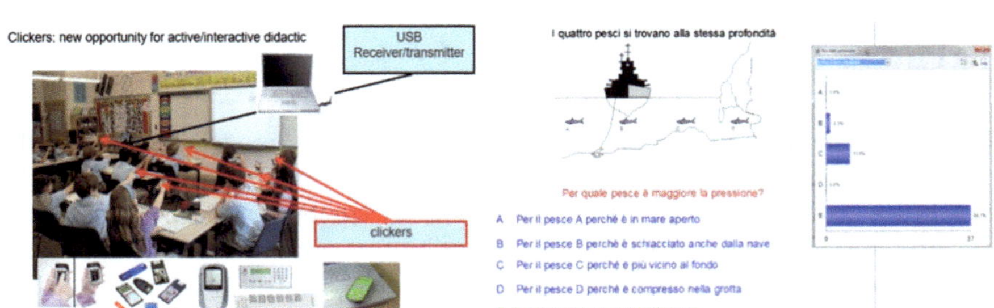

Fig. 16 Clicker in the classroom and Clicker question on Pascal principle (Michelini and Stefanel 2016)

The fact that they are dedicated tools makes clickers still more suitable for teaching, as documented in a large body of literature (Beatty 2004, 2011; Corrada-Emmanuel et al. 2007; Hobbs 2011; Lane 2011; Lindaas 2011; Challapalli et al. 2012).

The most effective strategies are those that encourage student interaction (Beatty 2011; Hobbs 2011; Challapalli et al. 2012), e.g., by alternating voting and peer discussion (Whitney 2011; Cheng et al. 2011; Stewart 2011). The design of the questionnaires (e.g., Fig. 16) and their use in the classroom must include the identification of the problematic issues to be addressed and the use of research-validated questions, the connection with the other parts of the teaching process, the definition of the timing of their use, the teaching strategy to be adopted and the methods of interaction and discussion of the student's answers (Stewart 2011; Kortemeyer 2011; Santi et al. 2014; Michelini and Stefanel 2016).

7 Games Based on Physics Concepts

Research in different fields highlighted the role of games in learning, in favoring the connection between physical abstract concepts and their formalization. Games designed with a learning purpose can be powerful research and educational tools in science education (McGonigal 2011). They can activate intuition to quickly solve complex computational problems (Carruthers and Stege 2013), reinforce motivations and active involvement, and ensure that the experience remains enjoyable with no fears of failure (McGonigal 2011). Recently, games based on quantum principles are developed and offered as apps for tablets, computers, and smartphones (QPlayLearn Platform 2022; van Nieuwenburg 2022; Wootton 2018). Research shows the effectiveness for students learning these games to visualize and simulate quantum mechanical phenomenology, as those related to the superposition principle and entanglement, which are relevant components in QM teaching/learning (Chiofalo et al. 2022).

8 App for Smartphone: New Opportunities for Learning Physics

The app for modern smartphones makes available in new modalities most of the opportunities for learning discussed in the previous section (Delgado 2021; Juskaite et al. 2019; Darmaj et al. 2019). The smartphones are equipped with different sensors or can be connected to sensors allowing the possibility to use it for managing measurements through apps available free on the web (Klein et al. 2015; Martin et al. 2010; Richtberg and Girwidz 2018; Delgado 2021; Buongiorno et al. 2018). For instance, the specific measurements illustrated in Fig. 3 can be performed using the light sensors incorporated in the smartphone (Monteiro et al. 2017). The accelerometer included in the smartphone allows the possibility to measure the motion. Sound intensity of magnetic field measurement becomes possible. Only measurements requiring special performances are problematic or impossible, as those illustrated in Fig. 4 that require too frequent and sensible acquisition system, not at disposal of ordinary smartphone equipment.

Based on the availability of such opportunities, a schoolwork activity was developed in which students carried out autonomous measures at home as a test development activity for apps and related innovative sample measures (Buongiorno et al. 2018).

Smartphones can also be used to make videos, implement activities with clicker (see Sect. 6), managed via the network with freely available apps, and implement the games discussed in Sect. 7. The possibility for students to use their smartphones in the classroom, in the laboratory, at home as well as in any other contest they find themselves in opens new opportunities for active teaching based on innovative approaches, both in content and in methods.

9 Long-Distance Teaching and Learning

Distance learning and in particular teaching/learning physics has long been a debated topic (Lambourne 2008). It has, however, affected groups of students, such as adults and working students, students with mobility difficulties or special geographical situations, such as in Australia or Canada. With the advent of the COVID pandemic, it has suddenly become central to the entire world population (Strubbe and McKagan 2021; Sokoloff 2022; Delgado 2021, 2022).

Some aspects were particularly problematic to address:

- Use of formalism and development of formal skills
- Development of teamwork/communicative–relational skills
- How perform experimental laboratory and development of related skills
- How support learning and provide feedback
- How to perform examination and assessment (Strubbe and McKagan 2021; Sokoloff 2022; Delgado 2021, 2022; Lambourne 2007, 2008; Ratnikova 2013; Pollock and Finkelstein 2013).

The numerous potentials offered by the ICT discussed in this paper and in particular their integration with the typical possibilities offered to synchronous and asynchronous interaction in the web made it possible to extend previous experiences (Braithwaite and Lambourne 1223) by realizing active/interactive distance learning in small as well as large groups (Strubbe and McKagan 2021; Sokoloff 2022; Stoica et al. 2011; Delgado 2021, 2022; Ametepe and Khan 2021; Bjurholt and Vetleseter Bøe 2023). In particular, experimental activities were possible almost in four different ways: use of RCL to have students carry out experiments remotely; RTL, managed by the teacher with the possibility of viewing the experiment, carried out in real time and/or filmed, and at the same time viewing the real-time graph and then discussed and analyzed by the students; pre-prepared kits designed by the researchers and sent to the students or self-produced kits also using smartphones to carry out experiments at home (Stefanel 2021).

10 Conclusion

The new information and communication technologies and especially the computer have opened new opportunities for physics education, some of the main ones being discussed in the present contribution for experimental, modeling, simulation, and active student involvement activities.

Sensors online with the computer or smartphone constitute a powerful apparatus for measurements of good precision and reliability that open up a whole world of possibilities, and disciplinary and methodological objectives. Its intrinsic multimedia nature allows for the measurements on films of experiments and phenomena (video analysis), connection to sensors locally and to experiments remotely, and integration with different hardware and software. This extends the experimental possibilities, and the spectrum of feasible experiments. The understanding of the most delicate and significant steps in a measurement process is made possible. The laboratory becomes an environment where students face interpretative challenges in experimental problem-solving, explore phenomenology, and make measurements. They also use computers to extract information about phenomena and construct phenomenological laws and interpretative models that they can implement in an appropriate software environment for comparison with experimental data. The strategy that enables the highest levels of formation to be achieved involves the involvement of students in the design activities of the experiment to be performed and the assembly of the apparatus to be used. In this way, the operational manual skills typical of the traditional laboratory are recovered. Modeling activities conducted following the stimuli of tutorials based on inquiry-type strategies offer students an understanding of the meaning of models in the various scientific fields, the role these models play in connecting with phenomenal reality, and the conceptual meaning of mathematical operators. Students understand the network of concepts underlying a formal model and do not look at it simply as a set of formulas providing a result. Computer simulations allow the experimental exploration carried out in the laboratory to be effectively extended. They are particularly important to the understanding of abstract concepts such as the field, or phenomenology's far removed from experience such as those in modern physics.

The real-time graphic visualization realized with the use of sensors connected online with the computer enables active teaching strategies in which students' learning difficulties are attacked, even in large groups, by carrying out experiments from the teacher's desk. It also allows that imaginative reduction of the physical phenomenon that activates the construction of scientific concepts, the connection between the observed phenomenon and its description through physical quantities. These educational opportunities are nowadays extended by the availability of apps that allow smartphones to be turned into excellent measurement tools that can be carried out both in and out of school. Smartphones are also an immediate tool for using games built for educational purposes, as has been done with games based on the principles of quantum mechanics that help students explore counterintuitive quantum behavior in effective operational contexts for learning the subtlest and most complex quantum concepts.

Computer-based experiments, modeling, and simulation activities naturally complement the use of interactive and multimedia whiteboards (interactive whiteboards). The intrinsic interactivity of interactive whiteboards constitutes their main added value for active and collaborative teaching strategies, which can be emphasized with the use of automatic responders (clickers) connected wirelessly to the computer and managed by the interactive whiteboard software itself.

The opportunities offered by new technologies have made it possible to implement active teaching, also in the experimental laboratory, not only in face-to-face activities but also in distance activities that have become commonplace with the COVID pandemic emergence.

References

F. Abd-El Khalick et al., Inquiry in science education: international perspectives. Int. J. Sci. Educ. **88**(3), 397–419 (2004)

M. Abu Baker Ilyas, A.M. Al-Tabtabaie, Improving knowledge delivery and information retention through 'smarter' interactive whiteboards, in *Proceedings of International Conference on Computers in Education* (2004)

W.K. Adams, Student engagement and learning with PhET interactive simulations. Il Nuovo Cimento **33C**(3), 21–32 (2010)

O. Akbaş, H.M. Pektaş, The effects of using an interactive whiteboard on the academic achievement of university students. Asia-Pac. Forum Sci. Learn. Teach. **12**(2), 1 (2011)

J.D. Ametepe, N. Khan, Teaching physics during COVID-19 pandemic. Phys. Educ. **56**, 065030 (2021)

L. Bao, K. Koenig, Physics education research for 21st century learning. Discip. Interdiscip. Sci. Educ. Res. **1**, 2 (2019). https://doi.org/10.1186/s43031-019-0007-8

I. Beatty, Transforming student learning with classroom communication systems. Educ. Center Appl. Res. **3**, 1–3 (2004). http://www.e-cause.edu/ECAR/TransformingStudentLearningwit/157511

I. Beatty, *Teaching with Clickers: How, for What, and with What Mind-Set?* (University of North Carolina Greensboro, Greensboro, 2011). http://ianbeatty.com/aapt-2011

R.J. Beichner, The impact of video motion analysis on kinematics graphs interpretation skills. Am. J. Phys. **64**, 1272–1277 (1996)

R.J. Beichner, Instructional technology research and development in a U.S. Physics Education Group. Eur. J. Eng. Educ. **31**(4) 383–393 (2006)

J. Benacka, Numerical modeling in secondary school physics and biology. Spreadsh. Educ. (eJSiE) **2**(3) (2008). http://epublications.bond.edu.au/ejsie/vol2/iss3/3

L. Benciolini, M. Michelini, A. Odorico, Formalizing thermal phenomena at 3–6 year old, in *Developing Formal Thinking in Physics*, ed. by M. Michelini, M. Cobal (Forum, Udine, 2002), pp. 391–396. ISBN: 88-8420-148-9

J. Bernhard, Does active engagement curricula give long-lived conceptual understanding?, in *Physics Teacher Education Beyond 2000*, ed. by R. Pinto, S. Surinach (Elsevier, Paris, 2001), pp. 749–752

J. Bernhard, Physics learning and microcomputer based laboratory (MBL), in *Science Education Research in the Knowledge-Based Society*, ed. by D. Psillos et al. (Academic Press, Kluwer, 2003), pp. 313–321

P. Bitzenbauer, Effect of an introductory quantum physics course using experiments with heralded photons on preuniversity students' conceptions about quantum physics. Phys. Rev. Phys. Educ. Res. **17**, 020103 (2021)

N. Bjurholt, M. Vetleseter Bøe, Remote physics teaching during the COVID-19 pandemic: losses and potential gains. Phys. Educ. **58**, 015004 (2023). https://doi.org/10.1088/1361-6552/ac96be

S. Bosio, V. Capocchiani, M. Michelini, L. Santi, Computer on-line to explore thermal properties of matter, in *Teaching the Science of Condensed Matter*, ed. by M. Michelini et al. (Udine, Forum, 1996), pp. 351–355

F. Bradamante, M. Michelini, A. Stefanel, The modelling in the sport, in *Teaching and Learning Physics in New Contexts*, ed, by E. Mechlova, L. Konicek (GIREP, Ostrava, 2004), pp. 206–208. ISBN: 80-7042-378-1

N.St.J. Braithwaite, R. Lambourne, The multimedia in physics teaching and learning community. J. Phys. Conf. Ser. **1223**, 011001 (2019). https://doi.org/10.1088/1742-6596/1223/1/011001

P. Bronner, A. Strunz, C. Silberhorn, J.-P. Meyn, Interactive screen experiments with single photons. Eur. J. Phys. **30**, 345 (2009)

D. Brown, W. Christian, Simulating what you see: combining computer modeling with video analysis, in *Proceedings of MPTL-16*, ed. by S. Divjak (MPTL, Ljubljana, 2011), pp. 15–17

D. Buongiorno et al., Alternanza scuola-lavoro nella prospettiva di ricerca con APP sul suono, in *Nuovi metodi e saperi*, ed. by G. Adorni et al. (AICA, Bologna, 2018), pp. 391–400. ISBN: 978-88-98091-47-8

S. Caravita, Costruzione collaborativa di prodotti e tecnologie della comunicazione. Ital. J. Educ. Technol. **3**, 6 (1995)

S. Caravita, O. Hallden, Reframing the problem of conceptual change. Learn. Instr. **4**, 89 (1995)

S. Carruthers, U. Stege, On evaluating human problem solving of computationally hard problems. J. Probl. Solv. **5**, 4 (2013)

CBE—Life Sciences Education (LSE), Describing teaching and learning at the intersection of biology and physics. CBE Life Sci. Educ. **12** (2013)

M. Ceberio, J.M. Almudi, A. Franco, Design and application of interactive simulations in problem-solving in university-level physics education. J. Sci. Educ. Technol. **25**, 590–609 (2016). https://doi.org/10.1007/s10956-016-9615-7

I. Cescon, A. Stefanel, Polarimetry measurement in a physics lab. Phys. Teach. **60**(2), 144–148 (2022). https://doi.org/10.1119/10.0009425

S.R.C.P. Challapalli, M. Michelini, A. Stefanel, Dall'esperimento al modello con la LIM nella scuola di base: il caso dell'ottica, in *Didamatica* (2012). http://mondodigitale.aicanet.net/2012-2/didamaica/PAPER/FULL/F153.pdf

S.R.C.P. Challapalli et al., L'uso dei Clicker per il personale coinvolgimento degli studenti di scienze della formazione nell'apprendimento della fisica, in *Tecnologie Informatiche per la Didattica* (2012). ISBN: 978-88-905406-7-7. http://mondodigitale.aicanet.net/2012-2/didamaica/PAPER/FULL/F143.pdf

K. Cheng, A. Pietan, M. Calglar, H. Dulli, Integration of computer-based pre-, in and post-lecture activities in physics, in *AAAPT 2011* (2011)

M. Chiofalo, C. Foti, M. Michelini, L. Santi, A. Stefanel, Games for teaching/learning quantum mechanics: a pilot study with high-school students. Educ. Sci. **12**, 446 (2022)

W. Christian, F. Esquembre, B. Mason, Easy Java simulations and ComPADRE OSP collection. Il Nuovo Cimento **33C**(3), 33–42 (2010)

W. Christian, F.-K. Hwang, *Magnetic Field from Loops Model* (2014). https://www.compadre.org/osp/items/detail.cfm?ID=8311

J. Clement, Model based learning as a key research area for science education. Int. J. Sci. Educ. **22**, 1041–1053 (2000)
P. Cooney, *Spreadsheet Physics Resources* (2003). http://sites.millersville.edu/pjcooney/ssresour.html
F. Corni, M. Michelini, L. Santi, A. Stefanel, Sensori on-line per la formazione insegnanti, in *Comunità Virtuale*, ed. by M. Michelini, M. Pighin, vol. 2 (Forum, Udine, 2005), pp. 1149–1161
A. Corrada-Emmanuel, I.D. Beatty, W.J. Gerace, *Group Discovery with Multiple-Choice Exams and Consumer Surveys: The Group-Question-Answer Model* (University of Massachusetts, Amherst, 2007)
D. Crăciun, M. Bunoiu, Boosting physics education through mobile augmented reality. AIP Conf. Proc. **1916**, 050003 (2017). https://doi.org/10.1063/1.5017456
D. Darmaj, D. Dwi Agus Kurniawan, A. Artha Lumbantoruan, S.C. Samosir, Mobile learning in higher education for the industrial revolution 4.0. Int. J. Interact. Mob. Technol. **13**(9) (2019). https://doi.org/10.3991/ijim.v13i09.10948
F. Delgado, Teaching physics for computer science students in higher education during the COVID-19 pandemic. Future Internet **13**(2), 35 (2021). https://doi.org/10.3390/fi13020035
F. Delgado, Post-COVID-19 transition in university physics courses. Educ. Sci. **12**, 627 (2022). https://doi.org/10.3390/educsci12090627
A.K. Dickerson et al., Mosquitoes survive raindrop collisions by virtue of their low mass. PNAS **109**(25), 9822–9827 (2012)
M. El Hadi et al., Real time free fall investigation for educational purposes using Arduino Uno board. Phys. Educ. **55**, 055027 (2020)
ESM, Educational studies in mathematics: bodily activity and imagination in mathematics learning. PME Spec. Issue **57**(3) (2004)
F. Esquembre, Computers in physics education. Comput. Phys. Commun. **147**, 13–18 (2002)
M. Euler, Physics and physics education beyond 2000, in *PhyTEB200*, ed. by R. Pintò, S. Surinac (Elsevier, Paris, 2001), p. 3
M. Euler, Quality development challenges to physics education, in *Quality Development in Teacher Education*, ed. by M. Michelini (Forum, Udine, 2004), pp. 17–30
M. Euler, A. Müller, Physics learning and the computer, in *Proceedings II International ESERA Conference*, ed. by M. Komorek et al. (1999)
European Union, *Recommendation of the European Parliament and of the Council of 18 December 2006 on Key Competences for Lifelong Learning* (2006), https://eur-lex.europa.eu/eli/reco/2006/962/oj
Final Report Hope Project WG2-New Competences for Physics Students (2017), http://hopenetwork.eu/content/final-report-wg2
N. Finkelstein, W. Adams, C. Keller, K. Perkins, C. Wieman, High-tech tools for teaching physics. Merlot J. Online Teach. Learn. (2006). https://jolt.merlot.org/vol2no3/finkelstein.pdf
W.P.S. Freitas et al., Arduino-based experiment demonstrating Malus's law. Phys. Educ. **53**, 035034 (2018). https://doi.org/10.1088/1361-6552/aab43d
L. Galliani, Formazione degli insegnanti e competenze nelle tecnologie della comunicazione educative. Ital. J. Educ. Res. **2–3**, 93–103 (2009)
A. Gallo, E. Mazzega, M. Michelini, Impiego della modellizzazione dinamica nell'insegnamento della TD. La Fis. nella Scuola **XXII**(S4), 50 (1989)
M. Gatti, L.G. González, M.G. Mereu, C. Tagliaferro, *L'impatto delle tecnologie dell'informazione e della comunicazione sulle competenze professionali e sulla formazione* (CEDEFOP, Salonicco, 1998), https://www.cedefop.europa.eu/files/5091_it.pdf; (2022), https://www.cedefop.europa.eu/en/publications/9130#group-downloads
M. Gervasio, M. Michelini, TERMOCRONO. Un semplice sistema economico e flessibile per misure di temperatura in tempo reale, in *Didamatica 2006*, ed. by A. Andronico et al. (AICA, Cagliari, 2006), pp. 522–529

M. Gervasio, M. Michelini, A USB probe for resistivity versus temperature and Hall coefficient measurements, in *MPTL14 Proceeding* (2009). http://www.fisica.uniud.it/URDF/mptl14/contents.htm
M. Gervasio, M. Michelini, Lucegrafo. A simple USB data acquisition system for diffraction experiments, in *MPTL14 Proceeding* (2009) CD-ROM. http://www.fisica.uniud.it/URDF/mptl14/contents.htm
G. Gervasio, M. Michelini, L. Santi, A. Stefanel, R. Viola, I progetti MOSEM e la superconduttività per una innovazione didattica a scuola. G. Fis. **55**(4), 363–373 (2014)
G. Giugliarelli et al., Tecnologie moderne e progettualità nell'attività sperimentale. La Fis. nella Scuola **XXVII**, 4 S, 113 (1994)
D. Glover, D. Miller, D. Averis, V. Door, The interactive whiteboard: a literature survey. Technol. Pedag. Educ. **14**(2), 155–170 (2005)
D.J. Grayson, Physics education for 21st century graduates. J. Phys. Conf. Ser. **1512**, 012043 (2020)
S. Grober, M. Vetter, B. Eckert, H.J. Jodl, Experimenting from a distance-remotely controlled laboratory. Eur. J. Phys. **28**, S127–S141 (2007)
S. Grober, M. Vetter, B. Eckert, H.J. Jodl, Experimenting from a distance RCL-experiments for teaching physics at high-school (2013). http://pen-physik.de/w_jodl/Docs/book-rcl_small.pdf
L. Grosslight, C. Unger, E. Jay, C.L. Smith, Understanding models and their use in science. J. Res. Sci. Teach. **28**(9), 799–822 (1991)
R. Hake, Interactive-engagement vs. traditional methods. Am. J. Phys. **66**, 64–74 (1998)
A. Heck, E. Kędzierska, T. Ellermeijer, Design and implementation of an integrated computer working environment. J. Comput. Math. Sci. Teach. **28**(2), 147–161 (2009)
P.R.L. Heron, D.E. Meltzer, The future of physics education research: intellectual challenges and practical concerns. Am. J. Phys. **73**, 390 (2005). https://doi.org/10.1119/1.1858480
D. Hestenes, Toward a modeling theory of physics instruction. Am. J. Phys. **55**, 440–454 (1987)
D. Hestenes, Modeling software for learning and doing physics, in *Thinking Physics for Teaching*, ed. by C. Bernardini et al. (Plenum, New York, 1995), pp. 25–65
R. Hobbs, Blurring the lines: ILD's (and other activities) in an integrated lecture-lab environment, WS, in *AAAPT 2011* (2011)
A. Hofstein, The laboratory in chemistry education. Chem. Educ. Res. Pract. **5**(3), 247–264 (2004)
A. Hofstein, V.N. Lunetta, The laboratory in science education: foundations for the twenty-first century. Sci. Educ. **88**(1), 28–54 (2004)
A.M. Hoskinson et al., Bridging physics and biology teaching through modeling. Am. J. Phys. **82**(5), 434–441 (2014)
http://modellus.fct.unl.pt/
http://www.compadre.org/portal/index.cfm
https://www.i2u2.org/
https://www.ises.info/index.php/en
https://phet.colorado.edu/it/simulation/legacy/band-structure
B. Huang, Open-source hardware—microcontrollers and physics education, in *122 ASEE*, Seattle, 14–17 June 2015
IP (2005), https://www.design-simulation.com/ip/simulations.php, and IP (2007), http://www.interactivephysics.com
A.H. Johnstone, A. Al-Shuaili, Learning in the laboratory; some thoughts from the literature. U. Chem. Ed. **5**, 42–51 (2001)
L. Juskaite, A. Ipatovs, A. Kapenieks, Mobile technologies in physics education in Latvian secondary schools. Period. Eng. Nat. Sci. **7**(1), 187–196 (2019). Available online at: http://pen.ius.edu.ba
E. Kedzierska et al., MOSEM 2 project: integration of data acquisition, modelling, simulation and animation. Il Nuovo Cimento **33C**, 3 (2010). https://doi.org/10.1393/ncc/i2010-10616-y
J. Kirstein, V. Nordmeier, Multimedia representation of experiments in physics. Eur. J. Phys. **28**, S115 (2007)

J. Kirstein, S. Haase, T. Mühlenbruch, V. Nordmeier, 20 Jahre Interaktive Bildschirmexperimente. PhyDidB Beiträge Frühjahrstagung **2016** (2016). http://www.phydid.de/index.php/phydid-b/article/view/739/872

P. Klein, J. Kuhn, A. Müller, S. Gröber, Video analysis in introductory mechanics physics courses, in *Multidisciplinary Research on Teaching and Learning*, ed. by M. Palgrave (2015)

R.D. Knight, *Five Easy Lessons: Strategies for Successful Physics Teaching* (Addison-Wesley, Boston, 2002)

A. Kohnle, C. Baily, A. Campbell, N. Korolkova, Enhancing student learning of two-level quantum systems. Am. J. Phys. **83**(6), 560–566 (2015)

G. Kortemeyer, The assessment continuum, in *AAAPT 2011* (2011)

Z.A.C. Krusberg, Emerging technologies in physics education. J. Sci. Educ. Technol. **16**, 401–411 (2007). https://doi.org/10.1007/s10956-007-9068-0

R. Lambourne, Laboratory-based teaching and the physics innovations centre for excellence in teaching and learning. Eur. J. Phys. **28**(3), S29–S36 (2007). https://doi.org/10.1088/0143-0807/28/3/S03

R. Lambourne, Physics and distance education, in *Connecting Research in Physics Education with Teacher Education*, ed. by M. Vicentini, E. Sassi (I.C.P.E. Book © International Commission on Physics Education, 2008). http://iupap-icpe.org/publications/teach2/Lambourne.pdf

W.B. Lane, Self-reported in-class emotional, in *AAAPT 2011* (2011)

B.V. Liengme, *Modelling Physics with Microsoft Excel* (Morgan & Claypool Publishers, San Rafael, 2014)

P.L. Lijnse, "Developmental research" as a way to an empirically based "didactical structure" of science. Sci. Educ. **79**, 189–199 (1995)

S. Lindaas, A research methodology for using clickers, WS, in *AAAPT 2011* (2011)

M.C. Linn, S. His, *Computers, Teaching, Peers: Science Learning Partners* (Lawrence Erlbaum Associates, Mahwah, NJ, 2000)

B. Martin, W. Brouwer, D. Austen, Creating digital learning objects to teach abstract ideas in modern physics and astronomy, in *CBLIS 2010*, ed. by C. Constantonou et al. (Oelizk, Warsaw, 2010), pp. 31–40

E. Martinez, V. Carbonell, M. Florez, J. Amaya, Simulations as a new physics teaching tool. Comput. Appl. Eng. Educ. **18**(4), 757–761 (2010). https://doi.org/10.1002/cae.20266

B. Mason et al., Report and recommendations on multimedia materials for teaching and learning QP. Il Nuovo Cimento **38C**, 105–116 (2015)

E. Mazur, Can we teach computers to teach? Comput. Phys. **5**, 31 (1991). https://doi.org/10.1063/1.4822968

L.C. McDermott, How we teach and how students learn: a mismatch? Am. J. Phys. **61** (1993)

L.C. McDermott, Oersted medal lecture 2001: physics education research—the key to student learning. Am. J. Phys. **69**, 1127 (2001)

J. McGonigal, *Reality Is Broken: Why Games Make Us Better and How They Can Change the World* (J. Cape, London, 2011)

A. Messeri, La questione delle competenze. Magellano. Riv. L'orient. **22**, 1–4 (2004)

M. Michelini, L'elaboratore nel laboratorio di fisica: alcune considerazioni di carattere generale. La Fis. nella Scuola **XXI**, 2, IR, 159 (1988)

M. Michelini, Modellizzazione ed esperimenti on-line. La Fis. nella Scuola **XXIV**, 4IR, 44–68 (1991)

M. Michelini, L'elaboratore nel laboratorio didattico di fisica: nuove opportunità per l'apprendimento. G. Fis. **XXXIII**(4), 269–294 (1992)

M. Michelini, Quale laboratorio per la formazione degli insegnanti, in *Uso del Laboratorio e Insegnamento della Fisica* (Garamond, Roma, 1999)

M. Michelini, Building bridges between common sense ideas and a physics description of phenomena, in *New Trends in STE*, ed. by L. Menabue, G. Santoro, vol. 1 (CLUEB, Bologna, 2010), pp. 257–274

M. Michelini, A. Stefanel, Hands-on sensors for the exploration of light polarization, in *Informal Learning and Public Understanding*, ed. by G. Planinsic, A. Mohoric (GIREP, Ljubljana, 2006), pp. 202–208. ISBN: 961-6619-00-4
M. Michelini, A. Stefanel, Clicker per l'apprendimento attivo della Fisica degli studenti dell'area Bio, in *Didamatica* (2016). ISBN: 9788898091447. http://didamatica2016.uniud.it/proceedings/dati/articoli/paper_106.pdf
M. Michelini, A. Stefanel, Il contributo delle ICT per l'apprendimento scientifico, in *Tecnologie per l'educazione*, ed. by P.G. Rossi, P.C. Rivoltella, Cap. 20 (Pearson, London, 2019), pp. 273–290
M. Michelini, R. Viola, V. Engstrom, Superconduttività: esperimenti e modelli, in *Informatica per la didattica*, ed. by A. Andronico et al. (parte I) (Laterza, Bari, 2008), pp. 110–114. ISBN: 978-88-8231-456-9
M. Michelini, L. Santi, A. Stefanel, Thermal sensors interfaced with computer as extension of senses in kindergarten and primary school. Il Nuovo Cimento **33C**, 3 (2010). https://doi.org/10.1393/ncc/i2010-10641-x
M. Michelini, L. Santi, A. Stefanel, Upper secondary students face optical diffraction using simple experiments and on-line measurements, in *FFP14*, ed. by E. Kajfasz, T. Masson, R. Triay (AMU, Marseille, 2014a). http://pos.sissa.it/archive/conferences/224/240/FFP14_240.pdf
M. Michelini, L. Santi, A. Stefanel, Teaching modern physics in secondary school, in *FFP14*, ed. by E. Kajfasz, T. Masson, R. Triay (AMU, Marseille, 2014b). http://pos.sissa.it/archive/conferences/224/231/FFP14_231.pdf
M. Michelini, L. Santi, A. Stefanel, JQM per affrontare nella scuola secondaria i fondamenti di MQ, in *Didamatica* (2016). ISBN: 9788898091447. http://didamatica2016.uniud.it/proceedings/dati/articoli/paper_96.pdf
R. Millar, I. Abrahams, Practical work: making it more effective. SSR **91**(334), 60 (2009)
M. Monteiro, C. Stari, C. Cabeza, A.C. Martí, The polarization of light and Malus' law using smartphones. Phys. Teach. **55**, 264 (2017). https://doi.org/10.1119/1.4981030
J.M. Ogborn, Dynamical modelling system, microcomputer software, in *ICPE (IUPAP) Communicating Physics* (Longmans, London, 1984), pp. 94–103
J.M. Ogborn, New technologies and the classroom. Commun. Phys. ICPE (IUPAP) 94 (1986)
Oprea, Arduino-based projects in physics education. eLearn. Softw. Educ. **2**, 107–113 (2018)
G. Planinsic, Experiments as building blocks of knowledge, in *Research and Innovation in Physics Education*, ed. by J. Guisasola, K. Zuza (Springer, Cham, 2020), pp. 1–13. https://doi.org/10.1007/978-3-030-51182-1_1
N.F. Podolefsky, K.K. Perkins, W.K. Adams, Factors promoting engaged exploration with computer simulations. Phys. Rev. Spec. Top. Phys. Educ. Res. **6**, 020117 (2010)
S.J. Pollock, N.D. Finkelstein, Impacts of curricular change: implications from 8 years of data in introductory physics. AIP Conf. Proc. **1513**, 310 (2013). https://doi.org/10.1063/1.4789714
C. Pontecorvo, A.M. Ajello, C. Zucchermaglio, *I contesti sociali dell'apprendimento* (LED, Milano, 1995)
QPlayLearn Platform. Available online: http://www.qplaylearn.com/. Accessed 09 Oct 2022
D. Ratnikova, Sold-out crowd examines distance learning in physics. APS News **22**(7) (2013)
S. Richtberg, R. Girwidz, Learning physics with interactive videos—possibilities, perception, and challenges, in *GIREP Congress 2018* (2018)
M. Riel, Educational change in a technology-rich environment. J. Res. Comput. Educ. **26**, 31–39 (1998)
Y. Rodríguez et al., Physics education through computational tools: the case of geometrical and physical optics. Phys. Educ. **48**, 621 (2013)
W.-M. Roth, Affordances of computers in teacher-student interactions: the case of interactive physics™. J. Res. Sci. Teach. **32**(4) 329–347 (1995). https://doi.org/10.1002/tea.3660320404
L. Santi, E. Mazzega, M. Michelini, Understand radiation interference by means of computer modelling, in *Light and Information*, ed. by L.C. Pereira, J.A. Ferreira, H.A. Lopes (GIREP-Universidade do Minho, Braga, 1993), pp. 372–380

L. Santi, S.P.R. Challapalli, G. Fera, M. Michelini, A. Stefanel, S. Vercellati, Clicker. La Fis. nella Scuola **XLVII**(S4), 78–85 (2014)
E. Sassi, Computer supported lab-work in physics education: advantages and problems, in *Physics Teacher Education Beyond 2000*, ed. by R. Pintò, S. Surinach (Elsevier, Paris, 2001), pp. 57–64
B. Sherwood, R. Chabay, VPythom: 3D programming for ordinary mortals. Il Nuovo Cimento **33C**(3), 59–63 (2010)
R. Simpson, R.K. Thornton, Computers bring new opportunity to science education. Comput. Phys. **9**(6) (1995)
C. Singh, Interactive learning tutorials on QM. Am. J. Phys. **76**(4), 400–405 (2008)
G.W. Smith, S. Puntambekar, Examining the combination of physical and virtual experiments in an inquiry science classroom, in *CBLIS 2010*, ed. by C. Constantinou et al. (Oelizk, Warsaw, 2010), pp. 153–164
D.R. Sokoloff, Active learning in lecture with interactive lecture demonstrations using microcomputer, in *Proceedings of MPTL11*, ed. by S. Divjak (MPTL, Ljubljana, 2011), pp. 27–34
D.R. Sokoloff, Active learning of introductory light and optics. Phys. Teach. **54**(1), 18–23 (2016)
D.R. Sokoloff, Exploring multimedia to adapt ILDs for home use, in *Physics Teacher Education—What Matters?*, ed. by J. Borg et al. (Springer, Cham, 2022), pp. 3–14. https://doi.org/10.1007/978-3-031-06193-6_1
D.R. Sokoloff, R.K. Thornton, P. Laws, Realtime physics: active learning labs transforming the introductory lab. Eur. J. Phys. **28**, S83–S94 (2007)
R.M. Sperandeo, M. Michelini, L. Santi, *Proposte didattiche su forze e movimento* (Forum, Udine, 2002). ISBN: 88-8420-075-X
A. Stefanel, Il contributo della multimedialità all'apprendimento attivo della fisica in DAD, in *Invited Talk, 107 Congress of the Italian Physical Society*, Milan (2021), pp. 13–17
A. Stefanel, C. Moschetta, M. Michelini, Cognitive labs in an informal context to develop formal thinking in children, in *Developing Formal Thinking*, ed. by M. Michelini, M. Cobal (Forum, Udine, 2002), pp. 276–283
J.C. Stewart, *Electricity and Magnetism Self-Testing and Test Construction Tool* (2011). http://physinfo.uark.edu/physicsonline
D. Stoica et al., The interactive whiteboard and the instructional design in teaching physics. Procedia Soc. Behav. (2011)
L. Strubbe, S. McKagan, I suddenly have to move my lab course online! What should I do? PhysPort—Expert Recommendation (2021). https://www.physport.org/recommendations/Entry.cfm?ID=119927
K. Swan, M. Miltrani, The changing nature of teaching and learning in computer-based classrooms. J. Res. Comput. Educ. **25**, 121–127 (1998)
V.D. Teodoro, Cognitive artefacts, technology and physics education. Interact. Educ. Multimed. **11**, 173–189 (2005). http://www.ub.es/multimedia/iem
A. Theodorakakos, E. Hatzikraniotis, D. Psillos, "PEC task explorer": a tool for ICT supported learning in science, in *CBLIS 2010*, ed. by C. Constantonou et al. (Oelizk, Warsaw, 2010), pp. 75–83
L.-J. Thoms, R. Girwidz, Symposium: virtual and remote labs in practice, in *GIREP Conference*, S. Sebastian, 9–13 July 2018
R.K. Thornton, D.R. Sokoloff, Learning motion concepts using real-time microcomputer-based laboratory tools. Am. J. Phys. **58**, 858–867 (1990)
M.H. Tinker, R.J.A. Lambourne, S.A. Windsor, The flexible learning approach to physics (FLAP): a review after the first two years. Int. J. Sci. Educ. **21**(2), 213–230 (1999). https://doi.org/10.1080/095006999290796
A.P. Titus, L.W. Martin, R.J. Beichner et al., Web-based testing in physics education. Comput. Phys. **12**, 117 (1998). https://doi.org/10.1063/1.168628
E. van den Berg, A.L. Ellermeijer, O. Slooten (eds.), *Modelling in Physics and Physics Education* (University of Amsterdam, Amsterdam, 2008)

E. van den Berg, F. Schweickert, R. van den Berg, Science, sensors and graphs in primary schools (2012). https://www.iederkindeentalent.nl/wp-content/uploads/2012/06/sciencesensors1.pdf

E. van Nieuwenburg, Quantum TiqTaqToe. Available online: https://quantumtictactoe.com/. Accessed 09 Oct 2022

M. Vendramini, M. Michelini, Sensori On-line nella scuola primaria per sviluppare il pensiero formale, in *Nuovi metodi e saperi per formare all'innovazione*, ed. by G. Adorni et al. (AICA-UniBO, Bologna, 2018), pp. 231–240

F. Wang, M.J. Hannafin, Design-based research and technology-enhanced learning environments. Educ. Technol. Res. Dev. **53**(4), 5–23 (2005)

H. Whitney, Development of active learning tools for a course on physics and music, in *AAAPT 2011* (2011)

C. Wieman, K. Perkins, Transforming physics education. Phys. Today **58**(11), 36 (2005)

J. Wootton, The History of Games for Quantum Computers (2018). Available online: https://decodoku.com/. Accessed 09 Oct 2022

D.A. Zollman, N.S. Rebello, K. Hogg, QM for everyone: hands-on activities integrated with technology. Am. J. Phys. **70**(3), 252–259 (2002)

Project-Based Learning in Secondary Science: Digital Experiences in Finnish Classroom

Anna Lager, Jari Lavonen, and Kalle Juuti

Abstract This chapter analyses the integration of digital tools and project-based learning pedagogy into secondary physics teaching and learning. The teachers' technological pedagogical content knowledge (TPACK) model is used as a framework for this analysis. TPACK combines Shulman's structure of pedagogical content knowledge (PCK), content or subject matter knowledge and knowledge and skills needed for the use of digital tools and environments. As a practical tool for planning a lesson, a modified content representation (CoRe) tool with emphasy to the use of digital tools (CoDiRe) is introduced. Two concrete examples of will be introduced and analysed. The first example introduces project-based learning and the use of digital tools, while students make sense of phenomena related to moving objects. The second example is from our collaborative research with physics teachers during the COVID-time, when physics teaching was organised in distance teaching mode and traditional laboratory activities was difficult to organise.

Keywords Technological pedagogical content knowledge (TPACK) · Content representation · Project-based learning · Scientific and engineering practices · Education technology

A. Lager · J. Lavonen (✉) · K. Juuti
Faculty of Educational Sciences, University of Helsinki, Siltavuorenpenger 5, 00170 Helsinki, Finland
e-mail: jari.lavonen@helsinki.fi

A. Lager
e-mail: anna.lager@helsinki.fi

K. Juuti
e-mail: kalle.juuti@helsinki.fi

J. Lavonen
Department of Childhood Education and Centre for Education Practice Research, University of Johannesburg, Soweto, South Africa

M. Streit-Bianchi et al. (eds.), *New Challenges and Opportunities in Physics Education*, Challenges in Physics Education, https://doi.org/10.1007/978-3-031-37387-9_15

1 Introduction

We will analyse in this chapter the use of digital tools in physics teaching and learning. We start the chapter by describing teachers' technological pedagogical knowledge or competence (TPACK), which they apply while plan, implement and assess their own teaching and students' learning. Second, we introduce project-based learning and give two case example, which outline how TPACK have been employed in the planning and implementing project-based learning in secondary physics classrooms.

2 Technological Pedagogical Knowledge (TPACK) as a Model for Physics Teachers' Knowledge Base

Teachers need knowledge and skills for the instructional design, including knowledge and skills needed for using digi-tools and platforms or educational technology in supporting various learners' learning, engagement and well-being. *Technological pedagogical content knowledge* (TPACK) was designed as a such knowledge base (Mishra and Koehler 2006). TPACK combines Shulman's structure of pedagogical content knowledge (PCK), content or subject matter knowledge and knowledge and skills needed for the use of digi-tools and -environments.

Shulman's original model divides teacher knowledge into subject matter (content) knowledge (CK or SMK), pedagogical content knowledge (PCK) and general pedagogical knowledge (GPK) (Carlsen 1999; Hashweh 2005). In addition to these three areas of knowledge, a teacher needs contextual and curriculum knowledge (Gess-Newsome and Lederman 1999). However, it is challenging to describe the use of teacher knowledge in different teaching and learning situations, because the work of a teacher is complex and a teacher utilises at the same time various domains of knowledge.

Subject matter knowledge (SMK) includes conceptual, factual and procedural knowledge in a certain SMK domain, here physics. A teacher needs to understand the nature of SMK, that is, the epistemological and ontological aspects of the subject matter (Shulman 1987). Because the physics SMK is broad, physics curricula in various countries have reduced and core-ideas and knowledge have been recognised and emphasised in the curricula. Core ideas and knowledge are significant and important across physics domains and could be used for planning experiments, explaining phenomena and solving problems (Krajcik et al. 2021). Core ideas and knowledge are also relevant in the personal, local and global contexts. On the other hand, physics teachers have to continuously update their SMK, because physics research produces continuously new knowledge as it is described in the first part of this book.

Pedagogical content knowledge (PCK) is the synthesis of the combined knowledge needed to teach a certain topic or an amalgam of SMK and knowledge of pedagogy (Carlsen 1999). PCK is "the knowledge that teachers bring forward to design and reflect on instruction" (Gess-Newsome 2015, p. 36) and includes, for example, the following areas of teacher knowledge: knowledge about (1) teaching or instructional strategies, assessment strategies and collaboration strategies (shortly teaching methods); (2) student interest, motivation and the learning of conceptual and procedural knowledge and skills; (3) learners, (mis)conceptions, experiences and thinking skills and cognitive and affective demands of the tasks and activities; (4) the resources available to support teaching and scaffold learning; (5) curriculum knowledge and goals for student learning (Loughran et al. 2008). Carlson and Daehler (2019) emphasise the complex layers of PCK and introduce collective PCK (cPCK), personal PCK (pPCK) and enacted PCK (ePCK). Because of this collective nature of PCK, it is important that physics teachers continuously discuss and reflect on their physics teaching and student learning.

Although PCK is a theory for teaching, it takes into account learning science research outcomes, which emphasise factors that support learners and groups in their engagement in learning (Sawyer 2015). For example, prior knowledge has been found to be one of the important factors for learning (Ausubel 1968). For example, Hattie and Donoghue (2016) argued that science inquiry promote learning only when the prior knowledge has been recognised. Students collaboration and interaction and contextualising of learning are examples of factors that support engagement in learning (Sawyer 2015).

An important characteristics to physics teaching methods is the students' interaction with nature and phenomena. In practice, a physics teacher guide students to make sense of phenomena through a demonstration or through engaging students in scientific and engineering practices. Scientific and engineering practices are similar to those of professional scientists, like reasoning, critical thinking and knowledge practices, such as questioning, observing, inferring, classifying, predicting, measuring, interpreting and analysing, as a part of learning (Krajcik and Merritt 2012).

The third main category of teacher knowledge is general pedagogical knowledge (GPK) (Gore and Gitlin 2004). Morine-Dershimer and Kent (1999) argued that general pedagogical knowledge consists of the following knowledge areas: (1) classroom management and organisation; (2) instructional models and strategies; and (3) classroom communication and discourse.

TPACK describes the knowledge base teacher needs for effectively teaching with technology (Mishra and Koehler 2006). The main idea of TPACK is stated as follows: *The basis of good teaching with technology requires an understanding of the representation of concepts using technologies; pedagogical techniques that use technologies in constructive ways to teach content; knowledge of what makes concepts difficult or easy to learn and how technology can help redress some of the problems that students face* (Mishra and Koehler 2006, pp. 1028–1029). Several researchers have characterised the seven domains of TPACK (Mishra and Koehler 2006; Lin et al. 2013), which are important for the versatile teaching and learning with digi-tools and platforms. Three domains, or SMK, PCK and GPK, were already introduced above.

Technological knowledge (TK) is knowledge about the use of digi-tools and digi-platforms or education technology. Digital tools are considered as tools, which process digital signals and are available in various environments and devices, such windows environment, laptops and mobile phone. Various tool applications are used for processing text, numbers, pictures, videos and music. Social media tools and digi-platforms or distance teaching and learning environments are adaptable for face-to-face, flexible, remote and mobile learning. In addition, digital learning materials such as learning games with interactive learning content are essential part of the learning environment. Furthermore, special digital tools are needed in various fields, like micro-computer laboratories and modelling tools in physics education. Robots, laser cutters and 3D printers are nowadays also used in physics education (Fuad et al. 2020). Technological content knowledge (TCK) is in turn knowledge about applying technology to represent CK, but this does not relate to its pedagogical purpose. One example of TCK is knowledge of using various laboratory tools and data-logging tools in a physics laboratory.

Technological pedagogical knowledge (TPK) is knowledge about applying various technologies in pedagogy for teaching and learning all subject domains rather than being focussed on specific content knowledge, such as using Zoom to organise students' lesson learning. Consequently, a teacher employs TPK or digi-pedagogy when he or she uses digi-tools or guides students to utilise digi-tools in learning. This TPK includes TCK or the skills needed for using digital tools, platforms and digital environments for teaching and learning, as well as the knowledge and skills needed to support students' engagement, learning and well-being in digital environments (Greenhow et al. 2020).

Consequently, TPACK refers to knowledge about the use of digi-tools in teaching and learning and in facilitating students' engagement, learning and well-being in a specific context (Greenhow et al. 2020). Although, previous views of TPACK seem teacher centred, TPACK emphasises teacher knowledge he/she employs in when she/he guides students to recognise their conceptions and experiences, work in a small group, interact with other students and be active in learning. Loughran et al. (2008) have suggested a list of eight questions, supportive for employing PCK in the planning of lessons and named the collection of questions as "the content representation (CoRe) tool", which could be used for structuring pedagogical content knowledge (PCK). In order to take into account the use of digi-tools in teaching and learning, we slightly modified this tool for better taking into account TPACK.

The modified CoRe or the content and digi-representation tool (CoDiRe) is:

- What do you want students to learn about the topic or what are the core ideas/big ideas/key concepts and models related to the topic? Do you have specific aims related to the use of digi-tools and platforms in learning?
- Why it is important (meaningful and relevant) for students to learn this topic (need-to-know)? Is it possible to support the development of interest through the use of digi-tools, for example, in the selection of appropriate context for learning?
- What else do you know about this topic—not going to teach students (the level of content)?

- What do you know about students' everyday experiences in the area of the topic? What experiences students have about the planned use of digi-tools (know based on previous studies or need to ask students during previous lesson).
- What do you know about students' conceptions/misconceptions related to the topic and how does it affect the teaching of the topic? Can you support students to recognise their conceptions through the use of digi-tools, for example, through online diagnosis test?
- How school context influences the teaching of this topic? (student, classroom and school context). What kind of digi-tools are available at school considering your aims? Do you need to book the tools beforehand?
- What kind of pedagogy you are planning to use, and how well the pedagogy suited for the topic? (Knowledge-in-use)? What kind of digi-tools support your pedagogy? Is the information easier available through the use of Web-browser or is it possible to support the observations to measurement through the use of digi-tools, such as data-logger, camera, video camera, thermal camera or microscope?
- How are you going to evaluate student learning (knowledge-in-use)? What kind of digi-tools support formative, summative and self-evaluation? Can you use, for example, Socrative, Kahoot or blog in evaluation?

3 Project-Based Physics Learning in Secondary Level

The idea of project-based learning (PBL) or project pedagogy has been suggested several times as an approach to a teaching reform and for engaging students in collaborative learning. On the other hand, the word "project" is used in various ways and all projects are not necessarily PBL in the way it is understood in this chapter. PBL is based on the ideas of John Dewey in the 1930s at the University of Chicago Laboratory School (1896–1903), where students engage in active and collaborative learning or project type of activities (Mayhew and Edwards 1965). However, based on Thomas's (2000) review on PBL studies, the studies lack common understanding, what project type of learning, such as PBL, means.

The PBL model presented in this chapter is based on the ideas of Blumenfeld, Krajcik and their colleagues (Blumenfeld et al. 1991; Krajcik and Shin 2015). In a PBL, students are engaged in a problem-oriented, meaningful learning in a small group, i.e. a project. The aim in the PBL is to support students to work in small groups to create artefacts that combine disciplinary core ideas or concepts with their previous knowledge. Artefact is a concrete output of learning, it is built by students, which can be, for example, a model, which describes a natural phenomenon based on the collected evidence. Artefacts are typically constructed with digi-tools, for example, with data-logging or modelling tools.

Learning sciences research has shown that students cannot learn physics SMK without engaging actively in disciplinary practices, here scientific practices, and they cannot learn these practices without learning the SMK through actively construct their understandings by working with and using ideas in real-world contexts. The key features of PBL (Blumenfeld et al. 1991; Krajcik and Czerniak 2013) are:

- PBL starts with a driving question, which contextualise learning and connect new ideas to previous ideas and experiences and guide learning process during the PBL (Greeno 2006).
- PBL focus on learning objectives/outcomes of the curriculum/standards that students are required to demonstrate mastery. Typically, the curriculum set learning objectives/outcomes to the learning of the scientific practices and use of technology. Consequently, these objectives/outcomes are also emphasised in PBL.
- Students explore the driving question through participating in scientific practices—processes of inquiry and problem solving that are central to expert performance in the discipline. Moreover, they use digi-tools in this exploring. As students explore the driving question, they learn and apply important ideas in the discipline. They investigate questions, propose hypotheses and explanations, argue for their ideas, challenge the ideas of others and try out new ideas.
- Students engage in collaborative activities to find solutions to the driving question. This mirrors the complex social situation of expert problem solving.
- Students create through the use of ICT tools a set of tangible products that address the driving question. These are shared artefacts, publicly accessible external representations of the class's learning.
- While engaged in the scientific practices, students are scaffolded in order to help them participate in activities normally beyond their abilities.

Consequently, in order to support students learning or in forming useable understanding, knowing and doing cannot be separated, but rather combined in planning, inquiring, problem-solving, decisions-making and explaining real-world phenomena situations. Learning is a kind of knowledge building, which refers to the process of creating cognitive artefacts, like concepts and models, as a result of common activity. Common activity means that students develop understandings through sharing, using, and debating ideas back and forth with others (Blumenfeld et al. 1991).

Finally, Krajcik and Shin (2015) emphasised the importance of cognitive tools, such as graphical representations in the computer screen, which help learners see patterns in data. Therefore, various digital tools could be considered cognitive tools because they allow learners to carry out tasks.

4 Examples of Lessons Where the Use of Digital Tools Have Been Integrated to Learning

4.1 Making Sense of Phenomena Related to Moving Objects Through Project-Based Learning and the Use of Digital Tools

The first example is written based on our research on students engagement and learning in the context of PBL (Inkinen et al. 2018, 2020; Schneider et al. 2020). The example are designed together with the CoDiRe tool with physics teachers (Juuti et al. 2021).

The teacher begins the lesson by introducing the topic of the lesson: "We will look at different movements, the change in movement and the reasons behind the change. We design experiments, model and discuss models. Experiments will be conducted with a video analysis software. A specific driving question is: Why do different objects take different times to fall when they are dropped from the same height? In order to understand the driving question, let us look at the drop of coffee filters. I have one filter in one hand and two nested filters in the other hand. What do you think, how do filters fall? Do they fall at the same time? Look closely at what is happening." Based on the teacher demonstration, it is found that a heavier object hits the ground first. The teacher continues the demonstration by doubling the masses of falling objects. After the demonstration, the teacher shows couple of video clips of a parachutist jumping.

Students are asked to summarise their findings in four-student group first independently and then combine the findings. The student report their findings to the online learning environment with two sentences. The summaries in the platform are analysed in a whole group discussion. The classroom recognised that the summaries focused to movement as such and to the reasons why a movement change or not change. The teacher says that the demonstration was the anchoring phenomenon of the upcoming study period, which introduces the students to the theme of the five lessons of the course.

The teacher guides the student to 4-student groups and asks them to draw up research questions on the basis of which the phenomenon can be studied and an answer to the driving question obtained. Questions were asked to write to an online learning environment. The teacher wrote supportive questions to online learning environment chat in order to help students to orient themselves in making the questions, for example, what do you already know about the topic and what do you want to find out by studying the phenomenon. Students formulate questions related to motion (e.g. how does velocity change during a fall? Is the speed of a falling object the same throughout the fall?) And questions related to the causes of motion change (e.g. how does the mass of a falling object affect the fall time? size (crumpled filter/ non-crumpled filter) affects the fall time?).

The teacher invites students back and asked students to classify the questions, posed in the learning environment in a meaningful way. The teacher says, "After you have classified the questions, your group will introduce them to the other group in order to discuss and compare the classification of other group. Make a common classification that you present to the whole class." The teacher asks students to choose questions that can be used to find the answer to the driving question. The groups present the classification criterion and examples of questions to the whole class and justify why the question is good for the phenomenon under consideration or takes the process forward.

The teacher says that next we start to study the anchoring phenomenon or similar phenomena based on the questions. First, a question or questions are selected to help investigate the falling motion (e.g. in what situation does the velocity of the falling object not change? What is the motion of the falling object then? What is the change in the speed of the falling object?). The reasons for the change in movement will be examined later. In this context, experiments related to the change of movement are not performed, but the movement itself is examined. The teacher demonstrates, how a data-logger or automated object tracking and video is used and data analysis done. Teachers shows how an app creates trajectory, position and velocity graphs for the object.

Next, the phenomenon was examined on the basis of movement-related questions. Students begin to design research in the direction of research questions in a small group. The teacher visited the groups and guides the use of mobile phone in capturing the movement. As students go further in measurement and modelling activities, the teacher submitted guides through the chat.

At the beginning of the next lesson, the group presents the results, such as graphical presentations, to another group. After the presentations, a joint discussion takes place, concluding that the movements can be grouped into two groups: a movement with constant velocity and movements in which the velocity changes. The students introduced their verbal and graphic patterns that described the studies movements. Under the guidance of the teacher, mathematical models describing the movements are also built and the use of the models in solving various problems is practiced.

4.2 Engaging Students in Scientific Practices in Remote Setting

The second example is from our collaborative research with physics teachers during the COVID-time, when physics teaching was organised in distance teaching mode and traditional laboratory activities was difficult to organise. The examples are only part of the PBL activities and include the assignments for distance working.

We have planned assignments collaboratively with physics teachers, which aim to engage secondary students in various scientific practices through the use of physics education technology (PhET) interactive simulations (http://phet.colorado.edu).

We have also investigated secondary students' use of scientific practices in remote setting. When working on assignments, students were free to use any digital resources and tools. Each assignment was completed in two sessions, and the duration of each session was between one and two hours. Students worked in small groups (2–4 students per group) and used shared online word processor (e.g. GoogleDoc) and messengers or telecommunications applications of their choice. The teacher was not present in the meeting sessions, but was available for online call if needed.

The following description is an example of assignment, designed together with physics teachers.

Assignment 1

(1) *Open the digital lab "flow"* (https://phet.colorado.edu/sims/cheerpj/fluid-pressure-and-flow/latest/fluid-pressure-and-flow.html?simulation=fluid-pressure-and-flow). *Investigate the phenomena by changing parameters: What have you observed? (write 2–3 observations).*
(2) *Choose one of your observations. For this observation: What could be possible further investigation? Formulate an investigation question.*
(3) *Let us follow the imaginary thin "slice of liquid" moving along the tube with different cross-sectional areas. In what way could we write the conservation of matter? Develop a model.*
(4) *Watch the video* (http://www.bozemanscience.com/ap-phys-098-equation-of-continuity) *or watch/read any other source on the topic.*
(5) *Explain the model given in the video/another source. Make conclusions about your model and the model presented.*
(6) *Does the presented model provide a background for any of your observations? Explain.*

Assignment 2

(1) *Investigate digital lab "Bouyancy"* (https://phet.colorado.edu/sims/density-and-buoyancy/buoyancy_en.html). *Observe what happens to the weight when block is under the water or oil. Write out your observations.*
(2) *Explain the reasons for differences in the observed weights. You can support your explanations with the graphic visualisation (e.g. net force). Use resources if necessary.*
(3) *Based on your digital lab observations and explanation, how could we find the density of unknown liquid? Write the plan of the experiment.*
(4) *Perform the experiment to find oil's density. Make a short report with collected data presented in tables/graphs and conclusions. Analyse the result and used models.*

At the beginning of the next lesson, the groups presented their assignments and graphical presentations to each other. After the presentations, a joint discussion took place. Under the guidance of the teacher, mathematical models describing the phenomena were discussed and the use of the models in solving various problems is practiced.

5 Discussion

In our discussion section, we analyse how the two examples met the CoDiRe questions and characteristics of PBL. PBL is characterised by the same features that are characteristic of working life projects. The project has aims and stages and ends with a concrete outcome, which can be, for example, a report, a video or a presentation. PBL cannot be defined by presenting a simple description of its progress as the PBL model is flexible. PBL is defined by its characteristics. The *first* characteristics are those of any given project, such as aims, stages and concrete outcomes (Krajcik and Shin 2015). All our examples were goal oriented, and there was a concrete outcome, a model that describe and explain the phenomena under investigation.

PBL design in both examples are based on the aims and objectives described in the curriculum. This is also the topic of the first question of the CoDiRe tool: it guides a teacher to think what a teacher want students to learn about the topic or what are the core ideas/big ideas/key concepts and models related to the topic. For example, the following core ideas were studied as part of the first PBL example: students analyse data on the motion and recognise when the object moves with constant or changing velocity; and, *second*, they analyse relationship among the net force on a macroscopic object, its mass and its acceleration. Moreover, in both examples they apply scientific and engineering practices in the design, evaluate and refine an experimental design which could be used for learning the core ideas. The CoDiRe tool ask the teacher to think what kind of conceptions and experiences the students have. For example, in the first PBL activity, students should have a basic understanding of the concepts of time, displacement, velocity, mass, acceleration and force.

The driving question guides PBL. It expresses the overall aim of learning. It gives students hints about the core ideas and practices that they will be working on during the learning period. It is a question, which help students to understand why the topic is important (meaningful and relevant) for students to learn. The driving question contextualises learning and shows the orientation or focus related to the phenomena under study. Asking a relevant driving question leads students to pose questions and design their studies. The driving question is thus the *third* key feature of PBL since the question constitutes an anchoring element through which students study the phenomenon. The driving question guides students to explore, leads them to ask additional questions and connects the lessons to each other.

Fourth, students are active in learning. This feature of PBL involves the idea of students' previous knowledge and experiences of the field of the phenomenon according to fifth and sixth questions of the CoReDi tool. The previous knowledge and experiences come up, for example, when students formulate research questions, make observations or internet searches and compile summaries. All of these activities are guided by the students' prior knowledge and values. It is crucial that the teacher elicits students' previous knowledge for review and offers possibilities for students to reflect on existential concerns and modern thinking dichotomies while constructing the outcomes. After all, it is well known that many of the beliefs of students are partly contradictory to the principles held by science. Moreover, this feature of PBL involves the idea that students actively process information and knowledge and reflect on their values through reading, making observations, discussing and creating artefacts. In PBL, students are active knowledge builders. Their activity also includes reflection on their own learning. Students are guided to analyse what they have learned in the direction of the driving question and what still needs to be learned. Therefore, PBL is an answer to the seventh question of the CoDiRe tool, which ask a teacher to think about the pedagogy.

Reflection is a general concept for those cognitive and affective functions by which an individual seeks to elucidate their experiences to construct knowledge or find new perspectives. With reflection, students make their thinking and learning visible to themselves and others. It is an exploration and awareness of the basics of one's thinking and action—first-hand insight. The reflexive process involves recalling experiences to the mind and recounting or telling others. During PBL, students use artefacts to present their own and group thinking and activities within and between groups. As the teacher tours class, they ask students what they have done up to that point and what they plan to do next. The questions are meant to support student reflection. The teacher's questioning helps students become aware of their own actions and develop their actions based on their own reflections and feedback. Consequently, the students are also able to apply their previous experience in new situations.

Fifth, students actively interact and collaborate during PBL. Students construct knowledge based on their previous understandings and experiences by interacting with other students, such as by asking questions, exchanging ideas, complementing the views of others, justifying their own views, linking concepts to other concepts and talking aloud about observations or conclusions. This interaction between students is similar to reflecting on different perspectives and testing claims or developing solutions based on information and data, which is part of the work of scientists and engineers. Furthermore, this collaboration and shared knowledge building enable students to reflect and understand different societal values and make connections between them. These issues belong to the fifth, sixth and seventh question of the CoDiRe tool, which ask a teacher to think about the pedagogy.

The *sixth* principle is that different learning tools are integrated into the learning, such as digital tools, which can acquire and process different information, datasets and models or simulate phenomena. This information can be represented and processed in a variety of ways. Macro- and micro-models explaining phenomena can be illustrated, and their dynamics can be elucidated through various simulations. An online learning environment or a common online document is suitable for taking notes and sketching patterns of phenomena and sharing information as it was emphasised in the second example. These issues belong to the fifth, sixth and seventh question of the CoDiRe tool, which ask a teacher to think about the students previous experiences with the digi-tools and how digi-tools are used for supporting students learning.

Seventh, working with concrete artefacts, texts, videos or patterns is integrated into PBL. Such artefacts include, for example, a list of possible research questions or a representation of a model describing the phenomenon under consideration. In all our example, the artefacts were models which describe and explain the phenomena the students were making sense. The purpose of producing artefacts is to inspire students to engage in processes similar to those that researchers are excited about when conducting research. Interacting with artefacts is a common thing conducive to learning and is emphasised in learning models that highlight situational learning. According to these models, learning occurs through interaction with social and cultural contexts and artefacts and participation in activities to create and invent such artefacts (Hakkarainen and Seitamaa-Hakkarainen 2022). Such models complement models of learning, which state that learning is a cognitive process within the mind. Artefacts also help the teacher evaluate students' learning process and progress as they make the students' thinking visible.

It is central to PBL that students' learning should be supported (scaffolded) to enable them to participate in activities in the proximal zone. These issues belongs to the eight question of the CoReDi tool. In the example above, there are several situations in which the teacher guides the learning. For example, in situations that the teacher knows to be challenging for students, they will provide instructions using a PowerPoint slide. The teacher's guidance entails questioning and supporting students' reflective thinking. The teacher instructs students to look at a phenomenon or matter being studied from different angles by raising the following questions: What is the information or data you are aiming to use while you conclude? What are you claiming? What do you find important? What kind of evidence would you need to convince others? What is your argument, and how does your evidence support it?

Acknowledgements We thank the Erasmus+ project "Schools Educating for Sustainability: Proposals for and from In-Service Teacher Education" for supporting the development of sustainable education in science. The study was supported by grants from the Strategic Research Council, Academy of Finland (345264, EduRescue; 312527, Growing Mind; 1340794, Clim Comp) and European Commission/H2020 (952470, SciCar; 4120113, Climademy).

Appendix

Examples of Assignments Related to Second Example

Assignments are to be completed in the electronic form using computers. Assignments are to be done in the small groups.

Assignment 1

(1) *Watch the video on Scientific method* http://www.bozemanscience.com/scientific-method.
(2) *You are about to investigate the force as a cause of change in motion. Have a look at the digital lab* https://phet.colorado.edu/sims/html/forces-and-motion-basics/latest/forces-and-motion-basics_en.html.

(3) *What can you investigate using this digital lab? Come up with an investigation question and write the plan of your investigation according to the steps of scientific method (research questions, identify variables, etc.).*
(4) *Perform the investigation, collect data and present it in a table.*
(5) *Using any computational program present the data as a plot/graph.*
(6) *Finalise the report and make a conclusion.*

Assignment 2

(1) *Use the simulation and find out, how the period (T) of a pendulum depends on a pendulum's length (L)?* https://phet.colorado.edu/en/simulations/pendulum-lab.
Set the friction equal to zero.
(2) *Using any computational programs plot the graph T(L), suggest a rule (formula) based on the graph.*
(3) *Find information in any resource on the pendulum's period and its formula. Analyse your answer to the question2 and make a conclusion.*
(4) *What happens if friction is not zero? Observe and explain* https://phet.colorado.edu/en/simulations/pendulum-lab.
(5) *We move the pendulum (period = 1 s) from the Earth to the Moon. How are the oscillations expected to change? Explain.*

Assignment 3

(1) *Watch the video* http://www.bozemanscience.com/voltage-current-resistance.
(2) *Explain in your own words the most important parameters of the electric circuit.*
(3) *Try out the digital lab* https://phet.colorado.edu/sims/html/circuit-construction-kit-dc-virtual-lab/latest/circuit-construction-kit-dc-virtual-lab_en.html.
(4) *How could one find out the unknown resistance of a resistor using this digital lab? Make a plan for such investigation.*
(5) *Based on your plan, use digital lab to determine the light bulb's resistance.* https://phet.colorado.edu/sims/html/circuit-construction-kit-dc-virtual-lab/latest/circuit-construction-kit-dc-virtual-lab_en.html.
(6) *Make a short report with collected data presented in tables and graphs, and conclusions. Analyse the used models.*

Assignment 4

(1) *Suggest, what gravitational force between 2 objects is dependent on. Why?*
(2) *Based on your assumptions, formulate one research question regarding how gravitational force is dependent on some parameter.*
(3) *Try out the digital lab* https://phet.colorado.edu/sims/html/gravity-force-lab-basics/latest/gravity-force-lab-basics_en.html.
(4) *Plan an investigation to answer your research question.*
(5) *Perform an experiment, collect data and present it in a table.*
(6) *Using any computational programs plot the graph, interpret data suggest a rule (formula) based on the graph.*

(7) *The mathematical expression for the gravitational force* $F_g = \frac{Gm_1m_2}{r^2}$. *At the same time, the gravity force in mechanics is usually considered* $F_g = mg$. *Explain, how those two are connected, the assumptions behind the expression mg, and limitations of that model.*

Assignment 5

(1) *Open the digital lab "flow"* (https://phet.colorado.edu/sims/cheerpj/fluid-pressure-and-flow/latest/fluid-pressure-and-flow.html?simulation=fluid-pressure-and-flow). *Investigate the phenomena by changing parameters: What have you observed? (write 2–3 observations).*
(2) *Choose one of your observations. For this observation: What could be possible further investigation? Formulate an investigation question.*
(3) *Let us follow the imaginary thin "slice of liquid" moving along the tube with different cross-sectional areas. In what way could we write the conservation of matter? Develop a model.*
(4) *Watch the video* (http://www.bozemanscience.com/ap-phys-098-equation-of-continuity) *or watch/read any other source on the topic.*
(5) *Explain the model given in the video/another source. Make conclusions about your model and the model presented.*
(6) *Does the presented model provide a background for any of your observations? Explain.*

Assignment 6

(1) *Investigate digital lab "Bouyancy"* (https://phet.colorado.edu/sims/density-and-buoyancy/buoyancy_en.html).
(2) *Observe what happens to the weight when block is under the water or oil. Write out your observations.*
(3) *Explain the reasons for differences in the observed weights. You can support your explanations with the graphic visualisation (e.g. net force). Use resources if necessary.*
(4) *Based on your digital lab observations and explanation, how could we find the density of unknown liquid? Write the plan of the experiment.*
(5) *Perform the experiment to find oil's density. Make a short report with collected data presented in tables/graphs and conclusions. Analyse the result and used models.*

Assignment 7

(1) *Go to the link and read about CERN and finding particles at Large Hadron Collider.*
(2) *Perform the assignment and find a particle! (just the way the physicists did it) (the task requires the google account sign-in)* https://colab.research.google.com/github/Freevolity/HST-2018/blob/master/Dimuon%20J_Psi%20for%20High%20School%20(Student%20Version).ipynb.

Assignment 8

(1) *Investigate the digital lab* https://phet.colorado.edu/sims/html/molecules-and-light/latest/molecules-and-light_en.html.
(2) *Choose 4 molecules in the digital lab, and make observations about them interacting with the waves of different frequencies. Write out your observations.*
(3) *Explain your observations using the theory of matter and radiation. You can use any sources.*
(4) *What models are presented in the digital lab, what limitations they have?*
(5) *For what purposes similar experiments are used?*

Assignment 9

(1) *Read the text about Nobel prize in physics 2018—"optical tweezers".* https://www.nobelprize.org/prizes/physics/2018/popular-information/.
(2) *Try out the digital lab which models the work of the optical tweezers.* https://phet.colorado.edu/sims/cheerpj/optical-tweezers/latest/optical-tweezers.html?simulation=optical-tweezers.
(3) *Make a short report about this discovery. In the report, explain the main physics law and models included, and practical applications.*

Assignment 10

(1) *Watch the video* https://www.youtube.com/watch?v=kdiHmSWI2Ks *and possible other sources to find information about Doppler effect (Doppler shift).*
(2) *Using* https://www.wolframalpha.com/input/?i=Doppler+shift+300Hz%2C+75mph&lk=3 *or other computational resource, calculate at what speed should move the guitarist to you, that the played by him note middle C starts to sound as middle A. Info about note frequencies here:* https://pages.mtu.edu/~suits/notefreqs.html.
(3) *Make a report about the Doppler shift and the reasons behind it.*

References

D.P. Ausubel, *Educational Psychology: A Cognitive View* (Holt, Rinehart, & Winston, 1968)

P.C. Blumenfeld, E. Soloway, R.W. Marx, J.S. Krajcik, M. Guzdial, A. Palincsar, Motivating project-based learning: sustaining the doing, supporting the learning. Educ. Psychol. **26**(3–4), 369–398 (1991)

W. Carlsen, Domains of teacher knowledge, in *Examining Pedagogical Content Knowledge: The Construct and Its Implications for Science Education*, ed. by J. Gess-Newsome, N.G. Lederman (Kluwer Academic Publishers, 1999), pp. 133–144

J. Carlson, K.R. Daehler, The refined consensus model of pedagogical content knowledge in science education, in *Repositioning Pedagogical Content Knowledge in Teachers' Knowledge for Teaching Science*, ed. by A. Hume, R. Cooper, A. Borowski (Springer Nature, 2019), pp. 77–92

M. Fuad, F. Ariyani, E. Suyanto, A.S. Shidiq, Exploring teachers' TPCK: are Indonesian language teachers ready for online learning during the COVID-19 outbreak? Univ. J. Educ. Res. **8**(11B), 6091–6102 (2020)

J. Gess-Newsome, A model of teacher professional knowledge and skill including PCK: results of the thinking from the PCK Summit, in *Re-examining Pedagogical Content Knowledge in Science Education*, ed. by A. Berry, P. Friedrichsen, J. Loughran (Routledge, 2015), pp. 38–52

J. Gess-Newsome, N.G. Lederman (eds.), *Examining Pedagogical Content Knowledge: The Construct and Its Implications for Science Education* (Kluwer Academic Publishers, 1999)

J. Gore, A. Gitlin, [Re]visioning the academic-teacher divide: power and knowledge in the educational community. Teach. Teach. Theory Pract. **10**(1), 35–58 (2004). https://doi.org/10.1080/1354060032000170918

C. Greenhow, C. Lewin, K.B.S. Willet, The educational response to COVID-19 across two countries: a critical examination of initial digital pedagogy adoption. Technol. Pedagog. Educ. (2020). https://doi.org/10.1080/1475939X.2020.1866654

J.G. Greeno, Learning in activity, in *Cambridge Handbook of the Learning Sciences*, ed. by R.K. Sawyer (Cambridge University Press, New York, 2006), pp. 79–96

K. Hakkarainen, P. Seitamaa-Hakkarainen, Learning by inventing: theoretical foundations, in *Invention Pedagogy—The Finnish Approach to Maker Education*, ed. by T. Korhonen, K. Kangas, L. Salo (Routledge, 2022)

M.Z. Hashweh, Teacher pedagogical constructions: a reconfiguration of pedagogical content knowledge. Teach. Teach. **11**(3), 273–292 (2005)

J.A.C. Hattie and G.M. Donoghue, Learning strategies: a synthesis and conceptual model. npj. Sci. Learn. **1**, 16013 (2016). https://doi.org/10.1038/npjscilearn2016.13

J. Inkinen, C. Klager, B. Schneider, K. Juuti, J. Krajcik, J. Lavonen, K. Salmela-Aro, Science classroom activities and student situational engagement. Int. J. Sci. Educ. **41**(3), 1–14 (2018). https://doi.org/10.1080/09500693.2018.1549372

J. Inkinen, C. Klager, K. Juuti, B. Schneider, K. Salmela-Aro, J. Krajcik, J. Lavonen, High school students' situational engagement associated with scientific practices in designed science learning situations. Sci. Educ. **104**(4), 1–26 (2020). https://doi.org/10.1002/sce.21570

K. Juuti, J. Lavonen, V. Salonen, K. Salmela-Aro, B. Schneider, J. Krajcik, A teacher–researcher partnership for professional learning: co-designing project-based learning units to increase student engagement in science classes. J. Sci. Teach. Educ. **32**(6) (2021). https://doi.org/10.1080/1046560X.2021.1872207

J.S. Krajcik, C.M. Czerniak, *Teaching Science in Elementary and Middle School: A Project-Based Approach* (Taylor and Francis, London, UK, 2013)

J. Krajcik, J. Merritt, Engaging students in scientific practices: what does constructing and revising models look like in the science classroom? Understanding a framework for K–12 science education. Sci. Teach. **79**(3), 38–41 (2012)

J. Krajcik, N. Shin, Project-based learning, in *The Cambridge Handbook of the Learning Sciences*, ed. by R. Keith Sawyer, 2nd edn. (Cambridge University Press, 2015), pp. 275–297. https://assets.cambridge.org/97805218/45540/frontmatter/9780521845540_frontmatter.pdf

J. Krajcik, E. Miller, B. Schneider, Science education through multiple literacies: project-based learning in elementary school, in *Transforming the Teaching and Learning of Science Through Project-Based Learning* (Harvard Education Press, 2021), pp. 1–16

T.C. Lin, C.C. Tsai, C.S. Chai, M.H. Lee, Identifying science teachers' perceptions of technological pedagogical and content knowledge (TPACK). J. Sci. Educ. Technol. **22**(3), 325–336 (2013)

J. Loughran, P. Mulhall, A. Berry, Exploring pedagogical content knowledge in science teacher education. Int. J. Sci. Educ. **30**(10), 1301–1320 (2008). https://doi.org/10.1080/09500690802187009

K.C. Mayhew, A.C. Edwards, *The Dewey School. The Laboratory School of the University of Chicago 1896–1903* (Routledge, 1965)

P. Mishra, M.J. Koehler, Technological pedagogical content knowledge: a framework for teacher knowledge. Teach. Coll. Rec. **108**(6), 1017–1054 (2006)

G. Morine-Dershimer, T. Kent, The complex nature and sources of teachers' pedagogical knowledge, in *Examining Pedagogical Content Knowledge: The Construct and Its Implications for Science*

Education, ed. by J. Gess-Newsome, N.G. Lederman (Kluwer Academic Publishers, Dordrecht, 1999), pp. 21–50

R.K. Sawyer (ed.), *The Cambridge Handbook of the Learning Sciences*, 2nd edn. (Cambridge University Press, 2015)

B. Schneider, J. Krajcik, J. Lavonen, K. Salmela-Aro, *Learning Science: The Value of Crafting Engagement in Science Environments* (Yale University Press, New Haven, 2020)

L.S. Shulman, Knowledge and teaching: foundations of new reform. Harv. Educ. Rev. **57**, 1–22 (1987). https://doi.org/10.17763/haer.57.1.j463w79r56455411

J.W. Thomas, A review of research on project-based learning. San Rafael, CA: Autodesk Foundation. https://www.autodesk.com/foundation

The New Methodologies in e-Learning and the Italian Experience in the Physics Teaching Field and STEM

Pierpaolo Limone and Giusi Antonia Toto

Abstract Background: Introduction on STEM learning, with updated national and European statistical data; brief background on the construct of motivation in the e-learning context. Research question: What e-learning strategies and best practices have been adopted in Italy for STEM teaching? Methodology: A literature review on the topic, with particular reference to the 2012–2022 period (decade of reference for the diffusion of MOOCs and online learning). Expected results: Reconnaissance of evidence-based experiences on the topic; realization of a theoretical framework for the study of the e-learning system in Italy, with specific reference to STEM teaching. Conclusions: Proposals to implement previously adopted models, e.g., integration of game-based learning in STEM online learning (gamified MOOCs, interactive microcredentialing, etc.).

Keywords STEM disciplines · Teaching · Digital innovation · Inclusion · e-Learning · Motivation · MOOCs

1 Introduction

This research paper aims to explore the e-learning strategies and practices adopted in Italy, in the last decade, with specific reference to STEM education from a pedagogical point of view according to a qualitative approach.

Starting from a literature review on the topic, evidence-based experiences will be reconstructed, reserving a special focus on STEM disciplines in Italy.

P. Limone (✉)
Developmental and Educational Psychology, "Pegaso" Telematic University, Centro Direzionale Isola F2, 80143 Napoli, Italy
e-mail: pierpaolo.limone@unipegaso.it

G. A. Toto
Didactics and Special Pedagogy, Department of Humanistic Studies, University of Foggia, Via Arpi 176, 71122 Foggia, Italy
e-mail: giusi.toto@unifg.it

M. Streit-Bianchi et al. (eds.), *New Challenges and Opportunities in Physics Education*, Challenges in Physics Education, https://doi.org/10.1007/978-3-031-37387-9_16

The enhancement of learning for STEM disciplines constitutes a today challenge to a global level for all educational systems (schools, universities, associations), in order to help students to have a broad and critical understanding of the present and to develop by themselves technical, creative, digital skills, communication and collaboration skills as well as competences in problem-solving, flexibility and adaptability to change.

Today, schools and universities are called upon to invest in STEM and to encourage educational innovation in curriculum and methodologies. Every learning environment, in the 3.0 world where we live, needs new technologies, resources and dedicated spaces where students can observe, build, create, learn and collaborate, using innovative educational and digital tools.

2 Teaching STEM Disciplines in Italy and Around the World

Among recent national and international research on teachers and teaching of STEM disciplines in Italy and around the world, a prominent place is deserved by the OECD TALIS (Teaching and Learning International Survey) 2018 (OECD 2019), the largest survey which aimed at investigating various aspects of the teaching profession, especially STEM teachers, in every order and degree school. It also aimed at developing international indicators, as Scippo et al. (2020) report in their study The Teaching of STEM Disciplines in Italy.

To initiate the survey, a questionnaire was developed for teachers in order to collect information related to a number of variables duly selected around four items: biographical, training and updating, teaching practices and school environment.

In addition to relevant data to the present study, the OECD TALIS 2018 (OECD 2019), which is a survey on training the motivation of teachers of these disciplines, shows that in Italy, in secondary schools of I and II degrees, there are many more teachers with a degree and trained in STEM discipline than teachers in elementary school where the percentages of duly trained teachers are much lower, adding that, for all degrees of education, the percentages are reduced if we narrow the focus to teachers in possession of an enabling title that also presupposes pedagogical–didactic training.

Then, with regard to continuing-education or in-service training, it is immediately apparent that in elementary school, the percentage of graduates is lower among colleagues in STEM disciplines than colleagues pertaining to humanities disciplines. In addition, in secondary school, Italian STEM teachers attend fewer formal programs and conferences in the educational field than colleagues from other disciplines and, in general, fewer teachers update on pedagogical skills, classroom management and inclusion, while more teachers update on the use of ICT (Information and Communication Technology). In secondary school, STEM teachers participate in more formal programs and teacher networks for professional development.

In addition to this, the landmark survey that offers an in-depth reflection on teaching practices in STEM disciplines is undoubtedly the PISA 2015 survey (OECD 2016), which focuses on learning outcomes and points attention to the relationship between teaching practices and performance on science tests, from students' perspectives.

The PISA 2015 (OECD 2016) survey distinguishes between four instructional approaches:

(1) teacher-directed instruction (*teacher-directed*), with structured explanations on a topic, classroom debates and student questions;
(2) instruction based on perceived feedback (*perceived feedback*), so as to inform students about their behavior, strengths and weaknesses, and how to improve their performance to achieve goals;
(3) flexible instruction (*adaptive instruction*), referring to how frequently the teacher adapts the lesson to the needs of the class or provides individual help to students with specific difficulties;
(4) inquiry-based instruction (*enquiry-based*), which involves students in experimentation and hands-on activities.

From the data collected, it would seem that, in order to achieve good results, unlike what is normally done, physics and science teachers need to wisely mix different teaching approaches so as to achieve meaningful learning based on experiential inquiry (Scippo et al. 2020).

Therefore, the strengths and weaknesses of the Italian teaching staff in science disciplines clearly emerge, and it is only recently, thanks mainly to the resources made available by the PNRR funds, that there is an opening to the possibilities offered by the new methodologies in STEM disciplines to develop in the younger generations transversal skills. It is also noted the greater seniority of the Italian teaching staff compared to their colleagues in other nations, with a higher percentage of male teachers in STEM disciplines, indicative perhaps of a gender stereotype still prevailing even in the teaching staff.

These characteristics of the Italian teaching staff, compared with what is happening in the various European and international scenarios, seem to point to the idea that it is sufficient to only know the subject in order to teach it. Moreover, solid pedagogical–didactic training is urgently needed (Moore and Smith 2014).

Educational research shows, how many teachers still teach according to a "traditional" methodology, making use of the classic frontal lecture in the same way it was taught years before and they usually clash with colleagues when the latter implement the learned innovative teaching strategies.

3 New Educational Paradigms

In light of the special attention to science disciplines, several projects have been funded in Italy and Europe to promote the interest of female students and encourage them to engage in these subjects.

For example, the ISEE project[1] (ISEE project Inclusive STEM Education to Enhance the capacity to aspire and imagine future careers, September 2016–August 2019) concluded in 2019:

> In our contemporary society of global uncertainties and social acceleration, our imagination of the future becomes problematic and source of anxiety [...] the young generation have difficulty in projecting themselves into the future, and in developing their potential as responsible and active persons [...] the goal of the I SEE project is to design innovative approaches and teaching modules to foster students' capacities to imagine the future and aspire to STEM careers (ISEE project).

Or again, the IDENTITIES project[2] (IDENTITIES project Integrate Disciplines to Elaborate Novel Teaching approaches to Interdisciplinarity and Innovate pre-service teacher Education for STEM challenges, September 2019–August 2022) which aimed to build specific interdisciplinary training modules and provide recommendations to policy makers:

> The IDENTITIES project designs novel teaching approaches on interdisciplinarity in science and mathematics to innovate pre-service teacher education for contemporary challenges [...]. Disciplinary IDENTITIES, their boundaries and forms of integration are enlightened through reflections on their epistemological and linguistics structures [...]. The project will lead to the construction of Open Education Resources for Blended modules and MOOCs, as well as recommendations for policy makers to promote interdisciplinarity and innovate prospective teachers' education for STEM challenges (IDENTITIES project).

The still ongoing FEDORA project,[3] also aims to develop future-oriented teaching methodologies and makes recommendations to policy (FEDORA acronym for Future-oriented Science Education to enhance Responsibility and engagement in the society of Acceleration and uncertainty, September 2020–August 2023):

> Scientific and technological development is driving rapid social changes, and educational systems are struggling to keep pace with these transformations [...]. Effective science education for future generations is an urgent need. The EU-funded FEDORA project aims to develop a future-oriented model for formal and informal science education to provide young people with skills of foresight, imagination and action [...] develop methodologies that target the main factors of the current misalignment between education systems and society. Recommendations will then be made for anticipatory policies for creating new and visionary attitudes on open education and institutional transformations (FEDORA project).

In Italy, with the implementation of the Ministry of Education Decree No. 147 of April 30, 2021, provided for under action #4 "Environments for integrated digital

[1] https://iseeproject.eu.

[2] https://identitiesproject.eu.

[3] https://www.fedora-project.eu.

teaching" of the National Digital School Plan (PNSD), the Ministry has promoted the creation of laboratory spaces and the provision of digital tools suitable to support schools' learning and teaching of STEM disciplines (Science, Technology, Engineering and Mathematics).

The National Digital School Plan envisaged in the "Buona Scuola" reform (Law 107/2015) is the policy document with which the MIUR intends to promote innovation and digitization of Italian schools and provides 35 actions financed for the purpose by drawing on resources from European Social Funds (PON Education 2014–2020) and from the funds of Law 107/2015.

Central in the same reform is the need to bring back to the center laboratory teaching, the concept of knowing and knowing how to do, which fosters a sustainable vision of school attention to the needs of teachers and students as well as the realities in which they are realized.

To do this, will be fundamental to consider classroom no longer as a closed place but an enabling and open environment, a "light" and flexible environment, the only capable of adapting to the digital world. Second, we need to overcome the system fragmentation whereby the concentration of investment on a few schools, in technology intensive environments case, has not produced system effects.

There is a need, today more than yesterday, for:

- classrooms "augmented" by technology for individual and collective use of the web and its content, for daily integration of digital into teaching, for interaction of diverse aggregations in learning groups, in wired and wireless connection;
- environments specifically dedicated to STEM teaching and alternative learning spaces, generally larger than classrooms which allow for continuous remodeling of settings consistent with the chosen teaching activity and capable of accommodating diversified activities for multiple classes or groups, spaces that, with these characteristics, can accommodate not only students, but also teachers in training;
- mobile laboratory spaces, mobile devices and tools in trolleys and mobile boxes available to the entire school, for various disciplines and laboratories, primarily those peculiar to STEM disciplines and physics;
- teaching equipment coding and educational robotics (educational robots of all sizes, integrated and modular sets programmable with apps, including with motors and sensors, drones educational programmable);
- programmable boards and educational electronics kits (programmable boards and expansion sets, smart electronic kits and modules and related accessories);
- tools for observation, scientific processing and three-dimensional exploration in augmented reality (educational kits for STEM disciplines, modular sensor kits, graphical-symbolic calculators, virtual reality viewers, 360° cameras, 3D scanners);
- 3D making and 3D creation and printing devices (3D printers, plotters, laser cutters, invention kits, tables and related accessories);
- innovative software and apps for digital STEM education.

Of course, space renovation does not remain an end in itself. This must be accompanied by adequate training of primary and secondary school teachers and increased interoperability, flexibility and inclusiveness of equipment to create an ecosystem of hardware and software devices that coexist with each other to accompany every teaching activity, cross-curricular, "hybrid," specialized, technologically augmented and consistent with the methodologies, age and different needs of students, especially for BES (English 2016).

4 Teacher Training

The challenges that are going to await our country are important and concern change management, introduced by COVID-19, in the social and professional world, besides the fact that transformation of countries toward new horizons in terms of digitization, ecological transition, sustainable mobility, education, social inclusion and cohesion, and health, all objectives of the development and investment lines identified by the EU for the revitalization of the economy and life in Europe. The PNRR give a look to people, even before technologies, as the engine of change and innovation. Technological progress can foster human development and help to create optimal conditions for the exercise of human rights. At the same time, its use and eventual misuse have broader implications to core values of democratic societies, including equality and fairness.

We cannot stop passively observing this. The complexity of today's society dominated by the development of artificial intelligence, big data technology and social networks (Morin 2016) is a challenge to be grasped and exploited to the fullest in the field of training, education and learning (Scarinci et al. 2022).

For almost a decade, European policies have been emphasizing the recognition of science disciplines in the declarations of educational objectives, the importance of the use of new technologies in different contexts, especially in the educational context involving physics and STEM disciplines, and encouraging the updating of teachers' professional profile, launching pathways that aim to overcome "digital educational poverty" (Pasta and Rivoltella 2022): suffice to think in Italy of the aforementioned 2015 National Digital School Plan, the PNRR funds, and then broaden our gaze to the more recent European program on digital education described in the Digital Education Action Plan 2021–2027 and the numerous training courses. However, the way is still far: teachers' total acceptance of new technologies for innovative teaching is "work in progress," because many remain anchored to a transmission of basic content.

An incentive in favor of a new digital literacy, the so-called digital literacy, and retraining of teaching staff has been given by the pandemic crisis that has swept through and invested in every sphere of life, but it has particularly touched educational institutions, universities and traditional learning environments that have had to reinvent themselves, adopting new communicative languages to meet the challenges of the moment. Smartphones, tablets, social platforms, moved the classroom into a virtual room where digital tools were the only mediators of knowledge, fostering everyone's involvement (Finestrone et al. 2022).

Content delivery methods and training settings have been touched, generating a change in knowledge acquisition and learning which has required to everyone, teachers and pupils of all grades, to possess specific digital and teaching skills, also to ensure performance of specialized teachers capable of fostering meaningful learning, computational thinking and critical thinking.

For the reasons just mentioned, digital literacy plays a major role in the pre-service training of teachers, particularly support teachers (Sailer et al. 2021), to integrate technological tools and digital media into education and training. It is necessary to create new dialogues, new laboratories and learning contexts: there must be a dialectical interchange between didactics and technology in order to experiment with new strategies and foster the growth of subjects and the enhancement of cognitive and relational styles, in order to qualitatively improve teaching processes and training settings.

In the age of Internet, the possibilities for web-based teaching have been expanded: there are MOOCs (Massive Open Online Courses), which have come from the United States, with the Open-Course-Ware, to Italy with the EduOpen platform. In a short time, MOOCs have gained great visibility, providing teachers in training, economic and time advantages and the opportunity to acquire language and computer skills, soft skills in the educational context.

The advantages of digital learning, in fact, are many: breaking down barriers, providing a teaching and Edu-training aid, simultaneously connecting individuals in different parts of the world through video chats, enabling online lessons that can offer the possibility of saving teaching material, correcting it and reviewing it, after careful consideration. While this is true for all teachers, greater benefits accrue to the support teacher, who, thanks to the use of instructional planning Web Apps (Learning designer, Edmodo), serious games and interactive Web Apps (Edpuzzle, LearningApps) and processing and production platforms (Padlet), can plan personalized or individualized instructional paths, meet the educational needs of pupils with BES, enhance differences, increase self-determination and develop soft skills useful in the world of work, making teaching strategies more appealing and improving their effectiveness, also to improve self-empowerment through metabolizing errors (Bernard et al. 2014). In such a scenario, institutions, in order to live fully in the present and contribute to the education of future world citizens, need to reconsider the curricula of degree programs with experienced teaching faculty and refresher courses to encourage "teaching with technologies." At the European level, this is determined by the eight key competencies arranged by the European Commission and Parliament in 2018 (which also include digital competence) and the Digital Competence Framework for Educators (DigCompEdu), which call for meta-competencies that mediate the relationship with students and support citizens' digital competence.

This and much more is learned, for example, by the support teacher during the specialization course, known as TFA, during which prospective teachers are asked to deepen their ICT skills, and through laboratory teaching, an active and engaging learning process is triggered, ranging from the formulation of questions to shared answers and the development of practical design skills, which underpin the teaching and inclusion process (Toto et al. 2022).

5 STEM Disciplines in Classroom

The idea of introducing STEM disciplines and enhancing them with the use of digital tools is not a new issue. Always throughout history, innovations have been introduced that have brought renewal in order to improve performance in instructional design and meet the needs of time. Today, targeting STEM disciplines as early as elementary school is possible thanks, for example, to the use of serious games, and there is nothing to prevent in continuing training in first and second grade secondary schools as well. In fact, the acronym STEM, which not everyone is familiar with, carries the initials in English for science, technology, engineering, art and mathematics, a quadrivalent of disciplines that many believe are increasingly important for the development of societies and, as a result, very expendable in the job market (Henderson et al. 2017).

Science and technology disciplines are crucial in many of future professions, so targeting projects in our schools which aim to provide an innovative teaching approach that can help new digital natives to acquire digital skills early from kindergarten through grade II classes is crucial.

Today, the novelty lies in making people learn the content of these disciplines by adopting various methods and tools (Spinelli 2022):

- world clouds, keyword clouds that remind the importance of concepts "to remember";
- earth speaker, a medium that invites young people to express their ideas on how to protect the planet and adults to listen to what the younger generation has to say;
- coding and robotics, the tool through which computational thinking, that logical process of breaking down a problem into smaller parts, finding a solution and developing it, is fostered in school; and introducing coding to school gives kids a chance to work on projects, encourage sharing and collaboration while learning from their mistakes;
- Nindendo Labo, to bring out kids' best skills and tell the story of how manual skills and technology can coexist and empower each other; and
- Minecraftedu, an immersive experience through which children and teens can learn by playing and experimenting, only so they can also make mistakes from their mistakes.

None of this is unfeasible, and concreteness of this comes from the evidence of projects launched in schools in southern Italy, for example.

From the programming of robotic arms to musical computing, from the search for innovative solutions for environmental sustainability to the use of tools to understand the mechanisms of cyberbullying and to promote local knowledge. These are just a few examples of the activities that will be launched in South Italy in 2023 thanks to the 7 projects selected by "Con i Bambini,"[4] as part of the Fund for Combating

[4] https://www.conibambini.org/2023/02/20/con-i-bambini-e-fondazione-cdp-7-progetti-per-le-steam.

Educational Poverty among Children, and by Fondazione CDP through the "Tools for Growth" call for proposals. The interventions will be supported with a total of €3 million and include pathways to promote the development of basic skills in STEM disciplines (science, technology, engineering, arts and mathematics) for children aged 11–17 living in southern regions.

The call was aimed exclusively in southern Italian regions where the percentage of pupils with sufficient math skills is lower than in other parts of Italy. The selected projects involve the adoption of methodologies capable to engage students particularly in didactic and educational paths, attracting their interest, stimulating their curiosity and increasing their knowledge of the world around them, also to develop active citizenship. In the declination of the individual educational proposals, the partnerships have taken into account the territorial specificities, structuring paths in line with the economic–social vocation of the territory and the context of the daily life of the children.

Thanks to the selected initiatives, actions will be launched in the provinces of Agrigento, Catania, Cosenza, Enna, Naples, Palermo, Salerno, South Sardinia, Taranto and Trapani. All projects include Steam laboratory activities integrated within the teaching for the expansion of digital skills through coding and educational robotics. In addition, the projects will offer students experiences in which they can put in practice what they have learned, so as to grasp its relevance in their daily lives. All initiatives include specific teacher training activities through the intervention of external experts, who will hold in-depth face-to-face lectures and be involved in side-by-side classroom activities to better support teachers in the use of new tools and methods.

6 Conclusion

In today's knowledge and communication society, the educational institution faces a new perspective and must broad its cultural horizons in light of a new Humanism, which requires creativity and dialogue to cope with the loss of the certainties of the past (Stoyanov et al. 2022).

It is precisely the pandemic crisis, mentioned above, that has turned the spotlight on a new style of education and highlighted the prerequisites in changing the course (Agrati 2021), showing how updating educational offerings and delivering teaching in an e-learning mode is essential for meaningful, interactive and lifelong learning, and to meet the educational needs of the new generations. Those are first steps of a journey in the making.

References

L.S. Agrati, L'emergenza da coronavirus come verifica delle competenze digitali dei docenti. Indagine sulla pregressa formazione in servizio. Form. insegnamento. Riv. int. Sci. dell'educ. form. **19**(2), 179–192 (2021)

R.M. Bernard et al., A meta-analysis of blended learning and technology use in higher education: from the general to the applied. J. Comput. High. Educ.**XXVI**, 87–122 (2014)

L.D. English, STEM education K-12: perspectives on integration. Int. J. STEM Educ.**3**, 1–8 (2016)

F. Finestrone, A. Scarinci, V. Berardinetti, F.P. Savino, Il cammino dell'inclusione e la sfida del digitale. Mizar. Costellazione Pensieri **2**(17), 96–108 (2022)

C. Henderson, M. Connolly, E.L. Dolan, N. Finkelstein, S. Franklin, S. Malcom, K. St. John, Towards the STEM DBER alliance: why we need a discipline-based, STEM-education research community. J. Geosci. Educ. **65**(3), 215–218 (2017)

T.J. Moore, K.A. Smith, Advancing the state of the art of STEM integration. J. STEM Educ. Innov. Res. **15**(1), 5 (2014)

E. Morin, *Insegnare a vivere: manifesto per cambiare l'educazione* (Raffaello Cortina Editore, Milano, 2016)

OECD, *PISA 2015 Results (Volume II): Policies and Practices for Successful Schools* (OECD Publishing, Paris, 2016). https://doi.org/10.1787/9789264267510-en

OECD, *TALIS 2018 Results (Volume I): Teachers and School Leaders as Lifelong Learners* (OECD Publishing, Paris, 2019). https://doi.org/10.1787/1d0bc92a-en

S. Pasta, P.C. Rivoltella, Superare la "povertà educativa digitale". Ipotesi di un nuovo costrutto per la cittadinanza digitale, in *La formazione degli insegnanti: problemi, prospettive e proposte per una scuola di qualità e aperta a tutti e tutte* (Pensa MultiMedia Editore, 2022), pp. 600–604

M. Sailer, F. Schultz-Pernice, F. Fischer, Contextual facilitators for learning activities involving technology in higher education: the Cb-model. Comput. Hum. Behav. **121**(106794), 1–13 (2021)

A. Scarinci, M. Di Furia, G. Peconio, Ambienti di apprendimento digitali innovativi: nuovi paradigmi. *Form. Lav. Pers. Anno* **XII**(36), 22–38 (2022)

S. Scippo, M. Montebello, D. Cesareni, L'insegnamento delle discipline STEM in Italia. Ital. J. Educ. Res. **25**, 35–48 (2020)

A. Spinelli, Potenzialità e responsabilità della scuola per favorire l'inclusione nelle discipline STEM, in *La nascita del sistema scolastico* (2022), p. 130

S. Stoyanov, T. Glushkova, V. Tabakova-Komsalova, A. Stoyanova-Doycheva, V. Ivanova, L. Doukovska, Integration of STEM centers in a virtual education space. Mathematics **10**, 744 (2022)

G.A. Toto, M. Rossi, D. Lombardi, Il digitale e la formazione dei docenti di sostegno. Form. Lav. Pers. **XXII**(36), 39–52 (2022)

Applications of Technology to Promote Active Learning with Examples from Acceleration and Gravity

David R. Sokoloff

Abstract Student interest in studying physics has continued to grow, stimulated in part by contemporary discoveries like the 2015 detection of gravitational-wave signals by LIGO. Yet many students are still taught in a traditional, passive manner, and leave the introductory course with significant gaps in their conceptual understanding, even about Newton's laws of motion, acceleration, and gravity. Active learning strategies have been developed over the last 40 or so years in parallel with and, in many cases, enhanced by developing computer technologies. These technologies provide students with unprecedented tools to explore and analyze the physical world as part of their introductory physics experiences. This chapter describes some of these strategies and the computer-based tools supporting them, with gravity-related examples.

Keywords Technology · Active learning · Computers · Gravity

1 Introduction

In 2015—about a century after Albert Einstein published his theory of general relativity—we entered the age of gravitational-wave astronomy with the detection of the first signals detected by the LIGO Collaboration (Bailes et al. 2021). During these exciting times in astrophysical research, there has been growing student interest in studying physics, with high school enrollment in the U.S., for example, nearly doubling to over 1.5 million since the turn of the twenty-first century (White 2021a), and college/university enrollments also continuing to grow (White 2021b). However, despite this interest, many students are still being taught in a traditional manner and Physics Education Research (PER) has continued to demonstrate that they often leave the introductory physics course with the same incorrect ideas about physics concepts that they had developed in their previous years experiencing the physical

D. R. Sokoloff (✉)
University of Oregon, Eugene, OR 97403, USA
e-mail: sokoloff@uoregon.edu

M. Streit-Bianchi et al. (eds.), *New Challenges and Opportunities in Physics Education*, Challenges in Physics Education, https://doi.org/10.1007/978-3-031-37387-9_17

world. In particular, this includes conceptual ideas about Newton's laws of motion and acceleration, and even classical ideas about gravitational force and acceleration (McDermott 2001).

During the last 40 or so years, a number of research-validated active learning strategies have been developed to improve students' understanding of basic physics concepts, enhance their excitement and interest in physics, and prepare them for the study of contemporary topics (Sokoloff and Yüksel 2023; Meltzer and Thornton 2012). A number of these strategies have incorporated novel applications of developing computer technologies to assist students in making observations of the physical world, and to help engage students more effectively in the learning process.

This chapter will explore some of these uses of technology, such as computer-based data acquisition tools, interactive video analysis and simulations, with examples of how they have been used in conjunction with active learning curricular materials to enhance understanding. The examples presented will emphasize the study of classical ideas about gravitational acceleration.

2 Computer-Based Data Acquisition in Laboratory and Lecture

Beginning around 1986, new computer-based laboratory tools have become increasingly popular for the real-time collection, display, and analysis of data in the introductory laboratory. These tools consist of electronic sensors, a computer interface, and software for data collection and analysis. Available sensors measure motion (displaying position, velocity, and acceleration), force, sound, magnetic field, current, voltage, temperature, pressure, rotary motion, humidity, light intensity, and other physical quantities. These tools provide a powerful way for students to observe the physical world through real experiments. For example, students can discover motion concepts for themselves by walking in front of an ultrasonic motion sensor while the software displays position, velocity, and/or acceleration in real time. Students can see a cooling curve displayed in real time when a temperature sensor is plunged into ice water, or they can use a microphone to see how a sound pressure versus time plot changes as one of them sings. Collected data are often displayed graphically and can be analyzed quantitatively. Students' time in class can be spent examining, discussing, and thinking about the observations, while eliminating much of the drudgery of collecting the data.[1]

[1] In the U.S., such tools are available from Vernier Science Education (www.vernier.com) and PASCO Scientific (www.pasco.com).

2.1 *Active Learning in the Introductory Physics Laboratory*

For example, after examining graphical representations of their own fairly constant-velocity motion walking toward and away from a motion sensor, students doing *Real-Time Physics (RTP): Mechanics* (Sokoloff et al. 2007, 2011) laboratories examine the motion of a low-friction cart with a fan unit mounted on it (see Fig. 1a). With the fan applying a constant force toward the motion sensor, the graphs of velocity and acceleration versus time shown in Fig. 1b are collected and displayed, when the cart is given a short push away from the motion sensor and released. Before observing this motion, many students predict that the acceleration will be positive as the cart moves away from the motion sensor, negative as it moves toward the motion sensor and that both the velocity and the acceleration will be zero at the moment the cart reverses direction. They are very surprised by the constant negative acceleration they observe! Most surprising to students is the nonzero acceleration at the instant the cart reverses direction.

Of course, it is essential for introductory students to understand the significance of Newton's laws in examining the motions of objects under the influence of applied forces. Later activities in *RTP: Mechanics* explore these laws. Applied force can be measured and displayed using a force sensor mounted on the cart, as shown in Fig. 2a. Before making any observations, students predict whether applied force is related to velocity (the Aristotelian view held by many students entering an introductory course (McDermott 2001)) or acceleration (the Newtonian view). Figure 2b shows the graphs collected when the student's hand jerks the cart back and forth in front of the motion sensor. It is clear that the shape of the acceleration–time graph matches the force–time one, suggesting Newton's second law. Later observations with this equipment using a modified Atwood's machine setup to supply the force, establish that a constant applied force results in a constant acceleration.

But what about gravitational acceleration? Experiments in traditional introductory laboratories are typically of the confirmation variety. A common experiment for

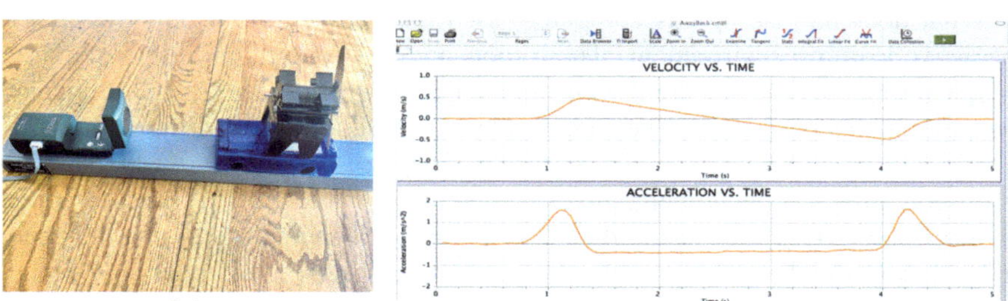

(a) (b)

Fig. 1 **a** A low-friction cart moving in front of an ultrasonic motion sensor, with the fan unit mounted on it blowing back toward the left (on the left). **b** Graphs of the data collected by the motion sensor as the cart moves to the right, slowing down, reverses direction, and then moves back to the left, speeding up (on the right). The graphs were collected using Vernier LoggerPro software

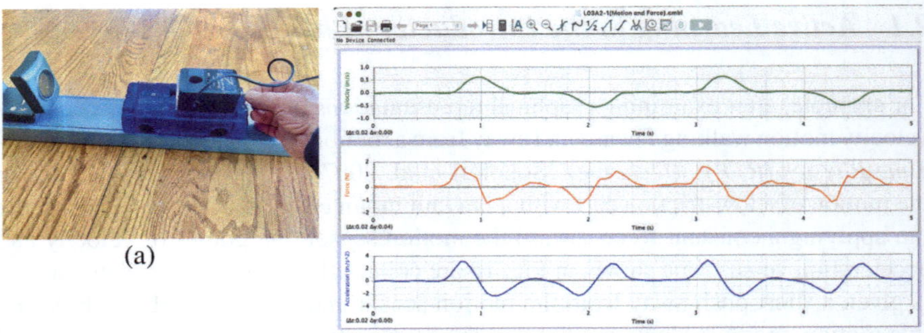

(a) (b)

Fig. 2 **a** Low-friction cart with a force sensor mounted on it (on the left). **b** Graphs collected when the force sensor is grasped, and the cart is jerked back and forth in front of the motion sensor (on the right). The middle graph is force–time, the top velocity–time and the bottom acceleration–time. The graphs were collected using Vernier LoggerPro software

gravity—of which there are many varieties—involves measuring and "confirming" the value for "*g*". Traditionally, no attempt is made in such laboratory experiments to emphasize the properties of the gravitational acceleration (and force) in the laboratory—close to the Earth's surface. However, in a later laboratory in *RTP: Mechanics*, students "discover" the gravitational force, and observe that it is constant in magnitude and directed downward in the location where they are making their measurements. Figure 3a shows the experimental setup consisting of a motion sensor on the floor and a soccer ball. When the ball is tossed in the air above the sensor, the graphs in Fig. 3b are collected. Since the students observe that these graphs are identical in shape to the ones in Fig. 2 for the cart moving under the influence of a constant applied force (and other results they have observed with the Atwood's machine), they conclude that there must be a constant downward force on the ball. The magnitude of the acceleration can be determined from the graph, but this is just a secondary objective of the laboratory!

What does the use of this computer-based data acquisition technology add to the active learning environment? These tools are easy to use, and do not require a long learning curve. They are flexible and versatile, designed to be independent of the experiments performed. They are usable in experiments with different levels of sophistication, with relatively high accuracy. They are designed to enable students to observe physical phenomena directly and clearly—often in real time, to appeal to the displayed results to justify their conclusions—often in discussions with peers, and to learn from their observations. Many of these observations could not be made without the computer-based tools. In summary, these tools enable the active learning pedagogy that helps students to learn from their observations and from each other.

PER studies have compared the conceptual learning in the *RTP* learning environment with learning in traditional, passive environments. For mechanics, studies using the research-based, multiple-choice *Force and Motion Conceptual Evaluation (FMCE)* assessing the learning in large enrollment introductory classes have

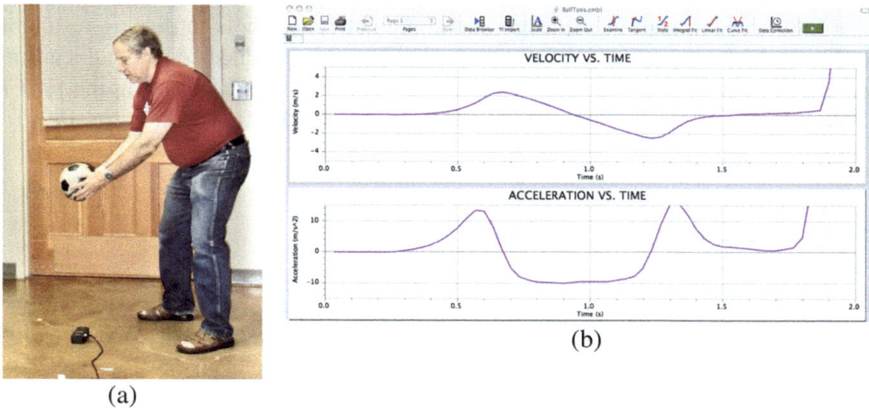

Fig. 3 **a** Photograph of a soccer ball being tossed upward above a motion sensor (on the left). **b** Velocity versus time and acceleration versus time graphs collected for the ball's up and down motion (on the right). The graphs were collected using Vernier LoggerPro software

found that very significant improvement in conceptual understanding in mechanics is achieved when *RTP* technology-supported, active learning laboratories replace traditional ones (Sokoloff et al. 2007; Thornton and Sokoloff 1998).

2.2 Active Learning in the Introductory Physics Lecture

Many introductory physics students spend the majority of their time in lecture, often a large lecture section with 100 or more students. When the information is flowing predominantly in one direction—from the lecturer to the students—this is a very passive experience. Can learning in a lecture class be made more active? One research-validated solution is the use of *Interactive Lecture Demonstrations (ILDs)* (Sokoloff and Thornton 1997, 2004).

With the eight-step *ILD* strategy, students are presented simple experiments and asked to make their own individual predictions about the outcomes. They then have small group discussions. Volunteers share their small groups' predictions with the entire class. Then, the experiment is carried out, and in many cases, the same computer-based tools are used to display the results on a large screen for the entire lecture class to see. Volunteers then describe and explain the results (often prodded by the instructor's leading questions) and suggest analogous physical situations or applications. The strategy is formalized through the use of Prediction Sheets on which students record their predictions (that will be collected at the end of class, but never graded).

ILDs in 28 introductory physics topic areas are available in the book *Interactive Lecture Demonstrations* (Sokoloff and Thornton 2004). With the combination of a robust active learning strategy and computer-based tools to display results in understandable ways, *ILDs* have been demonstrated to produce conceptual learning gains comparable to those achieved with *RTP* in research studies using the *FMCE* (Sokoloff et al. 2007; Thornton and Sokoloff 1998; Sokoloff and Thornton 1997).

3 Interactive Video Analysis

The existence of a constant gravitational force close to the Earth's surface is an important feature of projectile motion that is treated in a later *RTP* laboratory (Sokoloff et al. 2011). While 2D and 3D motion sensing systems have been developed, projectile motion can be examined productively using interactive video analysis. Figure 4 shows the experiment file for this laboratory, which includes a video of a tossed tennis ball of about 30 frames length.

The student marks the position of the ball in each successive frame, while the software plots graphs of *x*-position, *y*-position, *x*-velocity and *y*-velocity versus time. The distance scale of the video can be calibrated using the given height of the pile of books shown in the video. Figure 4 shows the *y*-velocity versus time that is a straight

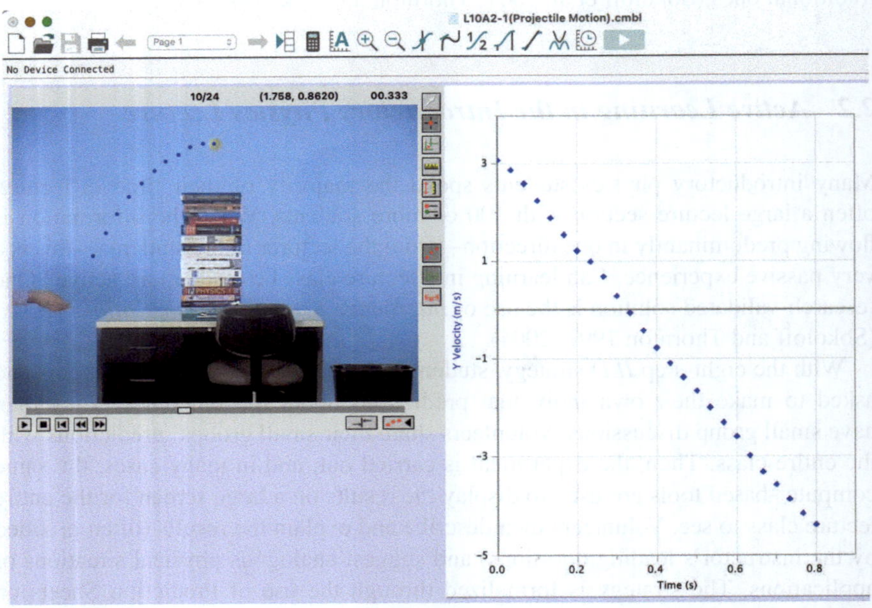

Fig. 4 Experiment file for the *RTP: Mechanics* projectile motion laboratory, showing a frame from the video of a tossed tennis ball on the left with the position of the ball on a subset of frames indicated, and a graph of *y*-velocity versus time plotted from these data points on the right. The video analysis and graphing were done using Vernier LoggerPro software

line with negative slope equal to the gravitational acceleration, as in the motion shown in Fig. 3. The students are, of course, asked probing questions about the meaning of the shape of this graph, the constancy of the x-velocity, and the independence of the x- and y-motions.

Interactive video analysis has wide application in the teaching of physics and has been incorporated into both of the major U.S. data acquisition software packages, Vernier LoggerPro and PASCO Capstone.[2] Also widely used is Tracker (https://physlets.org/tracker/), a free video analysis tool. In addition to *RTP*, video analysis is used in several *ILDs*. The LivePhoto Physics project has done extensive work in developing video analysis activities in mechanics and many other areas of physics (Physics with Video Analysis, https://www.vernier.com/product/physics-with-video-analysis/).

4 Simulations

The most common collections of physics simulations in the U.S. are PhETs (https://phet.colorado.edu/) and Physlets (https://www.compadre.org/Physlets/). Several PhETs deal with gravitational forces. *Gravity Force Lab* examines the dependence of the gravitational force on the masses and separation, i.e., Newton's Law of Universal Gravitation, while *Gravity and Orbits* looks at the orbital motion of objects under the influence of the gravitational force. While *Projectile Motion* includes displays of velocity and acceleration vectors with components, to illustrate the independence of the x and y motions, its focus is the effects of launch velocity, angle, and air resistance on the trajectory of the projectile. Figure 5 shows a screenshot from *Gravity and Orbits*. The Physlets chapter, *Gravitation*, also focuses on orbital motion and Kepler's laws.

5 Adaptations for Distance/Virtual Learning

Even before the COVID pandemic of 2020–23, there was a growing need for distance-learning materials to aid students studying at home to do meaningful laboratory work. During the pandemic, virtual learning became a necessity for many. This section describes two approaches using technology to meet these needs.

[2] See Footnote 1.

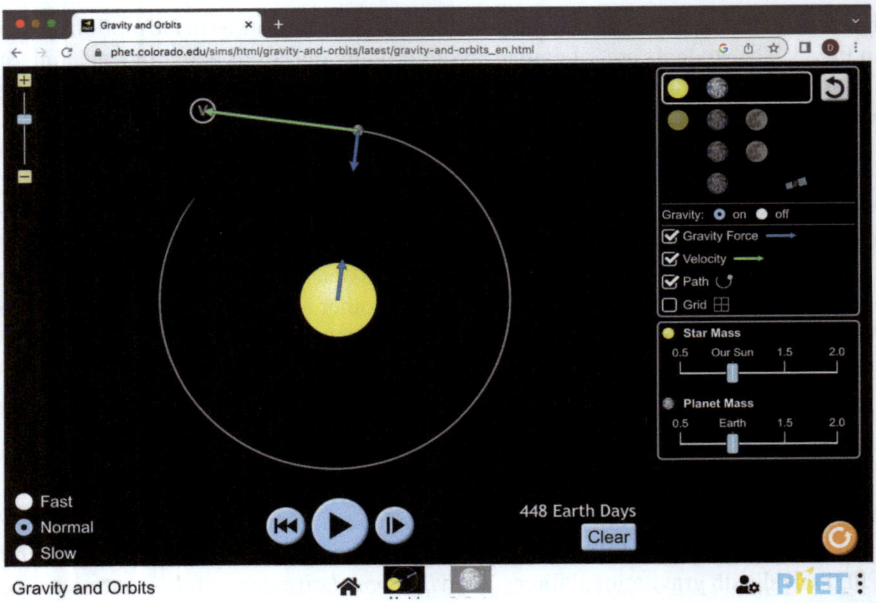

Fig. 5 Screenshot from PhET *Gravity and Orbits* showing the velocity (green) and gravitational force (blue) vectors for a planet orbiting around a star

5.1 Adaptation of RTP Mechanics for Use with the IOLab

Several years ago, researchers at the University of Illinois developed a relatively inexpensive, self-contained, computer-based data acquisition cart called IOLab (https://www.iolab.science/). The cart, pictured in Fig. 6a, rolls on three wheels and contains an optical encoder to measure displacement, from which velocity and acceleration can also be calculated and displayed. It also has a force sensor, an accelerometer, a gyroscope, and devices to measure a number of other physical quantities, e.g., light intensity. (Since that time, Vernier and PASCO have also developed their own versions of a "smart cart".[3]) Because of the growing need for distance-learning strategies, a version of *RTP: Mechanics*, adapted for the IOLab, was developed (Bodegom et al. 2019). The IOLab versions of *RTP* are available to download at: https://pages.uoregon.edu/sokoloff/IOLabInst32120.html. Figure 6b shows the velocity–time and acceleration–time graphs collected for an activity in which the IOLab was given a push up an inclined ramp and released (see Fig. 6a), moving up, reversing direction, and then moving down, all under the influence of the component of the gravitational force along the ramp. These graphs resemble those in Fig. 3b, although it is clear that the IOLab's wheels have significant friction and the force sensor significant extraneous noise.

[3] See Footnote 1.

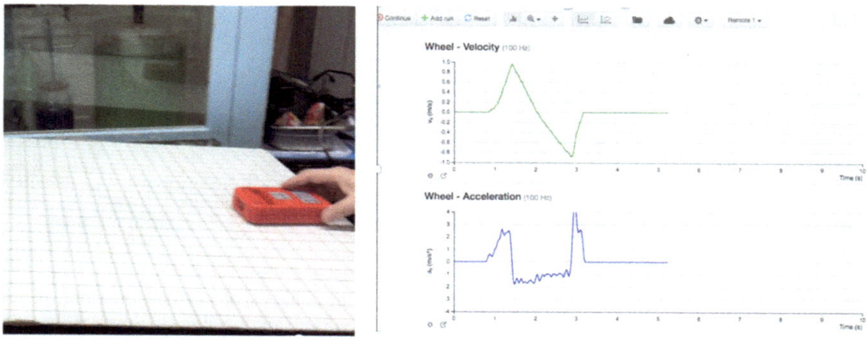

Fig. 6 **a** IOLab, self-contained computer data acquisition cart given a short push up an inclined ramp and released (on the left). **b** Screenshot of the graphs of velocity versus time and acceleration versus time collected with an optical encoder contained in the cart (on the right)

In a research study using the *FMCE*, distance learning with the IOLab version of *RTP* was assessed (https://www.iolab.science/). The conceptual learning gains by the IOLab groups—both in lab and at home—were consistently significantly better than the control groups that did traditional laboratories. Also, the students experiencing the laboratories, IOLab device, and IOLab software had generally favorable attitudes toward their experience, as indicated by their responses on end-of-term laboratory course evaluations (https://www.iolab.science/).

5.2 *Home-Adapted ILDs*

During the COVID pandemic, beginning in 2020, a project was launched to develop *Home-Adapted ILDs* that students could work on virtually, at home (Sokoloff 2020a). Student predictions on a Prediction Sheet were retained as the primary engagement mechanism, but in order to simplify implementation, student-to-student discussions were not included. After making predictions—on a downloadable Prediction Sheet (in Word) that could be sent to the instructor—student observations of the results of "demonstrations" were drawn from the wealth of available digital materials—videos, photographs, computer-based data acquisition, interactive video analysis, and simulations (Sokoloff 2020b). Figure 7 shows two examples of such observations: (a) a ball tossed straight upward and (b) the motion of a tennis ball projectile. In both cases, the graphs were collected using video analysis.

26 sets of *Home-Adapted ILDs* in different introductory physics topic areas, based on the in-class *ILDs* in the book *Interactive Lecture Demonstrations* (Sokoloff and Thornton 2004), are available for free use (Sokoloff 2020a).

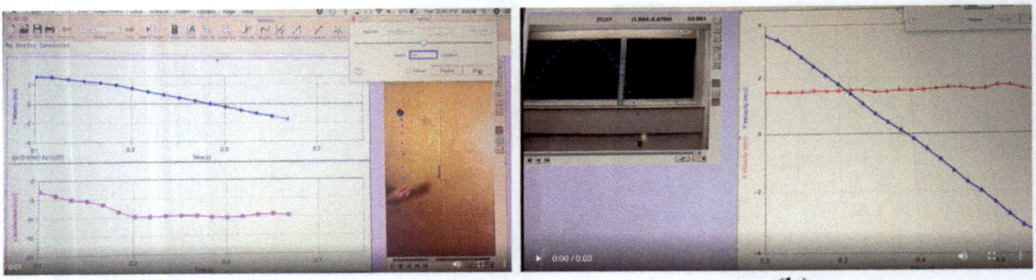

(a) (b)

Fig. 7 **a** Velocity–time and acceleration–time graphs from analysis of a video of a ball tossed upward, a frame of which is shown (on the left). **b** *x*-velocity–time and *y*-velocity–time graphs from analysis of a video of a tennis ball projectile, a frame of which is shown (on the right). The video analysis and graphing were done using Vernier LoggerPro software

6 Conclusions

In the last 40 or so years, PER has revolutionized the teaching of physics, inspiring major changes in the pedagogy used in the introductory physics course at the high school and university levels. Active learning curricula like *Real-Time Physics* and *Interactive Lecture Demonstrations* are used widely today. Curricula like these have been strongly supported by computer-based technologies for observing and simulating the physical world, making student observations more understandable. These two parallel developments have gone hand-in-hand to improve student's grasp of physics concepts, as demonstrated by other PER studies.

References

M. Bailes, B.K. Berger, P.R. Brady et al., Gravitational-wave physics and astronomy in the 2020s and 2030s. Nat. Rev. Phys. **3**, 344–366 (2021). https://doi.org/10.1038/s42254-021-00303-8

E. Bodegom, E. Erik Jensen, D.R. Sokoloff, Adapting realtime physics for distance learning with the IOLab. Phys. Teach. **57**(6), 382 (2019)

https://phet.colorado.edu/

https://physlets.org/tracker/

https://www.compadre.org/Physlets/

https://www.iolab.science/

L.C. McDermott, Oersted medal lecture 2001: physics education research—the key to student learning. Am. J. Phys. **69**, 1127 (2001). https://doi.org/10.1119/1.1389280

D.E. Meltzer, R.K. Thornton, Resource letter ALIP–1: active-learning instruction in physics. Am. J. Phys. **80**(6), 478–496 (2012)

Physics with Video Analysis, https://www.vernier.com/product/physics-with-video-analysis/

D.R. Sokoloff, *Home-Adapted Interactive Lecture Demonstrations*, self-published (2020a). https://pages.uoregon.edu/sokoloff/HomeAdaptedILDs.html

D.R. Sokoloff, Multimedia in physics education, in *Connecting Research in Physics Education with Teacher Education*, ed. by J. Guisasola, E. McLoughlin (Zenodo, 2020b), p. 3. https://doi.org/10.5281/zenodo.5792968

D.R. Sokoloff, R.K. Thornton, Using interactive lecture demonstrations to create an active learning environment. Phys. Teach. **35**(6), 340 (1997)

D.R. Sokoloff, R.K. Thornton, *Interactive Lecture Demonstrations* (Wiley, Hoboken, NJ, 2004)

D.R. Sokoloff, T. Yüksel, Physics education research and the development of active learning strategies in introductory physics, in *Handbook of Research on Physics Education, Volume 1: Learning Physics*, ed. by M.F. Taşar, P. Heron (AIP Publishing, Melville, New York, 2023), pp. 23-1–23-26

D.R. Sokoloff, R.K. Thornton, P.W. Laws, Realtime physics: active learning labs transforming the introductory laboratory. Eur. J. Phys. **28**, S83–S94 (2007)

D.R. Sokoloff, R.K. Thornton, P.W. Laws, *Realtime Physics: Active Learning Laboratories, Module 1: Mechanics*, 3rd edn. (Wiley, Hoboken, NJ, 2011)

The IOLab versions of *RTP* are available to download at: https://pages.uoregon.edu/sokoloff/IOLabInst32120.html

R.K. Thornton, D.R. Sokoloff, Assessing student learning of Newton's laws: the force and motion conceptual evaluation and the evaluation of active learning laboratory and lecture curricula. Am. J. Phys. **66**, 338–352 (1998)

S.C. White, Enrollments in high school physics in the U.S. Phys. Teach. **59**, 581 (2021a). https://doi.org/10.1119/10.0006467

S.C. White, Enrollments in introductory physics courses in physics departments. Phys. Teach. **59**, 204 (2021b). https://doi.org/10.1119/10.0003666

Future Competencies in Physics Education and Learning with Multimedia in Poland

Tomasz Greczyło

Abstract Availability of various multimedia and digital tools and procedures involving new technology for physics teaching and learning face Polish teachers with numerous educational challenges. Effective use of technology is not the only driving force for the creation of new learning paths and approaches. The key catalyst for new teaching ideas is the vision of knowledge, skills, attitudes, and values students need to transform society and shape the future for better lives. Both strongly and irreversibly drive educators to change their routines into innovative acting. The primary role in spreading new education approaches in physics teaching and learning in Poland is played by national universities which train new teachers and serve as centers of excellence. An important role is also played by regional teacher training centers and non-governmental organizations. They all conduct educational research and publish the results. They also deal with the transfer of findings and conclusions to school reality as well as offering ongoing support for in-service teachers. National annual teachers' meetings are the main grounds for discussions and the spreading of new ideas and examples on the integration of research results into everyday school practices. Noticeable sources of integration are also social networking sites and local groups operating around specific universities and institutions. National and international projects co-financed by European Union are also prominent sources of reports, new ideas, and solutions. During the realization of projects, materials are tested and made ready to implement in school environments. The chapter presents chosen results of national research and discusses the difficulties teachers encounter in implementing their conclusions and recommendations. Special attention is paid to describing the theoretical and practical background and proposing solutions supporting educators in facing obstacles and overcoming them.

Keywords Physics education · New education approaches · Teaching and learning processes challenges · ICT experience in Poland · From research to practice · Dissemination channels · Material for teachers

---

T. Greczyło (✉)
Institute of Experimental Physics, University of Wrocław, Pl. Uniwersytecki 1, 50-137 Wrocław, Poland
e-mail: tomasz.greczylo@uwr.edu.pl

M. Streit-Bianchi et al. (eds.), *New Challenges and Opportunities in Physics Education*, Challenges in Physics Education, https://doi.org/10.1007/978-3-031-37387-9_18

1 Introduction

Availability of various multimedia and digital tools and procedures involving information and communication technology (ICT) (UNESCO Institute for Statistics 2009) for physics teaching and learning face Polish teachers with various educational challenges. The most significant seems to be choosing and adjusting the way they organize teaching and learning processes in the direction that guarantees the increase of student outcomes and improves the realization of teaching objectives. In such circumstances, education researchers propose learning paths and strategies that take advantage of the new technologies and are simultaneously complying with national regulations.

The valuable use of technology is not the only driving force behind new learning paths and educational ideas. The key catalyst for new teaching and learning approaches is the vision of knowledge, skills, attitudes, and values students need to transform society and shape the future for better lives (European Commission 2019).

Both strongly and irreversibly drive educators to change their routines into innovative acting. Cognitive flexibility, digital literacy, and computational thinking along with a creative and innovative mindset are recognized as the most necessary competencies driving the change in physics teaching and learning toward multimedia-supported upshot.

2 From Key Competencies to Future Competencies

Factors differentiating teachers' competencies related to ICT and conditioning implementation of ICT methods and tools have been already investigated in Poland for decades (Baron-Polańczyk 2013). In this analysis, only the latest research conclusions and recommendations will be discussed and pictured in some detail.

The future competencies in multimedia-supported teaching and learning physics in Poland are rising from the broader list of ICT competencies that have been identified during research efforts (Wronka 2018). They result from the following actions recognized as a necessary element of educational processes conditioned using ICT:

- Making teaching methods more attractive using multimedia presentations, films, animations, photos, modeling, simulation, data acquisition, etc.
- Encouraging students to independently search for information, e.g., in a scientific database.
- Creating conditions when students are:
  - o Solving exercises, quizzes, and test to consolidate the acquired knowledge.
  - o Carrying out joint exercises and tasks with other individuals and schools, e.g., in eTwinning programs.
  - o Performing virtual experiments, which in normal school conditions are not possible or would be too dangerous or time-consuming.

  - o Performing joint group work, during which everyone is responsible for a part of the entire task (e.g., increasing the amount of available data by parallelizing work, etc).

- Making possible learning about the surrounding world with help of virtual reality technology, e.g., traveling in special glasses through the solar system, the inside of the body, and observation of the formation of substance compounds;
- Reinforcing students in the independent creation of thematically focused numerical models or programs;
- Speeding up reaching conclusions on a given topic thanks to the use of an interactive whiteboard;
- Motivating students to conduct self-preparation of demonstrations, experiments, and presentations.

The spectrum of cognitive skills related to the use of ICT is in Poland often referred to as computing thinking (Sysło 2014). Basic future competencies that are expected as crucial in multimedia-supported teaching and learning are in line with the global view of the issue (Sparks et al. 2016). In this framework, competencies are forming seven categories:

- Identification—grouping knowledge, abilities, and skills allowing to use of ICT tools to identify, define, and present information describing certain problems or topics;
- Access—grouping knowledge, abilities, and skills to find and retrieve necessary information describing certain problems or topics in the digital environment;
- Implementation—grouping knowledge, abilities, and skills to apply an existing organization scheme or classification of information the most appropriate to a certain problem or topic;
- Integration—grouping knowledge, abilities, and skills to interpret and present data—synthesizing, analyzing, and comparing data from different digital sources;
- Evaluation—grouping knowledge, abilities, and skills to evaluate data in terms of its adequacy, reliability, bias, timeliness, scope, and accuracy;
- Creation—grouping knowledge, abilities, and skills to generate information by adopting, designing, and creating information or data in the ICT environment;
- Communication—grouping knowledge, abilities, and skills to communicate, suitable to the context of the ICT environment, including directing information to the right audience and in the right place.

In Poland, the ongoing effort toward defining and supporting the growth of future competencies is nowadays considering all the aspects mentioned above and is structured by national documents developed by the Chancellery of the Prime Minister called Digital Competences Development Program 2023–2030. The document contains strategic directions for activities in digital competencies development (Republic of Poland 2023). An important section of the document is devoted to teaching and learning.

General and specific directions of activities listed in the program along with results of research work of which only chosen ones are described below were put together to form the following list of future competencies in multimedia-supported teaching and learning physics. These competencies are in three areas: cognitive, social, and technical—each area entails specific knowledge, skills, and attitudes (Włoch and Śledziewska 2019). The cognitive area includes **solving complex problems, critical thinking, creativity, and mental flexibility**. They are also colloquially known as thinking skills. The social area contains **cooperation, emotional intelligence, entrepreneurship, and people management**. They are essential in a work environment that requires human contact and teamwork. The technical area incorporates **basic digital competencies, advanced field-related digital competencies, and engineering competencies**.

3 Dissemination Channels

The main role in spreading new education approaches in physics teaching and learning in Poland is played by national universities which train new teachers and serve as centers of excellence (Bugaj and Szarucki 2019). An important role is also played by regional teacher training centers and non-governmental organizations (NGOs) (Karpińska et al. 2021). Dissemination tasks are carried out along with conducting educational research and publishing their results. Universities, training centers, and NGOs also support the transfer of findings and conclusions to school reality as well as offer ongoing support for in-service teachers.

National annual teachers meetings, e.g., Congress of Physics Teachers (Kongres Nauczycieli Fizyki), Congress of the Polish Physical Society (Zjazd Polskiego Towarzystwa Fizycznego), Congress of the Polish Association of Science Teachers (Zjazd Polskiego Stowarzyszenia Nauczycieli Przedmiotów Przyrodniczych), and Conference Computer Science at School (Konferencja Informatyka w szkole), are main grounds for discussions and spreading of new ideas and examples on the integration of research results to everyday school practices.

Noticeable sources of integration are also social networking sites (e.g., Facebook thematic groups) and local groups operating around specific universities and institutions (including district examination boards). The activities and problems being discussed there could be an interesting subject for a separate analysis. National and international projects co-financed by the EU are also noticeable sources of ideas and solutions tested and made ready to implement.

3.1 *Key Characteristics*

Disseminated materials, although they concern various aspects of teaching physics, all have common features. The materials, tools, and procedures facilitating the use of ICT by teachers and their students can be collectively characterized. Researchers and educators (Ostrowska and Sterna 2015) structure them in a way that supports:

- Focusing on the learning objectives, not on the use of ICT;
- Achieving learning goals faster than when ICT is not involved;
- Learning more (in terms of knowledge, awareness, and skills);
- Being more involved in learning and more intellectually active;
- Ensuring less time to master the same skills than when using traditional teaching methods.

Contemporary materials for multimedia-supported teaching and learning physics contain categories of teaching strategies (Pitler et al. 2015) that are supported by specific ICT tools:

- Setting goals and providing feedback;
- Motivating to make effort and rewarding achievements;
- Supporting collaborative learning;
- Formulating tips, questions, and introductory information and providing information in non-verbal form;
- Summarizing the material and taking notes, as well as doing homework and exercises;
- Recognizing similarities and differences along with formulating and checking hypotheses.

3.2 Key Sources

As the English language, mostly among teenagers and a population of young physics teachers, is no longer an impenetrable barrier in Polish education, all players in the arena are taking the advantage of platforms and tools widely available (e.g., Compadre repository, Phyphox application, Tracker program). Their use in specific educational contexts is researched abroad and is smoothly adopted into the Polish educational environment (Engstrøm et al. 2011; Sokolowska et al. 2019). Similarly, functioning tools are also developed in the national language.

Education in Poland is undergoing inevitable changes. No wonder, the world around us is constantly evolving, and the labor market and the needs of students are changing. In response to the new challenges in the year 2019, the Integrated Educational Platform (Zintegrowana Platforma Edukacyjna—ZPE) was created thanks to the realization of several projects that were co-financed by the European Union. The website contains a huge amount of multimedia-supported materials that were created with national and European Union funds under the European Social Fund (Operational Program Human Capital 2007–2013 and Operational Program Knowledge Education Development 2014–2020). ZPE was created in cooperation with leading national universities, educational institutions, and companies. The platform is a complex and comprehensive tool for teachers and students, supporting both stationary learning at schools and remote learning (Kulasa 2022). It is freely available and consists of materials considering current research conclusions which are described in the following text.

4 Students' Expectations of ICT Education

An important factor determining successful teaching is compliance of students' expectations with the methods used by their teachers daily. Knowing the expectations of primary, high school, and university students toward ICT education seems to be significant in the process of designing content for physics education and selecting multimedia.

The research conducted in the years 2018–2019 with a group of almost 700 primary, and junior high school students, high school students, and university students sheds some light on the problem of expectations in the Polish educational environment (Kuruliszwili 2020). The results of the work were elaborated in four areas:

- Concerning the declarative level of participants' digital competencies,
- Catalog of digital skills that the participants would like to gain/develop,
- Preferences concerning participation in additional activities making use of ICT,
- Satisfaction with the current education in ICT.

The findings allow concluding that only half of the students want to add, extracurricular, and improve their competencies in the field of ICT. They are most willing to improve their skills in the field of graphic design, film, and animation creation. It is important to emphasize the fact that the differences in choices between girls and boys were not significant.

The respondents more often choose the form of learning through electronic or blended training than traditional learning—only 20%. Students also declared that they learn most through self-education which is the basis of electronic learning making use of tutorials and materials available on the Internet in form of tutorials.

Satisfaction with ICT teaching declared in the study was not very high. Fewer than half of the participants were satisfied with their previous education and nearly half of them were dissatisfied with their schooling. Moreover, only 20% preferred to participate in classes using the traditional ways of teaching (Kuruliszwili 2020).

In the future, the analysis of our students' expectations should be considered an important factor in planning educational activities using multimedia in physics teaching and learning.

5 Examples of Research Work

The section presents chosen results of research—papers or reports published in national sources [e.g., "Foton" (Photon), "Edukacja" (Education), "TiK w edukacji" (ICT in education), "Fizyka w szkole" (Physics at school)] or materials being the outcomes of finalized projects and discusses the difficulties teachers encounter in implementing recommendations. The doctoral dissertations in physics education have been also included in the analysis. Some attention is paid to describing faced obstacles and proposing ways to overcome them.

Various aspects of future competencies in multimedia-supported teaching and learning physics have been investigated in Poland during the last decade. To create a reliable picture of the issue, the results of the analysis—theoretical and practical—conducted on relatively large groups of students have only been selected and presented. Each cited survey consists of information that allows looking at the list of future competencies from different perspectives.

5.1 *Current State and Needs in Teaching Physics in Polish Middle and High Schools*

Theoretical analysis of the current state and needs in teaching usually precedes actions aimed at implementing specific didactic solutions. It is particularly significant when educational activities include multimedia that support teaching and learning physics.

Crucial evidence brought the research and report which are co-results of the project titled "*SAT Project - From a screw to a satellite - good practices in teaching physics in schools*" (ERASMUS + PROJECT NO: 2015-1-PLO1-KA201-016801) realization (Szurek et al. 2017). The outcomes of teaching and learning processes judged on the results achieved by Polish students in international surveys [e.g., Programme for International Student Assessment (PISA), Trends in International Mathematics and Science Study (TIMSS)] are comparable to those of their peers from the most developed EU countries. Nevertheless, there remain many areas that require particular attention. One of these areas is the use of communication and information technology in education. Although in private life Polish students and teachers are proficient users of computers, smartphones, and applications, they do not transfer these skills to the school and professional sphere. The main reason for this is the low quantity and poor quality of ICT-related equipment in schools. It can be said that in this way a huge potential of teachers and students is wasted (Szurek et al. 2017). It mostly affects the development of future competencies in the technical area, particularly **advanced, field-related digital competencies,** and **engineering competencies**.

On the other hand, Polish teachers have developed an impressive amount of modern no technology-related teaching methods that are proving to be accessible and effective. A balance must be found between technology and non-technology involving methods of teaching and learning physics. Polish teachers should learn how to effectively use ICT so well that they can use it effortlessly. Technology should support active learning and should always be used for a specific purpose (Szurek et al. 2017).

Another important conclusion from this analysis is the need to share educational achievements between Polish teachers. During the research, it has been discovered that there is a huge number of good practices at schools of different educational levels (e.g., education projects' materials, innovative proposals, and curriculum materials) but they are not widely shared, published, or disseminated. Many teachers of physics and other science subjects have developed modern, creative, interdisciplinary, and effective methods and use them daily. The great challenge is to collect these good practices, reward their authors, and make them available for use by a wide range of teachers in a convenient way. It is also important to encourage and motivate teachers and educators to create and develop innovative educational practices. Such actions are shaping social areas' future competencies, especially **cooperation** and **entrepreneurship**.

5.2 *Remotely Controlled and Virtual Laboratories*

Remotely controlled (RCL) and virtual laboratories (VL) are recognized as valuable multimedia-supported use of ICT in physics education and are intensely and comprehensively investigated internationally (Gröber et al. 2007). This aspect of physics education is also explored in Poland (Chmielewski and Zielińska 2017), and several such experiments have been implemented in ZPE.

Nationally carried research confirmed the advantages of each type of laboratory which are namely:

- Support concept explanation—RCL and VL experiments are well anchored in a clearly defined subject matter—this supports the development of future competencies in the cognitive area specifically: **solving complex problems** and **critical thinking**.
- No place restrictions and high accessibility—as the RCL and VL are accessible via the Internet network, the only need from users concerns technical support which is in line with shaping future competencies in the technical area, mainly **basic digital competencies**.
- Medium or low costs.

Parallelly disadvantages of remotely controlled and virtual laboratories are confirmed:

- Lower flexibility;
- Indirect or no interaction with equipment;
- Results are idealized data or have limited credibility.

It is observed that the complexity, accuracy, and consequently costs of traditional laboratory equipment are increasing, so the standard approach to designing and arranging school laboratories has changed. On the other hand, the steady tendency of cost reduction and instantaneously increasing amount of knowledge, especially in the field of ICT, make remotely controlled and virtual laboratories alternative substitutes for traditional laboratories.

The time of COVID-19 brought the universality of remote communication and tamed learning methods that did not require face-to-face contact. As one of the consequences, students started looking for experiments regardless of their location and time. They started to be not only spectators of work done by others but also were willing to supervise, modify, and play with some virtual or remotely controlled experiments. Recently the support of these laboratories by artificial intelligence is being implemented. All such aspects are creating a new region for innovation that perfectly interacts with shaping competencies in the cognitive area especially **creativity** and **mental flexibility**.

5.3 Application of Computer Methods in Teaching Physics

The issue of the use of computational methods is an extremely capacious term. It includes both displaying content, which is the least interactive use, and supporting experimentation and data processing which seems to be the most sophisticated. This term is also related to the use of systems that allow simultaneous collection of data monitoring information about changes in perception of the problem during teaching and learning processes. Such interactive pilot systems' so-called clickers are used in Polish education, and their impact on teaching and learning outcomes has been researched (Binek 2018).

The results of a study conducted on a group of more than 600 high-school students show that this kind of multimedia affects teaching and learning outcomes. The possible increase in knowledge resulting from the use of an interactive system of pilots along with the subjective feelings concerning students' reception of teaching and learning supported by the system have been investigated (Binek 2018).

Checking the increase in knowledge was based on a traditional written pre-test and post-test consisting of different types of physics tasks (among others: open questions, test tasks, and problems). The judgment of student perception was based on an anonymous questionary.

The findings allow making claims that:

- Entry tests showed that every year the young people who started their education in schools participating in the research showed a comparable level of knowledge and skills but those who took part in physics lessons with pilots learned significantly more;
- Within the classes participating in the study, the knowledge of each certain class rose compared to its previous year's increase;
- Students expressed a positive attitude to learning with the use of each of the interactive forms of support;
- Students indicated that lessons on which they additionally perform experiments are rated the highest when it comes to building motivation.

Widely understood computer methods allow for an increase in the level of individualization of the teaching process. They give the lesson an interactive character, as they predestinate the formation of immediate feedback. They contribute to increasing the attractiveness of lessons, which helps students to build a positive attitude to physics. The use of computer methods in teaching physics brings measurable benefits in terms of achievements and motivation. However, in the art of teaching and learning, they should not be assigned a leading role. Although they give a lot of possibilities and have many advantages, the main role should always be assigned to students and a teacher (Binek 2018). In this approach, computer methods allow for shaping the competencies in the social areas, primarily **cooperation** and **emotional intelligence**.

5.4 Use of Smartphones and Tablets in Teaching Physics

Smartphones and tablets have opened new possibilities in multimedia-supported teaching and learning physics and are subjects of common investigations (Hochberg et al. 2018). Also in Poland, some actions are aimed at determining how smartphones and tablets could be used in teaching and indicating the educational effectiveness of those tools (Rochowicz 2019).

Nationally recognized research in this area has mainly participatory character and is conducted on small student groups and in very diverse conditions (Kawecka 2018). Certain solutions or specific educational tasks are proposed as the main result of such investigations. They all point to the advantages of smartphones and tablets which:

- Are tools that can support the work of teachers at various school subjects;
- Are relatively readily available tools;
- Make it easier to work with the flipped classroom method;
- Contain sensors that allow to perform experiments, collect data, and elaborate them;
- Positively affect self-confidence;
- Consolidate the group and facilitate the performance of tasks within it.

The articles presenting research results also point to the disadvantages of using this type of ICT tools among which there are:

- Ease of using tools to perform tasks not related to the main task;
- Treating the tool as a kind of black box whose mode of operation and functionality is not understood;
- Requires funds to buy them.

According to a survey conducted in the year 2017 by the National Library (Biblioteka Narodowa 2018), 93% of teenagers own smartphones and use them to search for information, communicate with friends, listen to music, and play digital games. The use of smartphones or tablets allows you to change the school environment into a place where modern educational aids encourage students to learn physics. In such circumstances, the future competencies which are specifically developed locate in two areas—cognitive: **mental flexibility** and technical: **basic digital competencies** and **advanced field-related digital competencies.**

5.5 Didactic Effectiveness of Multimedia-Supported Teaching and Learning

Multimedia supporting teaching and learning physics, apart from the benefits rising from the possibility of diversifying ways of reaching the interest of students, have the potential of increasing the effectiveness of educational efforts. The effectiveness of teaching is also a key element of feedback necessary for the appropriate organization of teachers' workshops in which ICT is a crucial element (Karpińska et al. 2021).

Interesting results of research on various elements that make up multimedia support for teaching and learning physics were carried out on a group of junior high school students (Kamińska 2009). The forms of teaching and learning included both basics ones such as multimedia paths, multimedia textbooks, and multimedia encyclopedias all containing films, sounds, simulations, animations, and sophisticated interventions such as models of phenomena, computer-aided measuring instruments, numerical models, and everyday objects (e.g., kitchen equipment, toys). These various elements may be properly associated to create so-called multimedia packages (Kamińska 2009).

In two of the three hypotheses, significantly higher didactic effects due to the applied multimedia forms of support were observed. The significant differences were demonstrated in terms of assimilating the content of their understanding and the ability to apply the content in new situations—problem solving. The least effective was the method in which students were free to use multimedia forms. A comprehensive analysis of results allows concluding that the best way to use different forms of multimedia support among junior high school students turned out to be the one where they are acting as one of many different teaching aids.

However, it is important to point out that the specific order of educational path should be followed. In one lesson, students should have the opportunity to conduct a real experiment, then simulate it and analyze similar phenomena occurring in computer animation. The research confirmed a very important conclusion—multimedia-supported forms of physics teaching and learning should be used in every physics lesson in junior high school, but they cannot be arbitrary. They must be strictly selected by the teacher based on current feedback (Kamińska 2009). From this perspective, the diversity of tools subordinated to a specific educational path is conducive to shaping future competencies in the cognitive area: **mental flexibility**, as well as in all aspects of the technical area.

5.6 Impact of Multimedia-Supported Teaching and Learning on Students' Motivation Toward Science

Motivation plays an important role in teaching and learning processes and could be stimulated with the use of multimedia-supported materials especially when novel and attractive topics are the subject of teaching (Cepic 2017). Between the years 2019 and

2021, the project titled *"NanoDay—a day with nanotechnology and nanoscience"* (POWR. 03.01.00-00T154/18) has been carried out to enable students of Polish primary and high school schools to broaden their knowledge and skills related to nanotechnology and nanoscience (Kaczmarek and Greczyło 2023).

The total number of research participants was more than 600 pupils: around 2/3 from primary schools and 1/3 from secondary schools. Before the start of the set of activities, they completed an entry evaluation sheet, made available online. Then, the students participated in all the forms of educational events that were richly supported by multimedia: lectures, exhibitions, and hands-on workshops. Finally, they filled in the evaluation sheet, also made it available online. Ex-ante and ex-post evaluation was the main source of quantitative and qualitative data. Analysis of the answers provided in the evaluation survey allows us to clearly state that:

- The comprehensive educational activities brought measurable benefits to the students participating in them—both the average score in terms of knowledge and the area covered by declarative questions increased;
- The introduction of up-to-date scientific findings done with the use of ICT significantly increases students' motivation toward science;
- Offering different forms of teaching and learning increases didactic effectiveness;
- Facilitating learning by doing increases accessibility;
- A selection of topics rich in practical contexts stimulates interest.

The design and implemented teaching cycle based on topics in nanoscience and nanotechnology can only be realized in the form when there is versatile multimedia support (Kaczmarek and Greczyło 2023). Unfortunately, such support requires additional funding for the equipment along with an extensive and comprehensive teacher training course. If implemented undeniably supports shaping competencies in the social area: **cooperation** and **people management** as well the technical area especially **basic digital competencies** and **engineering competencies**.

The above-presented examples are only a proprietary choice selected from many studies of various scopes and scales done recently in Poland, but they all form a valuable and coherent perspective. The overall view on different areas of work facilitating multimedia-supported teaching and learning in physics clearly shows that the list of future competencies presented above although done on rather a broad level of detail is valuable and helpful. On one side, it supports the development of educational materials and procedures but still leaves space for creativity and novel approaches.

6 Perspectives

The future competencies can only be effectively shaped within a broad and flexible educational ecosystem focused on lifelong learning.

Conclusions from the quantitative and qualitative study conducted in the years 2018–2019 in seven main Polish cities on a group of more than 1000 responders and participants of Design Thinking workshops show that students and graduates of

Polish universities are aware of the importance of future competencies in the context of legal functioning on the changing labor market (Włoch and Śledziewska 2019). They also claimed that the Polish educational system in its current structure and with the present goal is not conducive to the development of future competencies. Therefore, additional actions should be taken to involve the university in cooperation with other entities of the emerging educational ecosystem, especially with employers, public institutions, and non-governmental organizations. This could be done by including them in the process of creating the content, form, and purpose of the university's didactic mission (Poziomek 2022).

The actions undertaken nationally on the way toward future competencies in multimedia-supported teaching and learning physics are in line with global efforts, e.g., OECD projects Education 2030—The future we want (OECD 2019).

It should also be emphasized that the use of information and communication technologies in multimedia-supported teaching and learning process requires teachers to spend a significant amount of time, for example, on familiarizing themselves with the software and preparing classes. Therefore, a very important action that should be taken by school authorities and institutions supporting their work is comprehensively motivating teachers to use ICT. This support should go hand in hand with the creation of conditions in schools for the preparation of materials and tools by teachers, as well as their collection and duplication, carried out, for example, by providing teachers with additional workspace (back room and subject room), technical means (science laboratories, teaching aids), and technical support in the person of a supporting teacher (Greczyło 2017).

7 Recommendations

All actions supporting the development of future competencies in media-supported teaching and learning physics should begin with changing the knowledge transfer model based on the use of ICT, moving toward greater activation of students and their motivation. The experience of many schools in Poland shows that it is not enough to equip schools with innovative ICT solutions, but it is also necessary to parallelly change the model of conducting classes. Our effort as educators should lead to a situation where ICT tools are used not only by teachers, but above all by the student. This requires a different than traditional arrangement of educational activities during which the student is playing a central role and ICT is considered a valuable set of instruments (Dębski and Bigaj 2019). The coherent use of ICT is possible only when future competencies are possessed.

Due to the rapid development of new technologies especially in the educational sphere, there is an urgent need for regular supplementation of training courses for teachers, allowing them to prepare to use ICT tools during lessons and as educational tools also at home (Bochenek and Lange 2019). Important actions aligned with this recommendation are promoting good practices and grassroots-free initiatives supporting the use of ICT in educational contexts. There are schools in Poland [e.g., working in the net of "Szkoła ćwiczeń" (Exercise school) (ORE 2018)] that can be serving as examples of responsible use of ICT in education. They can be a role model for others. These institutions should have the privilege of presenting their experiences in a form of demonstration lessons or study visits and sharing their ready-to-use example during training and regional conferences.

All kinds of interactions showing the possibility of using new technologies, applications, and procedures (including those used by students daily) to implement the core curriculum should be supported both by national and local government institutions, as well as units supporting the school, in particular centers of excellent teachers. The implementation of this recommendation can contribute to increasing the attractiveness of classes conducted by teachers, as well as reducing the distance between students and teachers in terms of the ability to use new technologies (Plebańska et al. 2017).

Above all important insides into future competencies, we should pay special attention to supporting the culture of sustainable development in all areas of human activity.

References

E. Baron-Polańczyk, Factors differentiating teachers' information competencies and their implementation of ICT methods and tools (research report). Eduakcja Ustawiczna Dorosłych **43**, 129–145 (2013)

S. Binek, *Zastosowanie metod komputerowych w nauczaniu fizyki. Praca doktorska* (Uniwersytet Śląsk, Katowice, 2018)

M. Bochenek, R. Lange, *Nastolatki 3.0 Raport z ogólnopolskiego badania uczniów* (NASK Państwowy Instytut Badawczy, Warszawa, 2019)

J. Bugaj, M. Szarucki, Doskonałość naukowa oraz doskonałość dydaktyczna jako kluczowe kompetencje uczelni publicznych w Polsce. *Przegląd Organizacji* **2**(949), 7–14 (2019). https://doi.org/10.33141/po.2019.02.01

M. Cepic, Introducing current research results into education: experiences from the case of liquid crystals, in *Key Competences in Physics Teaching and Learning, Springer Proceedings in Physics*, vol. 190, eds. by T. Greczyło, E. Dębowska (Springer, New York, 2017), pp. 41–54

T. Chmielewski, K. Zielińska, Survey of remotely controlled laboratories for research and education. Appl. Comput. Sci. **13**(1), 85–96 (2017)

M. Dębski, M. Bigaj, Młodzi cyfrowi. Nowe technologie. Relacje. Dobrostan (Gdańskie Wydawnictwo Psychologiczne, Gdynia, 2019)

V. Engstrøm, T. Greczyło et al., *MOSEM^2 Teacher Guide, Modelling and Data Acquisition for Continuing Vocational Training of Upper Secondary School Physics Teachers in Pupil-Active Learning of Superconductivity and Electromagnetism Based on Minds-on Simple Experiments* (Simplicatus Research and Development AS, Lillestrøm, 2011)

European Commission, *Directorate-General for Education, Youth, Sport and Culture, Key Competences for Lifelong Learning* (Publications Office, 2019), https://doi.org/10.2766/569540. Accessed 7 March 2023

T. Greczyło, Zestawy materiałów dla nauczycieli szkół ćwiczeń: przyroda, Zestaw 5 Zeszyt 3: Wykorzystanie technologii informacyjno-komunikacyjnych w edukacji fizycznej. *Ośrodek Rozwoju Edukacji* (2017)

S. Gröber et al., Experimenting from a distance-remotely controlled laboratory (RCL). Eur. J. Phys. **28**, S127 (2007)

K. Hochberg, J. Kuhn, A. Müller, Using smartphones as experimental tools: effects on interest, curiosity, and learning in physics education. J. Sci. Educ. Technol. **27**, 385–403 (2018). https://doi.org/10.1007/s10956-018-9731-7

II C, *Stan czytelnictwa w Polsce w roku 2017, Biblioteka Narodowa* (2018). https://www.bn.org.pl/download/document/1535441771.pdf. Accessed 14 March 2023

M. Kaczmarek, T. Greczyło, Kształtowanie zainteresowania przedmiotami ścisłymi z wykorzystaniem o zagadnień z obszaru nanonauki i nanotechnologii, paper sent to quarterly „Eduakcja" (2023)

A. Kamińska, Efektywność dydaktyczna multimedialnych form nauczania fizyki. Praca doktorska. (Uniwersytet Mikołaja Kopernika, Toruń, 2009)

A. Karpińska, M. Zińczuk, K. Kowalczuk, *Nauczyciel we współczesnej rzeczywistości edukacyjne* (Wydawnictwo Uniwersytetu w Białymstoku, Białystok, 2021)

E. Kawecka, Smartfony i tablety na lekcjach fizyki. Meritum **4**, 30–37 (2018)

T. Kulasa, Rola Zintegrowanej Platformy Edukacyjnej MEiN w szkole przyszłości (od zpe.gov.pl do edukacja.gov.pl), in *Materials from 18th Conference on Informatics in Education, 28–30 June 2022, Toruń* (2022)

S. Kuruliszwili, The expectations of primary, high school and university students towards ICT education. Edukacja Ustawiczna Dorosłych **3**, 43–56 (2020)

OECD, *OECD Future of Education and Skills 2030, Conceptual learning framework, Concept Note* (2019)

M. Ostrowska, D. Sterna, Technologie informacyjno-komunikacyjne na lekcjach. *Przykładowe konspekty i polecane praktyki, Warszawa* (2015)

H. Pitler, E.R. Hubbel, M. Kuhn, *Efektywne wykorzystanie nowych technologii na lekcjach, Warszawa: Ośrodek rozwoju edukacji w Warszawie* (Centrum Edukacji Obywatelskiej, 2015)

M. Plebańska, M. Sieńczewska, M. Szyller, *Polska szkoła w dobie cyfryzacji. Diagnoza 2017. Podsumowanie wyników badania zespołu badawczego Wydziału Pedagogicznego Uniwersytetu Warszawskiego i PCG Edukacja, Warszawa* (2017)

Poradnik. *Jak skutecznie wdrożać „Model szkoły ćwiczeń"* (Warszawa, Ośrodek Rozwoju Edukacji, 2018)

U. Poziomek, Science education in Poland: diagnosis and challenges. Studia BAS **2**(70), 119–146 (2022)

Republic of Poland, *Resolution on the Establishment of a Government Program Called "Programme for the Development of Digital Competences"* (2023). https://www.gov.pl/web/premier/uchwala-w-sprawie-ustanowienia-programu-rzadowego-pod-nazwa-program-rozwoju-kompetencji-cyfrowych. Accessed 11 March 2023

K. Rochowicz, Smartfon jako pomoc w nauczaniu fizyki. Fizyka w Szkole **5**, 36–37 (2019)

D. Sokolowska et al., Physlety Dla Polskich Nauczycieli. Foton **145**, 23–38 (2019)

J. Sparks, I. Katz, P. Beile, *Assessing Digital Information Literacy in Higher Education: A Review of Existing Frameworks and Assessments With Recommendations for Next-Generation Assessment: Assessing Digital Information Literacy in Higher Education*, 2016, ETS Research Report Series (2016)

M.M. Sysło, Computational thinking. A new perspective on digital competence. *Inform. Technol. Educ.* **11**, 15–32 (2014)

K. Szurek, K. Wilk, R. Gabryszewski, Analiza stanu obecnego i potrzeb w nauczaniu fizyki w polskich gimnazjach i szkołach ponadgimnazjalnych. *SAT Project* (2017)

UNESCO Institute for Statistics, *Guide to Measuring Information and Communication Technologies (ICT) in Education* (UNESCO Institute for Statistics, 2009). https://unesdoc.unesco.org/ark:/48223/pf0000186547.locale=en

R. Włoch, K. Śledziewska, *Kompetencje przyszłości. Jak je kształtować w elastycznym ekosystemie edukacyjnym?* (DELab UW, Warszawa, 2019)

S. Wronka, Modern technologies in teaching science, in *Competences of the Future*, ed. by S.M. Kwiatkowski (Fundacja Rozwoju Systemu Edukacji, Warszawa, 2018)

Digitalization in Early School Education in North Macedonia

Marina Vasileva Connell and Darko Taleski

Abstract This chapter reports on the various initiatives, projects and challenges faced by the introduction of digitalization in the classroom in the Republic of North Macedonia. The Ministry of Education and Science with the help of National experts in setting up the Program for development of ICT in education gave recommendations for implementing changes in the educational process and suggested the development of strategies for the sustainability of such essential transformation. The good reactions to, and active participation of, young students and teachers to innovative projects supported by many institutions and external financial help allowed the easy implementation of ICT also in schools in rural areas. These led to several prizes being awarded by Microsoft in the international context to digital programs developed and used by students in primary schools. The further challenges represented by the distance learning imposed by the recent pandemic were linked to the lack of available digital platform and have been overcome thanks to the clear guidelines received from the Ministry of Education, the creation of video and the support from private initiatives and the dedication of Macedonian experts from abroad the country. The importance of Friends of Education meetings to informing, and doing workshops for, teachers facilitating the implementation of STEM digital innovative programs in the classroom is also outlined.

Keywords Application of ICT · Digital tools and digital projects · STEM · Robotic · Distance learning

M. V. Connell (✉)
I Am Learner; NGO Friends of Education, Edinburgh EH47 9JB, Scotland, UK
e-mail: marinaconnell@iamlearner.net

M. V. Connell · D. Taleski
Secondary School Teacher, NGO Friends of Education, Prilep, Republic of Macedonia
e-mail: dt155278@schools.mk

M. Streit-Bianchi et al. (eds.), *New Challenges and Opportunities in Physics Education*, Challenges in Physics Education, https://doi.org/10.1007/978-3-031-37387-9_19

1 Introduction

The Republic of North Macedonia, after independence (1991), faced new social and economic relations that imposed the need to take note of modern European and world global trends and technologies and to incorporate them as much as possible into educational activity. Hence, schools should supply knowledge commensurate with the environment in which the students are expected to live and work, which implies implementing modern technologies and giving more weight to STEM contents in teaching.

The intensive and rapid development of information and communication technologies (ICT) leads to an inevitable massification of their use in daily life, to the diversity in their use, need of readiness for constant and rapid innovation for prompt and adequate reaction to changes in societies.

It is worth to remember that North Macedonia is a multiethnic and multicultural country. The population by end 2022 was 2.08 million persons. Primary and lower secondary school students in 2021/2022 were 186,649 and in the upper secondary school 71,018 (State Statistical Office, Republic of North Macedonia). The teaching is in 4 languages: Macedonian (official language of the country), Albanian, Turkish and Serbian. The country has 21 higher education institutions and 7 Universities.

This chapter attempt to explain how the Republic of North Macedonia, since its independence, has sought to adapt its education system not only to economic and social changes but also how reacted to digital challenges.

2 Events Related to the Computerization and Application of Digital Technologies for Education in the Republic of North Macedonia

Computerization and digitization in the Republic of North Macedonia developed intensively after 2002, when a Chinese donation was received which enabled a certain degree of mainstreaming of ICT in primary and secondary schools. This donation encouraged state institutions to start thinking about the need for sustained and adequate systemic reforms and transformation of education, to respond to the demands of competitiveness, competence, participation and connectivity, imposed by global and European societies, and the modernization of their educational systems. In that sense, the changes in education that referred to the intensive introduction of ICT in the educational process required the re-definition and strategic restructuring of the educational system in the Republic of North Macedonia. They were intended to lead to social and educational development, and at the same time encourage the creation and incorporation of new programs, as well as initiatives for equipping schools, and above all for the effective and innovative use of ICT in the educational system. As a result, in 2005, the National Program for the Development of Education 2005–2015, the Draft Program for the Development of ICT in Education (2005–2015) was prepared by the relevant institutions and working groups of experts and began to be implemented.

According to the report of the Foundation for Sustainable Information Solutions, Metamorphosis, issued in June 2010, on the topic "*The use of computers in the educational system in the Republic of North Macedonia*", the chronology of the events has been the following:

- 1986
  - Informatics as a subject was introduced in secondary schools.
- 2002
  - The government of the Republic of Macedonia received a donation from the Chinese government of 6000 computers for primary and secondary schools.
  - Microsoft supplied 6000 licenses for using Windows on school computers.
- 2003
  - The USAID e-School project, at the initiative of the Government of the Republic of Macedonia, deployed 2000 computers in 91 secondary schools from the Chinese donation.
  - Immediately after that, initial training for using them began.
- 2004
  - The Ministry of Education and a group of experts prepared the National Education Development Program 2005–2015.
  - The National Project Modernization of Education began to be implemented.
  - By the end of the 2004, all secondary schools received a computer laboratory with 20 computers through the e-School project.
- 2005
  - The Commission for Information Society prepared a National Policy for Information Society.
  - In April, the Information Society Commission and a working group of experts prepared the National Strategy for the Development of the Information Society.
  - In July, the Draft Program for the Development of ICT in Education (2005–2015) prepared by the Ministry of Education and a working group of experts was published.
  - FIOOM provided computer laboratories with an additional 225 computers in 45 primary and 180 computers in 18 secondary schools
  - Through the "Macedonia Connects" project, 460 primary and secondary schools were provided with internet until 2008 (WLAN in 360 primary schools and LAN in 100 secondary schools) (Metamorphosis Foundation 2010).

3 Application of ICT in Education in the Republic of North Macedonia

In 2004, the National Program for Development of Education was prepared by the Ministry of Education and Science and the National experts group. A part of this program was the Program for development of ICT in education which put more stress on implementing digital technology in education, giving recommendations for changes in the educational process as well as the development of strategies for sustainability of transformation in the educational system. The vision of this program was digital literacy for all teachers and implementation of ICT in education, digital literacy of students starting from 4th grade, networked schools with fast internet connections and multimedia computers, support services for the development of educational multimedia content in mother tongue. It also prepared an action plan for digital literacy in the primary, secondary and high schools following the National program for the development of education. Before 2002, the use of ICT in teaching was carried out in the context of the school subject Informatics on the basis of several computers per school. Informatics was implemented for the first time in high schools in 1986. By 2002, the intensive computerization of education began at the initiative of the then president, Boris Trajkovski, with the provided donation of 6000 computers. In addition to that donation, Microsoft also donated 6000 licenses for the use of the Windows operating system.

However, the provision of ICT equipment was not sufficient for the final use in primary and secondary schools; therefore, the United States Agency for International Development (USAID) got involved as a strategic partner in the realization of the effective integration of ICT in education. From that year, USAID began to implement programs such as "*e-School*" (2003–2008), "*Macedonia Connects*" (2004–2007), and later the "*Primary Education Project*" (2006–2011) in order to provide continuous support for computerization in education. Over these 8 years, these programs enabled the provision and installation of ICT equipment, software and internet infrastructure, as well as the provision of training for teachers on the integration of ICT in teaching.

In September 2003, the "*e-School*" project deployed 2000 donated computers in 91 secondary schools and began implementing teacher training. At the start, a computer laboratory was piloted in each of three high schools, and by the end of 2004, all high schools received a computer laboratory with 20 computers.

In 2005, an additional 3000 donated computers were installed in 360 central and remote district primary schools. Computer laboratories in primary schools were provided with 5–20 computers depending on the number of students (Draft Program for the Development of ICT in Education, 2005) (Hathaway 2005). In 2007, an additional 1500 computers arrived in primary schools. During that period, the Open Society Foundation Macedonia (FIOOM) and USAID additionally donated another 400 computers (180 in 18 secondary schools and 225 in 45 primary schools) (Draft program for the development of ICT in education, 2005). Also at the beginning of the school year 2005–2006, computers in both primary and secondary schools were connected to the internet through the *"Macedonia Connects"* project, implemented by USAID, the Government of the Republic of North Macedonia, the Education Development Agency (AED) and the internet provider ONNET. In total, 460 primary and secondary schools were provided with internet for three years by this means.

Within the *"e-School"* and *"Macedonia Connects"* projects, a series of trainings was provided for primary and secondary school teachers on basic ICT skills, as well as trainings for the integration of specific software solutions in interactive teaching Windows tools, Toolkid (ToolKid, etc.), web site development training (Mambo) and the use of the internet for search, collaboration and communication.

The government project *"Computer for every child"* began to be implemented at the end of 2006 in 366 primary and 93 secondary public schools in the Republic of North Macedonia, and was based on the National Program for the Development of Education (2005–2015). Computers were planned for every student in primary and secondary schools.

This project took place in several, interdependent segments:

- Procurement, installation and maintenance of equipment;
- Creation of local networks and internet connection;
- Training of teachers on the use of equipment, software tools and e-content;
- Development of learning management environment (e-obrazovanie.mk);
- Development of electronic contents and electronic textbooks (skoool.mk, e-ucebnici.mk).

The implementation of all segments was started, and depending on the project plans, each of them had a different level of progress.

Edubuntu, the Linux distribution, was chosen as the operating system for the purchased equipment. Along with the operating system, the Ministry of Information Society identified over 120 educational tools from the subjects: informatics, mathematics, physics, chemistry, geography, musical art and Latin language. From them, the experts from the Education Development Bureau selected 43 that best correspond to the educational goals in question. The applications were localized in Macedonian and Albanian language (Ministry of education and science of the Republic of North Macedonia 2021).

In 2021, the Ministry of Education and Science of the Republic of North Macedonia adopted a new Concept for Primary Education. The new Concept describes the need for further digital integration in primary education in the Republic of North Macedonia, emphasizing the need for a combined approach to learning through digital platforms and distance learning, as well as the use of digital technology to support the inclusion of students with disabilities in regular teaching.

Within this concept, national standards are defined, including the competencies that students should acquire by the end of primary education, across eight areas. The fourth area is digital literacy, which refers to using ICT as a source of information, using it skillfully and effectively for problem solving, sharing ideas, communication and collaboration within the school and outside it, creating digital content, as well as the ethical and safe use of digital technology in everyday life. Digital competencies are intended to be acquired through their inclusion in several compulsory and optional teaching subjects. In the content of the curriculum for each subject, the national standards are implemented through the competencies that students should acquire, which are related to the specific program, as well as the transversal competencies that should be included in most curricula, one of them being digital competencies, i.e., digital literacy (Ministry of education and science of the Republic of North Macedonia 2021).

4 International Successes of the Teachers from the Republic of North Macedonia in the Innovative Use of Digital Technologies in Education

In the process of digitalization of education in North Macedonia, as well as popularizing the use of digital technologies in education, the Microsoft team in North Macedonia made its contribution in the period from 2009 to 2017 through the Partners in Learning program. Within the framework of this program, several events were organized with the aim of encouraging teachers to use and integrate digital tools in the teaching of various teaching subjects.

Some of the activities that were carried out during this period are the following:

- Partners in Learning monthly newsletter for teachers to learn about the latest Microsoft tools that have applications in education
- Microsoft Innovative Educators Forum—a competition for teachers for an innovative way of using digital tools in education. The first forum organized in December 2009 supported by USAID
- The forum for children's safety on the internet
- Competition in entrepreneurship for secondary schools in cooperation with the British Council
- Microsoft Hour of Code week—the week of coding is an international event promoting programming for students as a way to solve problems, develop logical and critical thinking and creativity (Blog post 2017).

Teachers from Macedonia have achieved distinguished results from participating in the Microsoft Forum of Innovative Teachers, for several years in a row. The **Grandma's Games** project was winner out of 80 other projects from Europe at the European Forum of Innovative Educators in Moscow in 2011, thus taking it forward to the world Microsoft Innovative Educators forum 2011 in Washington DC where they were 1st Runner Up in the category Educators' Choice.

The Grandma's Games project was implemented as an interdisciplinary approach to learning in which students, through physical activities involving playing, interacting, competing, creating digital learning resources, and using the latest technology for programming, were able to develop key twenty-first century skills through all the project cycles. The project emerged as a multicultural bridge between students and teachers of different ethnic backgrounds as well as from different geographic regions across the country and beyond (Vasileva 2011a). As the founder and author of this project, Dr. Marina Vasileva Connell was given a series of awards, first at the national level, then onto the Microsoft Education Olympiad for Innovative Teachers where she achieved first place at the European level in Moscow, and finally when she won at the global level in Washington DC.

The project activities were realized within a curriculum that applied interactive playing and gaming methods, acquiring knowledge from outside the classroom walls, from the school playground, with open access that allowed the implementation of various workshops using different devices and technologies, including a pioneering application of robotics. Regular lectures were conducted according to a predetermined program, for which, the author of the project developed a special manual for teachers (Vasileva 2011b).

Different methods were implemented to achieve the students' learning objectives including, the "web-quest" method, "learning by doing," critical thinking, role playing, use of community resources (e.g., organizing a visit to school by grandparents and creating events with traditional grandma food recipes in different communities), learning from each other (video conference classes between schools from different regions and countries), Grandma's Games Idol (competition for the best lyric, best music for the anthem and best singer), PRES method (Point, Reason, Example, Summarize), enquiry based learning, etc. At the same time, the project endeavored to show maturity in critical cultural heritage and traditional values connected to life and technology, children's rights and democratic values.

The key benefit emerging from the project was a recognized need to find ways to deal with the negative impact of the increased time spent playing video games on the physical and mental development of children, while at the same time discovering its potential as a tool for digitalization of the cultural heritage.

4.1 Grandma's Games Activities

The project activities, in line with the curriculum, enabled students to develop an interest in the use of both their native language and English as essential means of communication in the world, to learn the power of public oral skills and modern presentational skills using ICT, to come to know old traditions as a characteristic feature of their national identity. While collecting the Grandma's Games, playing the role of "the journalist," they created survey questionnaires and developed listening, writing, communication and comprehension skills (Malinovski et al. 2014). They enhanced their abilities in oral expression while explaining the games to their peers via video conference classes involving more than 150 schools (Malinovski et al. 2015). In addition, students learned to apply the standard linguistics norms (phonetic, morphological, syntactic, lexical and spelling in their mother language and in the English language).

In art classes, students developed their sense of color, line, forms in open space; they broadened their art literacy through the illustration of the games; they developed

Fig. 1 Students' activities

their creative abilities and acquired experience for coherence in composition; they gained new technology skills such as drawing in MS Paint (Fig. 1).

Mathematical concepts were adjusted according to develop cognitive skills and abilities of the students. For example, the game "Lady (Dama)" is correlated with other subjects and develops perception ability (particularly visual and tactile), orientation in space and time, process thinking, ability to analyze and synthesize abstractions and generalization in deciding next steps in the game or solving specific problems.

Grandma's Idol was directly linked to Music Education. In terms of the curriculum, the games offered an opportunity to encourage intuitive positive behavior, empathy and physical development, enhance self-esteem, and develop perseverance/responsibility, accuracy and good work habits. Students were collected, classified, compared, presented and interpreted data.

The logical thinking involved in games strengthened behavior as well as competitive skills among students (Vasileva et al. 2014).

The collaboration took place in the following key directions:

- Student–student of different age (mentors), regions and countries through MSN, e-mail, video conference links;
- Teacher–teacher from different regions and countries to exchange project ideas and develop activities;
- Students–other people from different professions to achieve joint products.

Through the games, students were gradually introduced to problematic situations, encouraging them to develop their critical thinking over a number of steps:

1. First, to allow them to consider how they can influence their peers with an idea to reduce the time spent in front of an IT device, to develop creative research and to conduct a survey to collect the most popular Grandma's Games. This also allowed them to develop further thinking on not only how to play and have a fun but also on how to learn and develop new skills (Stojanovska and Vasilev 2014).
2. Secondly, students grasped their own learning while playing Grandma's Games, establishing logical links between the games and making generalized conclusions about the set tasks as a challenge.
3. Thirdly, they were able to interpret the findings, self-evaluate with justification both of their evaluation and of their ability to ask critical questions and lead productive discussions.

The project introduced new ways of learning which focused on "learning by doing" and all participants of the project were driven by the motto *"Let's play games, let's be friends, let's learn together!"*.

4.2 The Technology Context

A mixture of MS applications and other tools were used: designing and reporting research/interview survey (MS Word and Excel); illustration of Grandma's Games drawing in MS Paint; writing the games rules in MS Word; presenting the games using MS Power Point; exchange of collected information in between schools via Windows Live MSN (@live account and videoconference); programming a robot Pippin to draw shapes, such as rectangles and squares and to recreate the design of the old game known as Hopscotch; video recording and processing the games played in the school yard with Movie Maker; publishing the calendar and poster with 12 Grandma's Games using MS Publisher; presenting the games pictures using Auto Collage and Photo Synth; and a final Grandma's Games portfolio published in MS OneNote. Some of the technologies such as Movie Maker, Skype, MS OneNote and Publisher were introduced for the first time to this primary and lower secondary age group of students. This led to organizing student mentorship groups who were in charge of performing various sets of activities based on these applications and then transferring the knowledge to other students. For the first time, MS Kinect was introduced to play the game "Zavor," where children not only had fun playing one of the Grandma's Games in Kinect (X-Box) but also learned how to convert measurements in innovative ways. Students achieved curricular learning objectives using ICT in different ways, for example, online MSN and Skype video conference links (Vasileva 2010). Students wrote the game rules and survey reports about the games on a computer, and they then created Bing maps about the geographical origin of each game. They drew their own designs for games, inserted the data in MS Excel tables, developed surveys in MS Word, recorded videos with a Flip camera and then published each game. They created presentations in Power Point, took pictures of activities and made an auto collage of images. Using MS Songsmith, students wrote a huge number of poems and lyrics for the Grandma's Song competition, composed an anthem for Grandma's Games, and then selected the best singer to perform the anthem. The impact of the games in comparison to the usual mode of accessing technology enabled more ecological and healthier patterns to the child's learning and their physical development and well-being.

4.3 Stop-Motion Animation

In 2012, the project "*Fun education stop-motion animation*" as a state representative at the European Forum of Innovative Teachers in Lisbon, Portugal, was winner in the Teachers' Choice category. By winning this competition, the project was taken forward to participate in the Global Forum of Innovative Teachers 2012, which was held in Prague, Czech Republic, where it was 1st runner up in the Teachers' Choice category.

The leader of this project was the co-author of this chapter, Darko Taleski, an art teacher. He, together with Sofia Grabuloska, also an art teacher, Zorica Trajanoska and Dragica Zdraveska both English language teachers, and Mirjana Trompeska, a mathematics teacher, conceived and developed the project. In the second phase of the project, representatives from the Faculty of Information Sciences and Computer Engineering Bojan Kostadinov and Nikola Mijanovic were also involved. As result, more than 45 animations for 12 teaching subjects from students aged 8 to 14, in total more than 500 students and 30 teachers from North Macedonia were involved in the realization of this project.

The challenge was making stop-motion animations from all types of school subjects in primary and secondary education spanning from STEM to humanities. This project proved that stop-motion animation is adaptive to all school subjects, age and levels.

The aims and learning outcomes of the projects can be summarized as follows:

- Use of stop-motion animation in education.
- Wide-ranging new working classroom method capable of reaching students with different motivations and interesting them in science subject and technology, economically affordable anywhere independently of resources.
- An opportunity for cross-disciplinary teaching and learning, adaptive to more than one style of learning and teaching outcome (like one video = multipurpose storytelling, quizzes, interactive lesson, feedback on learning outcomes, competition).
- Develop critical thinking and skills in practical learning, allow the students to play a very active role in the two levels of the learning process: the content and the technology (Fig. 2).

In addition, the students found making stop-motion animation very stimulating as they learned the subject and developed their animation story. They first learned to do photography prior to computer processing which added to their interest and

Fig. 2 Students' activities about healthy food, "Fun education, stop-motion animation" project

motivation and encouraged them to modify, change and adapt the subject to obtain the wanted result. The process of creating stop-motion animations helped the students to visualize and make models for certain phenomena in the field of sciences, making them more understandable. Stop-motion animation was introduced in education for the first time in North Macedonian.

4.4 This is My Voice

Two years later in 2014 at the next Microsoft Teachers Experts Forum held in Barcelona, Spain the project "**This is my voice**", conceived and lead by Darko Taleski and Sofija Grabuloska, was the winner in the category Collaboration.

Teachers and students between the ages of 13 and 15, from 2 rural primary schools in 5 different villages, participated and gave voice to the project. They used digital technology producing powerful videos, animations, posters, and photo stories that provided the young people the possibility to explore and express their opinion on issues that their peers are facing in the rural community. Two distinct topics *Premature marriage*, often used to get away from family, and *Cultural heritage* and one joint media project on *Bullying* were worked on by the two schools. They created 16 targeted media products, including 3 animations, 3 videos, 4 photo stories and 6 posters providing their vital viewpoint on the issues that their peers are facing. Student art exhibits were organized in each school, and in a final exhibition the students spoke out for changes in their own and other people's lives (Taleski 2013).

5 Digitalization in the Time of Pandemic

The changes that occurred with the introduction of quarantine drastically changed everyday life and had a great impact on the education system, one of the most affected areas. As a consequence of this, the need to use digital tools for teaching by teachers in North Macedonia was rendered inevitable. Teaching staff were forced into activities for which most of them were not prepared. The closure of schools and the transition to online teaching in the Republic of North Macedonia was introduced at the beginning of March 2020 by a decision of the Government. A basic problem that appeared in the transition to online teaching was the lack of an online learning platform for primary and lower secondary school in Macedonian that teachers could use. At the beginning, according to the guidelines of the Ministry of Education, each school was allowed to choose a free platform that they could use to connect with students, share digital content, and create electronic tests. Google Classroom, Edmodo and social networks were the first platforms that were most accessible to teachers. They also faced a lack of digital educational content for most of school subjects in the mother tongue. This deficiency was more noticeable among teachers who teach in a language other than the Macedonian language (Metamorphosis Foundation 2020).

The project "Setting up an environment that enables the improvement of the quality of teaching and learning through co-creation and innovation," was started in March 2020 during the pandemic with the Eduino platform. The Eduino portal is implemented by SmartApp—the Laboratory for Social Innovation, funded by the Government of the United Kingdom through the British Embassy in Skopje with the support of the Education Development Bureau as an institution responsible for pre-school, primary and secondary education, the Ministry of Education and science, the Ministry of Labor and Social Policy of the Republic of North Macedonia, and UNICEF (InnovationLab 2020a). The project aimed to encourage innovation in education and create an EDUINO community, a group for professional cooperation and sharing, with the aim of collaboratively creating resources and strengthening the skills of the teaching staff. Through this project, teachers had the opportunity to be actively involved in the creation of digital teaching content for various teaching subjects by recording video lessons that were shared on this platform. The portal was enriched with a large number of resources, i.e., video lessons for teachers and students who had the opportunity to apply them during online teaching. Within the framework of the project, webinars were organized for teachers on various topics aimed at helping with the implementation of online teaching. By the end of 2020, 2,925 videos for primary education were created, 13 webinars were held, attended live by a total of 1,700 teachers, and a total of 155 published educational activities (InnovationLab 2020b).

The support of teachers and students with online content included private initiatives such as the portal Science for Children, organized by the *Kantarot foundation*—a foundation for science, culture and mental fitness founded by Nikola and Dragana

Stikovi. Nikola has a bachelor's degree, a master's degree and a doctorate from Stanford University, and currently works as a professor at the Polytechnic Faculty of the University of Montreal. Dragana has a bachelor's degree in management from UKIM and a master's degree in educational administration from Santa Clara University (Qantarot 2019). The *Science for Kids* portal aims to bring together scientists, teachers and media experts in one place, and that in times of a pandemic when the need for digital educational content had become great! Interactive digital contents from the fields of: mathematics, physics, chemistry and biology have been placed on the portal. On this portal, interactive digital content in mathematics has been uploaded, a total of 39 content for students 11–14 years old, physics a total of 24 content for students 13–14 years old, chemistry a total of 49 content for students aged 13–14, biology a total of 32 content for students aged 12–14 years old. In addition to the interactive content for these teaching subjects, other activities have been published that include science from different areas, given as an opportunity for website visitors to create and share them (Nauka za deca 2020). In order to popularize science among children, the first "*Science for Children*" conference was organized in March 2021, which includes workshops for children where they had the opportunity to get acquainted with holograms, brought them closer to the world of magnets, activities to encourage children's curiosity and get to know the scientific method, discussions were held on the topic of play and learning, technology and the future of education.

At the beginning of the 2021/2022 academic year, the Ministry of Education and Science of the Republic of North Macedonia, in cooperation with the Faculty of Information Technologies and Computer Engineering, launched the national online learning platform for primary and secondary schools www.schools.mk. In order to interact and implement live online teaching, the platform was connected to the Microsoft Teams communication platform, which is part of the Microsoft 365 cloud service that was provided to all teachers and students in primary and secondary schools in North Macedonia (Ministry of Education and Science of the Republic of North Macedonia 2020).

The well-known Macedonian physics teacher, Aida Petrovska, puts a lot of effort into creating digital online resources for both teachers and students. Her statement for this matter…. "*I am glad that lately there is a space opening for the introduction of STEM classes in our educational system, which implies interdisciplinary learning of science through technology and engineering. For me, STEM is a perfect model no matter how rejected it is in our school system, the excuse such as pocketing funds for equipment that STEM requires, is not a big issue, in reality, it is not large funds, to be honest, we spend much more irrationally funds and aimlessly. Education and motivation of the physics teaching staff is needed, and the students will be more motivated by the very novelty—taking a break from chalk and blackboard. Students through practical examples get to know the ways of making life easier and dealing with many life challenges and problems*" (Petrovska 2020).

Fig. 3 First Educonference for teachers and IT experts 2015

6 Other Initiatives for Using Digital Tools in Education in the Republic of North Macedonia

The non-governmental sector has also made its own contribution to the development of digitization in education in North Macedonia. One organization that has had a big impact on promoting digital tools for education is Friends of Education[1] a nonprofit organization for promoting the modernization of education through the use of digital resources while simultaneously embracing a holistic approach to teaching and learning. It was established by a group of teachers in 2014 as a result of the international successes they had in the innovative use of digital tools in education.

The Association Friends of Education become known in North Macedonia for organizing the First Educonference for teachers and IT experts in 2015 which was the biggest conference for education organized in North Macedonia (Friends of education 2015) (Fig. 3).

The following years were organized 5 more conferences. On conferences took place many international and domestic experts in science, IT, robotics and pedagogy. Various topics were covered like STEM education, robotics in education, prosocial values, open education resources, professional and career development of teachers, digital tools for teaching, science and art in education, virtual and augmented reality in education, inclusion in education and gamification in education. The last Sixth Educonference for teachers and IT experts was held in September 2022. It was dedicated to the Robotics versus Bullying, Erasmus + Project KA3—Support for Policy Reform 612872-EPP-1-2019-1-IT-EPPKA3-PI-FORWARD, in partnership

[1] https://www.friends-of-education.org/en/.

with organizations and institutions from Europe, Polo Europeo Della Conoscenza (Italy), Friends of Education (North Macedonia), CLEMENTONI S.p.A, Ministry of Education (Spain), University of Burgos (Spain), Ushak Provincial Directorate of National Education (Turkey), Panevezys District Education Center (Lithuania), Kino Information Technology Education (Bulgaria), MAKE IT BETTER Association for Innovation and Social Economy (Portugal), Association ARID (Poland), ASOCIATIA SINAPTICA (Romania).

The project, Robotics versus Bullying, an Erasmus + Project KA3—Support for Policy Reform, started in 2021 and will last until the middle of 2023. Its goal is to prevent bullying from the early age of school through the use of cutting-edge educational pedagogical strategies built on the cooperative use of robots in a co-constructivist setting. Combining peer education as a technique for inclusion, learning by doing in the field of educational robotics, socializing, learning and teamwork with innovation, learning and education. Learning by doing helps students become aware of bullying and cyber bullying. By providing teachers with courses that advanced their knowledge, the project worked in both the classroom and higher education sectors.

The values of this project are innovation and cross-disciplinarity, requiring a more innovative approach from teachers in their teaching while recognizing the challenges of social inclusion. At the conference were presented educational contents and practices for the implementation of STEM education—robotics, which allows students aged 6–12 to take their first steps in science, technology and engineering in a fun way and through social inclusion. The outcome is not only to increase students' digital skills, but above all the skills in problem-solving, critical thinking, experimentation and teamwork which are crucial in a child's present and future life (Friends of education 2022a). Teachers from North Macedonia and partner countries took active role on the conference by organizing 20 workshops covering the topics of innovative use of robots in the classroom including: Coding and robotics tools for inclusive and interdisciplinary approaching in the classroom; RobotArt-Drawing while programming; how to implement Roby's educational games in class; Robotics in Scotland's Primary schools—Bringing STEM to Life, the 4th Education Revolution (Friends of education 2022b).

Another initiative is the project "Digitalization of services in the education sector" (CUP 2021) funded by the Government of the United Kingdom, through the British Embassy in Skopje, and is implemented by the nonprofit, non-governmental organization Development Association, Center for Change Management (CMP).

This project's main objective is to further digitalize educational services, which are frequently used and can have a significant influence on both users and the Manufacturing Execution System (MES) as service providers. The project includes the complete digitalization of these services (reaching transactional level in accordance with EU benchmarks), support for restructuring internal MES administration processes for these services, development of MES administration capacities for services administration, and creation and delivery of promotional campaigns to encourage the target user groups to use the e-services. The implementation of the project was in the period from October 1, 2021 to March 31, 2022.

7 Conclusions

After describing the chronological development of events related to the computerization and digitalization of education in the Republic of North Macedonia, limited to primary and lower secondary school, we showed how in practice teachers created many innovative examples of the use of technology through interdisciplinary project activities.

Today, more than ever, ICTs define society not only in terms of access to information but dictate its transformation into a society of knowledge, ability and skills. The real transformation of society is achieved only if education responds with effective changes through the introduction and application of new concepts known as *Teaching for the twenty-first century, Learning for the twenty-first century and Skills for the twenty-first century*. Following these concepts, ICT is clearly not only a means for the realization of educational goals but also a significant factor in the complete restructuring of the educational system, the introduction of new interactive and participatory models of teaching, new educational pedagogy, continuous and lifelong learning.

Annex 1: Links to Digital Applications for Teachers

Grandma's Games Manual for teachers. https://www.academia.edu/9753118/GRANDMAS_GAMES_Teachers_Manual_Marina_Vasileva

Fun education stopmotion animation teacher manual https://www.academia.edu/s/2901997916

Eduino, North Macedonia web platform for digital educational content, collaboration and professional development of educators https://www.eduino.gov.mk/

Nauka za deca (Science for children) is under the patronage of the civic initiative Kantarot, with the desire to promote scientific culture and mental fitness among the youngest https://naukazadeca.mk/

References

Blog post, *Microsoft Partners in Learning North Macedonia* (2017). https://pilmacedonia.wordpress.com

CUP, *Дигитализација на услуги во образовниот сектор (Digitalizacija na uslugite vo obrazovniot sector, Digitization of services in the education sector)* (2021). https://cup.org.mk/proekti/digitalisation-of-services-in-the-education-sector

Friends of education, *First International Educonference* (2015). https://www.friends-of-education.org/info-3/

Friends of education, *EduConference2022a* (2022a). https://www.friends-of-education.org/info-6/

Friends of education, *Agenda EduConference 2022b* (2022b). https://drive.google.com/file/d/1hyPgIS4RfVCeXzaAESNntP7F1Xm_Pafb/view

D.M. Hathaway, *Educating Educators in Macedonia: The Next Step* (2005). http://mason.gmu.edu/~dhathawa/Portfolio/Coursework/macedonia873.htm

M.K. InnovationLab, *EDUINO* (2020a). https://www.eduino.gov.mk/za-inicijativata/

M.K. InnovationLab, *Report* (2020b)

T. Malinovski, M. Vasileva, T. Vasileva-Stojanovska, V. Trajkovik, Factors influencing students' experience during integration of classroom games in primary school program. J. Exp. Educ. **1**(789), 1–39 (2014)

T. Malinovski, T. Vasileva-Stojanovska, M. Vasileva, D. Jovevski, V. Trajkovik, Adult students' perceptions in distance education learning environments based on a videoconferencing platform: QoE analysis. J. Inform. Technol. Educ. **14**, 1–19 (2015)

Metamorphosis Foundation, *Употребата на компјутери и интернет во образовниот систем на РМ (Upotreba na kompjuterite i internet vo obrazovniot system na RM, The use of computers and the Internet in the educational system of the Republic of Macedonia)* (2010). https://metamorphosis.org.mk/wp-content/uploads/2014/10/upotreba-na-kompjuterite-vo-obrazovanieto.pdf

Metamorphosis Foundation, *Состојбите и предизвиците за спроведување онлај настава во основните училишта (Sostojbite i predizvicite za sproveduvanje onlajn nastava vo osnovnite ucilista, The conditions and challenges for implementing online teaching in elementary schools)* (2020). https://metamorphosis.org.mk/wp-content/uploads/2020/09/oor_istrazuvanje_2020.pdf

Ministry of Education and Science of the Republic of North Macedonia, *Националната платформа достапна за учениците и родителите (Nacionalna platform dostapna za ucenicite i roditelite, The national platform available for students and parents)* (2020). https://mon.gov.mk/content/?id=3356

Ministry of education and science of the Republic of North Macedonia, *Koncepcija za osnovno obrazovanie (Concept for Primary Education)* (2021). https://mon.gov.mk/stored/document/Koncepcija%20MK.pdf

Nauka za deca, *Science for Children* (2020). https://naukazadeca.mk/index.php#fuseschool

A. Petrovska, *Blog* (2020). https://aidapetrovska.blogspot.com/

Qantarot, *KANTAROT* (2019). https://qantarot.substack.com/about

V. Stojanovska, M. Vasilev, Digitized children's games from the past in function of the realization of the mathematics curriculum in primary education. Pract. Theory Syst. Educ. **9**(3), 236–244 (2014)

D. Taleski, *This is My Voice* (2013). https://thisismyvoice.wixsite.com/thisismyvoice

M. Vasileva, Electronic portfolio as an alternative model. Educ. Magaz. Sci. Magaz. Educ. Quest. **157**, 155–161 (2010)

M. Vasileva, V. Bakeva, T. Vasileva-Stojanovska, T. Malinovski, V. Trajkovik, Grandma's games project: bridging tradition and technology mediated education. TEM J. **3**(1), 13–21 (2014)

M. Vasileva, *Grandma's Games, Education and Technology, Innovation in Learning and Cognitive Developed, Annual Scientific and Methodological Magazine* (Year I. Burgas, University prof. Dr. Zlatarov, 2011a), pp. 271–272

M. Vasileva, *Grandma's Games Manual* (2011b). https://www.academia.edu/9753118/GRANDMAS_GAMES_Teachers_Manual_Marina_Vasileva

Digital Transformation for Vietnam Education: From Policy to School Practices

Duy Hai Tuong, Trinh-Ba Tran, and Duc Dat Nguyen

Abstract The Vietnamese Government is enhancing the implementation of the digital transformation plan to make Vietnam one of the leading digital nations. In the digital transformation process, education is prioritized with the mission of applying new models and teaching methods to implement STEM education, contributing to student career orientation to meet the Fourth Industrial Revolution. This chapter presents the current situation of digital transformation in education and the practice with examples of using COACH software to implement STEM education based on the digital transformation trend, which shifts from traditional teaching to multimedia teaching, combined with software and hardware exploitation such as traditional experimental equipment and sensors for data collection and automatic control.

Keywords Digital transformation in education · STEM education · COACH software · Vietnam education · Science teaching

1 Introduction

Digital transformation is the process in which digital technologies create disruptions that prompt organizations to strategically alter their methods of value creation, while also managing the accompanying structural changes and organizational obstacles that can impact the process positively or negatively (Vial 2019). Recent years have seen a strong correlation between digital transformation and significant changes occurring in society and industries due to the use of digital technologies (Agarwal et al. 2010;

D. H. Tuong (✉) · T.-B. Tran · D. D. Nguyen
Faculty of Physics, Hanoi National University of Education, 136 Xuan Thuy Street, Cau Giay District, Hanoi, Vietnam
e-mail: haitd@hnue.edu.vn

T.-B. Tran
e-mail: trinhtb@hnue.edu.vn

D. D. Nguyen
e-mail: ducdat@hnue.edu.vn

M. Streit-Bianchi et al. (eds.), *New Challenges and Opportunities in Physics Education*, Challenges in Physics Education, https://doi.org/10.1007/978-3-031-37387-9_20

Tolboom 2016). In the context of education, digital transformation is implemented in academic organizations to simplify the learning experience by addressing a range of obstacles, including limitations related to time and proficiency that are typically encountered in traditional modes of education (Alhubaishy and Aljuhani 2021). Vice versa, the provision of suitable digital literacy training to human resources through education is crucial in driving the implementation of sustainable development of society (Reddy et al. 2020). However, it should be noticed that educational technologies are an important but not exhaustive part of the digital transformation of education systems which involves both digital educational technologies and non-technology initiatives (Qayyum 2023).

In 2020, research by Brunetti et al. (2020) proposed four strategic measures for digital education, which included promoting digital learning paths to enhance logical and computational skills, creating digital lifelong learning programs, establishing a digital platform for skill-based educational plans, and investing in e-learning by educating individuals on the use of digital media and technology while ensuring effective teacher-student interactions through a combination of digital and analog methods (Brunetti et al. 2020).

Vietnam, with the vision of implementing digital transformation plans to make Vietnam one of the leading digital nations, also prioritized education with the mission of applying new models and teaching methods to implement STEM education, contributing to student career orientation to meet the Fourth Industrial Revolution. In terms of policy, to what extent has Vietnam progressed in this endeavor in, and what particular measures have been implemented to facilitate this progression? In this chapter, the current situation of digital transformation in education will be illustrated, followed by the strategies influenced by the policies including using the COACH platform to implement STEM education.

2 The Policy of Digital Transformation for Education in Vietnam

The digital transformation on the national scale, in general, and in education, in particular, has been reinforced by Vietnam through policy direction to specific actions in recent years, from educational management activities to teaching and student assessment. Vietnam's digital transformation strategy is always associated with STEM education because those fields contribute to career orientation and competence development to adapt to the current 4.0 Industrial Revolution.

Since 2017, the Vietnamese government has directed the implementation of information and communications technology (ICT) in teaching and carried out digital transformation in education by forming a shared digital learning resource repository for the entire sector, including lectures, e-textbooks, simulation software, and other learning materials to support teaching activities in schools. The goal by 2025 is to increase the level of application of ICT in managing and supporting teaching and scientific research activities to reach an advanced level in the ASEAN region, meeting the fundamental and comprehensive goals of education and training innovation. Therefore, ICT has become a driving force for innovation in management, content, teaching methods, and assessment in education and training (Vietnamese Government 2017a).

In 2018, the Vietnamese government issued a career orientation strategy for high school students with the goal of innovating the content, methods, and forms of vocational education in high schools, strengthening education activities in the direction of integrated STEM education in programs that are suitable for the country's vocational development trends, meeting the labor market's demands, and preparing human resources to meet the requirements of the Fourth Industrial Revolution. The strategy emphasizes the need to implement STEM education in some representative localities for economic regions, focusing on training teachers on integrated vocational education and STEM education as well as supporting teaching equipment to support vocational education and STEM education in some high schools (Vietnamese Government 2018).

In 2019, the Vietnamese government continued to make decisions to include digital literacy in the national high school education program until 2030, when nationwide coverage of 5G mobile networks and affordable access to high-speed Internet. The teaching and learning methods were innovated based on digital technology by encouraging new education and training models based on digital platforms. In these models, STEM professions were identified as a priority resource for implementing national research programs on prioritized and focal technologies, such as information and communication technology, electronics, new technologies in the energy sector, artificial intelligence, biotechnology, and biomedical electronics (Politburo of the Communist Party of Vietnam 2019).

In 2020, the Vietnamese government issued a national digital transformation strategy with the goal of making Vietnam a stable and prosperous digital nation, pioneering new technologies and models by 2030. Specifically, Vietnam aims to be among the top 30 countries in terms of information technology and cybersecurity, as well as to provide widespread 5G and fiber optic broadband services. The government has prioritized education as one of the critical areas for digital transformation because it has the most significant social impact, is closely related to people's daily lives, can bring about quick changes in perceptions, is cost-effective, and provide efficiency. In the national digital transformation strategy, Vietnam has prioritized STEM fields and STEM education, including the implementation of an integrated education model that incorporates science, technology, engineering, mathematics, arts, business, and industry (STEAM/STEAM/STEAME education), English language training, and information technology skills, as well as ensuring information security at all levels of education. Career guidance and training will also be provided to prepare students with the necessary skills for the digital environment (Vietnamese Government 2020).

For the education sector, the national digital transformation strategy requires the development of a remote teaching and learning support platform, the sufficient application of digital technology in managing, teaching, learning, digitalizing teaching materials and textbooks, building a platform for sharing teaching and learning resources in both direct and online forms, and developing educational technology aiming at personalized training. By 2030, the goal is to have 100% of educational institutions implement remote teaching and learning, including testing training programs that allow students to study online at least 20% of the curriculum, and to apply digital technology to assign homework and check students' preparation before class (Vietnamese Government 2020).

In 2022, the Ministry of Education and Training in Vietnam set the goal of digital transformation for the education sector by developing an interactive smart classroom system, modern laboratory/practice rooms (Lab), simulation Labs, and applying virtual reality (VR) and augmented reality (AR), machine learning, big data analysis, and artificial intelligence technologies in researching and practicing. They will also implement and pilot advanced digital teaching and learning models through a blended learning approach (smart classrooms, interactive study groups, self-learning with virtual assistants). As for data and learning materials, priority will be given to developing shared digital learning resources, open educational resources, including e-lectures, TV teaching lectures, multimedia digital learning materials, e-textbooks, simulation software, and other learning materials. They will also develop an online question bank system for all subjects in general and continuing education (Vietnam Ministry of Education and Training 2022).

Digital transformation in education and career orientation should always be carried out according to the STEM, English, and computer science education direction through implementing an integrated STEM/STEAM education model, developing programming thinking, and implementing appropriate computer science programs. They will introduce common skills in digital literacy, online safety and security, open platforms, and open-source software in the primary education program to establish the necessary skills for digital citizenship (Vietnam Ministry of Education and Training 2022).

Digital transformation and STEM education aim to approach the Fourth Industrial Revolution and create significant changes in education policies, contents, methods, and vocational training to generate a workforce capable of receiving new manufacturing technology trends. They will focus on promoting science, technology, engineering, mathematics (STEM), foreign languages, and computer science training in general education programs. They have piloted these changes at some high schools since the academic year 2017–2018 to strengthen fundamental skills, creative thinking, and adaptability to the Fourth Industrial Revolution requirements (Vietnamese Government 2017b).

Therefore, digital transformation in education is currently a national strategy of Vietnam, encouraging the implementation of new educational models, including integrating STEM education for career orientation in STEM fields for students. The following content will present STEM education in the implementation of digital transformation for career guidance of students.

3 STEM Education in High Schools in Vietnam

3.1 The Orientation of Implementing STEM Education in High Schools in Vietnam

The concept of STEM education was first introduced in Vietnam in 2016 through a pilot project titled "Applying the UK's STEM approach to the Vietnamese context in 2016–2017" by the British Council in Vietnam and the Ministry of Education and Training of Vietnam. The pilot project involved 15 public and private middle and high schools in 5 northern provinces, including Hanoi, Hai Duong, Nam Dinh, Hai Phong, and Quang Ninh (Nguyen Van et al. 2019). The project aims to enhance STEM education competence for teachers and school leaders to improve students' academic performance and practical skills. Prior to this, STEM education activities in schools were often organized for a small group of voluntary students to participate in science and technology competitions such as Robotcon and Intel ISEF, where students could pay for necessary materials/tools and instructors could be science or computer teachers in the school or experts, scientists from universities, research centers, and institutes. The pilot project results have contributed to promoting STEM education in Vietnam. In 2018, the Ministry of Education and Training included STEM education in the secondary school curriculum through subjects such as Mathematics, Natural Sciences and Social Sciences, Science, Technology, Computer Science, Physics, Chemistry, and Biology (Vietnam Ministry of Education and Training 2018).

In the general education program, from primary to secondary level, the essence of STEM education is defined as an interdisciplinary education model that helps students apply scientific, technological, engineering, and mathematical knowledge to solve practical problems in specific contexts. Mathematics education establishes connections between mathematical ideas, mathematics and reality, mathematics and other subjects and educational activities, especially with Science, Natural Science, Physics, Chemistry, Biology, Technology, and Computer Science, to implement STEM education. In the general education program, STEM education is emphasized in each subject. For example, in Natural Science, STEM education helps students gradually develop natural science competencies through observation and experimentation, synthesis of knowledge, and skills to solve problems in life. In Technology and Computer Science, STEM education plays a leading role in preparing students to search, receive, expand knowledge, and be creative in the era of the Fourth Industrial Revolution and globalization; provides strong support for students' self-learning; establishes a solid foundation for the application of digital technology to serve the development of new knowledge content, new educational methods, and modern education activities for all subjects. Analyzing the essence of STEM education in the subjects of the general education program shows that the STEM education goals for each level of education are expressed in the following Table 1.

Table 1 Characteristics of STEM education for different levels of education in mainstream subjects

Primary school level	Middle school level	High school level
STEM education is implemented in subjects such as science, computer science, and mathematics to provide opportunities for students to integrate and apply knowledge and skills from different fields of study and subjects, such as mathematics, computer science, and technology, to solve real-life problems at a level appropriate to their abilities	STEM education in natural science, computer science, and technology subjects aims to meet the demand for young human resources for the industrialization and modernization of the country	STEM education helps guide career orientation in various subjects and subject groups as follows: **Physics, Chemistry, Biology:** Implementing STEM education to develop for students to integrate knowledge and skills in researching and solving practical situations **Technology:** STEM education contributes to developing students' competencies, qualities, and career guidance **Computer Science:** Computer Science plays a central role in connecting other subjects, promoting STEM education, and fostering student creativity to create digital products with high levels of ICT **Mathematics:** Mathematics provides opportunities for students to apply mathematics to solve interdisciplinary and practical problems, contributing to the scientific basis for STEM education

3.2 *Technologies for Implementing STEM Education in High Schools in Vietnam*

Alongside guidance for implementing STEM education in high schools, the Ministry of Education and Training also proposes a minimum list of equipment for teaching each subject and implementing STEM education for each grade. In particular, devices and technologies for implementing STEM education are exploited from the minimum list of equipment that is generally regulated for all subjects and encourages the use of devices and easily processed, easy-to-find, inexpensive materials for students, especially by enhancing the use of software and digital resources to actively engage students in learning (Vietnam Ministry of Education and Training 2020, 2023). Devices and technologies for implementing STEM education are classified into three categories in terms of application directions, as shown in the following diagram (Fig. 1).

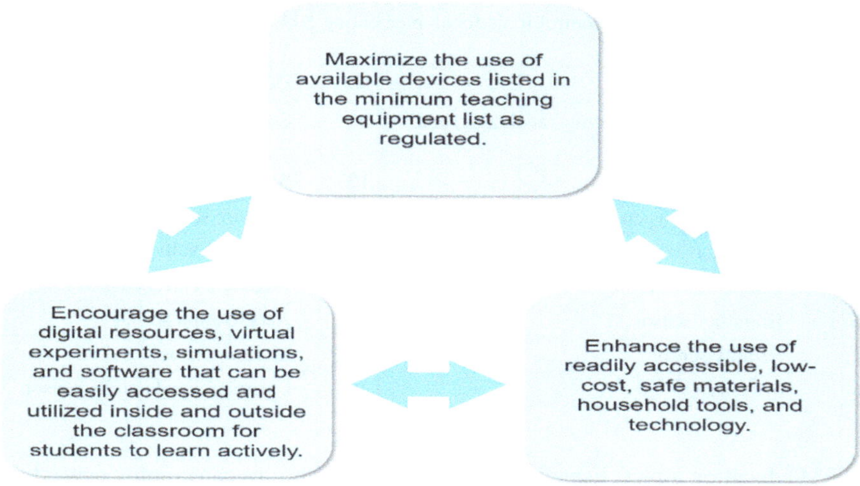

Fig. 1 Directions for applying technology to implement STEM education in schools

The Ministry of Education and Training also issued a circular on the specific equipment list to implement STEM topics in subjects and scientific and technical research activities in schools. Supporting software and data collection tools are specified for each level of education. For primary education, students are equipped with application software, teaching software, e-learning resources, search software, graphics software, interactive programming software, and natural science simulation software (Vietnam Ministry of Education and Training 2021a). For secondary education, students use data analysis software, sensor connections, 3D sound simulation software, molecular simulation software, atomic simulation software, optical simulation software, electrical simulation software, magnetic simulation software, and software for the structure and function of organisms, mind mapping software, and video design software (Vietnam Ministry of Education and Training 2021b). For high school education, the proposed list of software includes more complex simulations such as 3D simulations of the solar system, Earth and Moon, tides, eclipses, gravitational fields, electronic circuits, technical drawing software, robot control programming software, project management software, and 3D graphics software (Vietnam Ministry of Education and Training 2021c). In addition to the software, the following list includes connectivity devices such as data collection sensors (Table 2), execution devices, motherboards, educational robots, tooling, mechanical processing, and 3D printers to implement STEM topics with technology.

Attached to the list of sensors is a list of mechanical processing equipment and consumables by school year as follows (Table 3).

In addition to the minimum equipment categories specified above, educational institutions may supplement or provide additional equipment while implementing

Table 2 List of sensors and control tools for implementing STEM education in high schools in Vietnam

No.	Device	No.	Device
1	Equipment for collecting, analyzing, and displaying data	9	Stepper motor
2	CO_2 sensor	10	Sound sensor
3	Dissolved oxygen sensor	11	Pressure sensor
4	O_2 sensor	12	Electrical voltage sensor
5	Temperature sensor	13	Electrical current sensor
6	Humidity sensor	14	Acceleration sensor
7	Salinity sensor	15	Motion sensor
8	pH sensor	16	Electrical conductivity sensor

STEM education topics. Especially, educational institutions can employ new educational models that align with career-oriented and digital transformation goals, as outlined in the Ministry of Education and Training's initiative, including the implementation of STEM education, development of programming mindset, relevant computer science programs, and integration of content on digital skills, cybersecurity, open platforms, and open-source software into primary-level teaching to early cultivate essential skills for digital citizens (Vietnam Ministry of Education and Training 2022).

3.3 COACH Software in STEM Education

The COACH software for STEM education (Science, Technology, Engineering, and Mathematics) was developed by the Dutch Computer Technology Center (CMA), and a system of educational materials and STEM equipment was also built for teachers and students to use in teaching and learning.

STEM. This software is available in 7 languages, including Vietnamese, making it easy for Vietnamese students and teachers to use and organize their learning. The research team based their work on the COACH software to build a bank of STEM education topics according to the Vietnamese national curriculum for students from primary to high school levels, ensuring that teachers and students can carry out STEM topics during their 45- or 90-min class periods as specified in the Vietnamese school schedule. The criteria for the lesson plans are designed to closely follow the students' curriculum and build a teaching process, study materials, and instructional videos for teachers and students to explore and use the software to implement STEM topics with the aim of only using phones, tablets, or computers with the COACH software wirelessly connected to sensors for data collection.

Table 3 List of actuators and consumables for implementing STEM education in schools

Set of equipment	Name of components
Set of materials	Formex plastic sheet (A3 size, 3 and 5 mm thick)
	Acrylic sheet (A4, transparent, 3 mm thick)
	Hot glue stick (10 mm diameter)
	Threaded screws and M3 nuts
	Wood screws (several types)
	Drill bit (diameter 3 mm)
	Wheels (diameter 65 mm, axle 5 mm)
Set of mechanical equipment	Leaf ruler (300 mm long)
	Mechanical caliper (common type)
	Marker head (common type)
	Protractor (common type)
	Level gauge (common type)
	Paper knife (common type)
	Acrylic plastic cutter (common type); Small vise (50 mm aperture)
	File (flat, round) one each
	Hand saw (common type)
	Universal screwdriver set (common type)
	Small wrench (common type)
	Square beak pliers (common type)
	Glue gun (type 10 mm, power 60 W)
Set of manufacturing machines	Small size 3D printer (Printing technology: FDM, Layer resolution: 0.05–0.3 mm, Print head diameter: 0.4/1.75 mm, Printing material: PLA, ABS, Making size Maximum work: (200 × 200 × 180) mm, Connection: SD Card, USB Port)
Set of electrical materials	Hand-held electric drill (using batteries), quantity 03 pcs
	Lithium battery type 3.7 V, 1200 maH, 9 batteries;
	Lithium battery base (triple type)
	Black, red power cord (diameter 0.3 mm)
	Double-headed alligator clip (300 mm long)
	Heat shrink gene (2 and 3 mm diameter)
	Insulation tape
	Single-sided copper key (A4, 1.2 mm thick)
	FeCL3 salt, 500
	Soldering tin coil (100 g type)
	Turpentine 300 g

(continued)

Table 3 (continued)

Set of equipment	Name of components
Set of electrical equipment	Lithium battery charger (double charging tray, charging current 600 mA)
	Digital multimeter (common type)
	Electrical tester (common type)
	Wire stripping pliers (common type)
	Sharp beak pliers (common type)
	Pliers (common type)
	Electric screwdrivers (common type)
	Tin soldering iron (AC 220 V, 60 W), with soldering iron (common type)

The learning activities with COACH are designed on an open platform based on practical problems that help students explore and create solutions suitable for the situations and contexts they commonly encounter.

These learning activities are demonstrated based on the main functions of COACH, such as collecting data through sensors, processing data using analytical tools that include zooming in and out, reading values, finding slopes, finding the area of a graph segment, processing tools that include selecting and deleting data, smoothing out graphs, calculating new variables using mathematical functions, curve fitting, frequency spectrum analysis, and statistical tools that include finding descriptive statistics and creating statistical charts. Video analysis helps students bring the real world into the classroom so that students can analyze the fascinating events of nature because some phenomena and processes occur quickly, making it difficult to observe details directly with the naked eye, or some phenomena and processes are difficult to reproduce in the classroom. Programming with COACH allows students to build and create their own control models. These activities help students orient their careers and improve their ability to apply technology to the real world, solving practical problems. Modeling helps students access scientific and technological research models and build computer models before designing and implementing them in practice. The modeling software function of COACH helps students represent experimental data and models using animated graphic objects with attributes (position, size) assigned to this data.

In particular, for K12 education and higher education, the COACH library has provided sample activities and devices for students (from 8 years old and up) to create a series of STEM educational activities based on exploration and discovery. The COACH software with these functions has been developed by the research team into a library of hundreds of lessons for students and teachers to use (Fig. 2).

When teachers or students download and install the software, they can select from a library of available STEM lessons to study. The COACH software has promoted innovation in Vietnamese education through STEM education as part of the digital transformation model to meet the demand for human resources in the age of Industry 4.0.

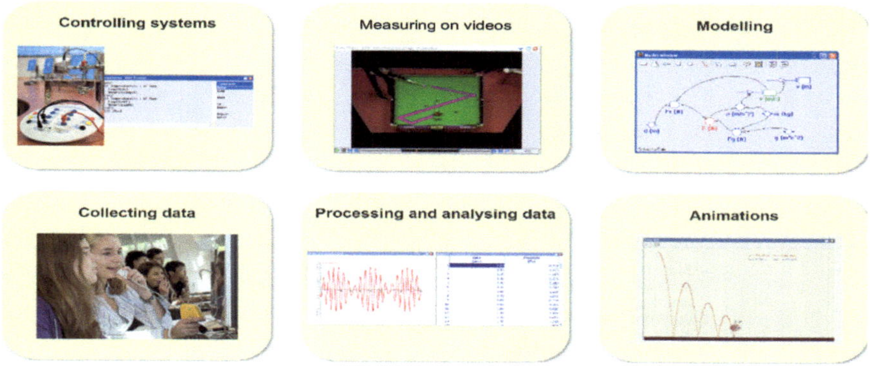

Fig. 2 STEM activities in COACH software

Specifically, the COACH software was developed by a research team to meet the needs of teachers and students in Vietnamese education with the following orientations:

- Exploiting the COACH data system, built on modern educational methods such as exploration, technical design cycles, and creating space for student creativity.
- Directing activities for teachers through a rich database of learning surveys that are regularly updated.
- Accessing the global education system through open data sources that are regularly updated by COACH subject teachers, experts, and education scientists from participating countries in COACH development.
- Building a system of diverse sample activities for teachers to implement in the classroom, without the need to create new ones from scratch and adjustable to different localities in Vietnam.
- Creating a learning survey system for students to self-study and conduct research using the exploration and technical creativity models associated with real-life situations.
- Supporting textbooks, exercise books, and teacher book authors in developing lesson models based on diverse COACH platforms, such as analyzing videos based on available lesson models and learning activities.
- Supporting the teacher community in building shared learning materials and experience-sharing lesson plans.
- Supporting educational scientists in researching and developing lesson models by providing direct feedback and transferring data files to practicing teachers.
- Creating a community that connects textbook authors, teachers, and students to support learning and self-study activities.

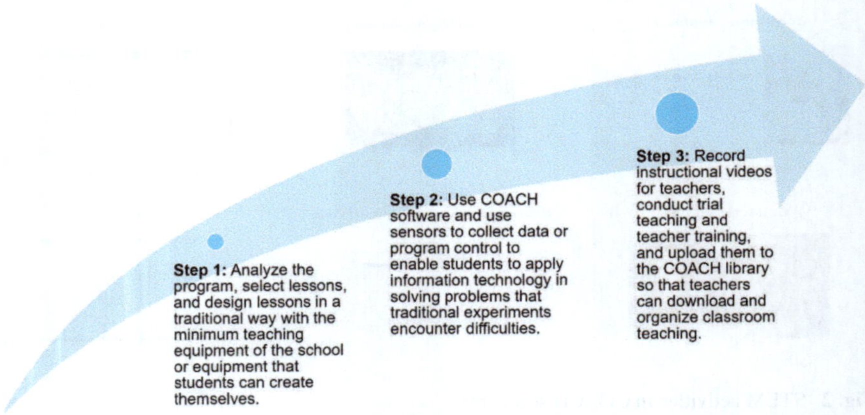

Fig. 3 Process of building a STEM lesson library helps teachers implement digital transformation in schools

3.4 Building the COACH Library in the Digital Transformation of STEM Education in Vietnam

Starting with the educational program content of science, technology, mathematics, and computer science subjects in Vietnam, topics with practical applications and control to design lessons according to the engineering design process or scientific inquiry process are determined (Fig. 3).

With the above process, STEM education has been implemented with the role of connecting ICT on COACH software for students to solve practical problems based on traditional experiments.

For example, in the 6th-grade technology program, there is a topic on the use of electric fans in the household. It includes an electric fan model, which can integrate the knowledge of control programming with temperature sensors, distance sensors, humidity sensors, and COACH software to help students control smart electric fans. To enable students and teachers who are not computer science majors to understand and implement this STEM topic in class for 45 or 90 min, an explanation of the commands in the lesson plan and instructional videos for teachers and students should be publicly uploaded on YouTube. Therefore, computer science is integrated into the technology subject, and technology is integrated to help students learn computer programming to meet the requirement of the computer science subject, which is to solve problems with computer support, as shown in the Figs. 4 and 5.

To support teachers in effectively utilizing the STEM library, each topic includes a list of devices with accompanying images uploaded onto the COACH software. There are also direct or online training activities for teacher development and sharing of software links in Vietnamese for easy installation and access (Figs. 6 and 7).

Based on the contributions of teachers during training and practical teaching, the STEM topics are being improved to closely follow the high school education

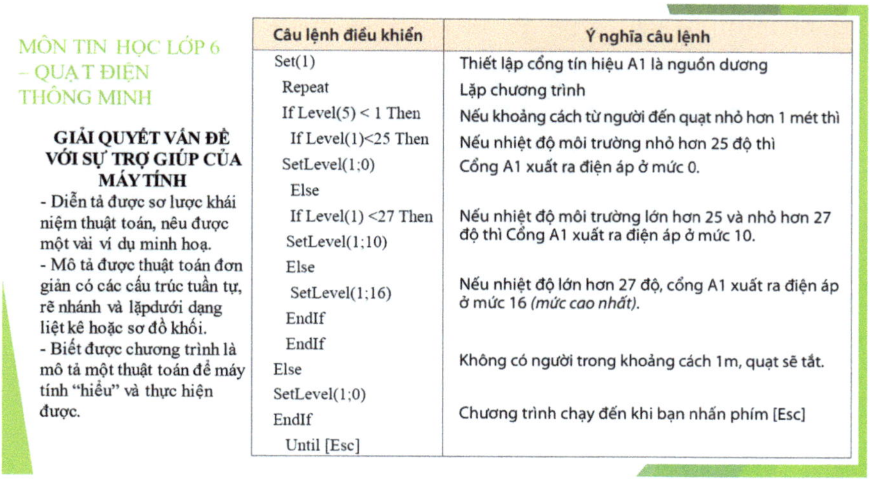

Fig. 4 Integrated COACH for programming control with commands in smart electric fan topic

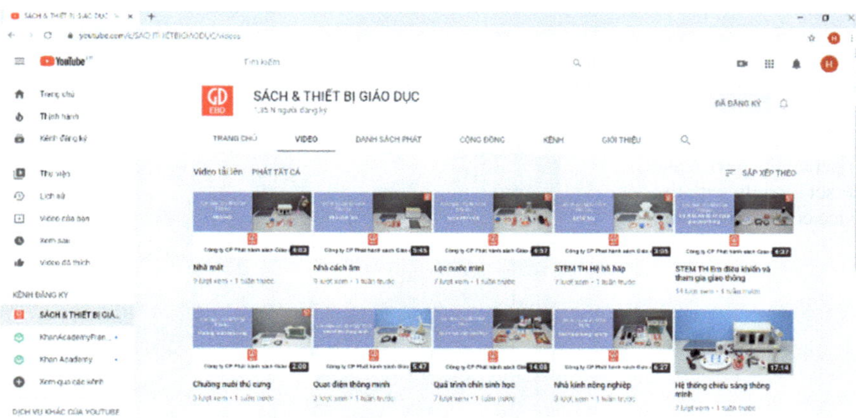

Fig. 5 List of instructional videos on STEM topics for teachers and students posted on YouTube[1]

curriculum, and teachers can use them for teaching within the time frame of each subject. The digital transformation in education is supported by the STEM topic library divided into each grade and subject, which helps students to self-study at home through video clips and tasks assigned on the software. STEM products that students create can be controlled or adjusted in the classroom. With the exploration of 68 STEM topics in English and 89 STEM topics in Vietnamese through the

[1] Link to the instructional videos: www.youtube.com/c/SÁCHTHIẾTBỊGIÁODỤC/videos

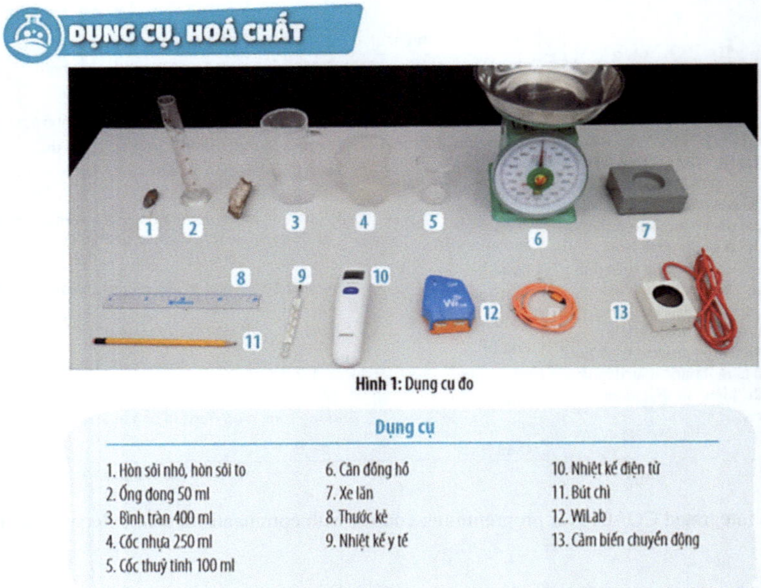

Fig. 6 Catalog of traditional equipment with sensors for teachers to observe, and organize teaching according to the topics designed in the library

Fig. 7 Teachers are trained to operate the device and interact directly with the device control software

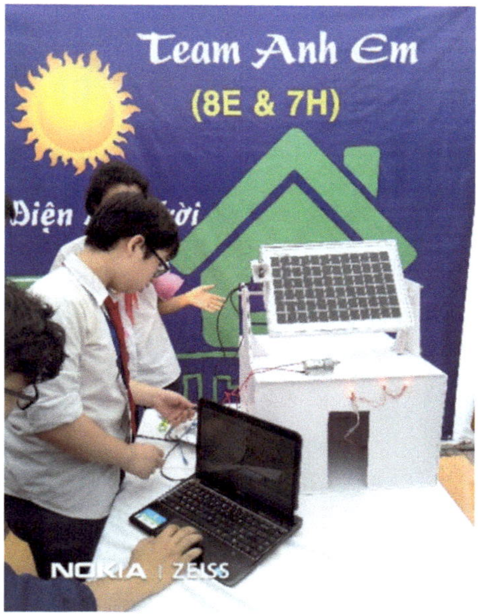

Fig. 8 Students implement the topic of controlling a solar powerhouse with COACH software

COACH software, teachers and students have many opportunities to learn and practice, contributing to the rapid digital transformation in education (Figs. 8, 9 and 10).

4 Conclusions

Vietnam has prioritized digital transformation and STEM education to adapt to the Fourth Industrial Revolution. The government has implemented various policies, such as integrating ICT in teaching and forming shared digital learning resources, to reach an advanced level in the ASEAN region by 2025. Career orientation strategies and the inclusion of digital literacy in the national high school education program until 2030 have also been implemented. The 2020 national digital transformation strategy prioritizes STEM fields and education, including STEAM education, English language training, and information technology skills. The Ministry of Education and Training has set goals for 2022, such as developing an interactive smart classroom system, modern laboratory/practice rooms, and shared digital learning resources. Digital transformation and STEM education aim to create significant changes in education policies, contents, methods, and vocational training to generate a workforce capable of receiving new manufacturing technology trends. One example of this implementation is the use of COACH software in Vietnam to facilitate STEM education.

Fig. 9 Students assemble and control smart agricultural greenhouses with pre-designed commands

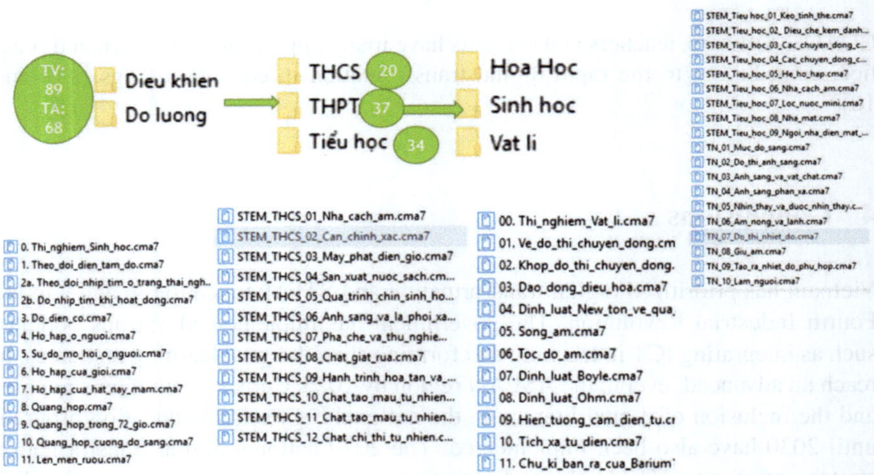

Fig. 10 List of STEM topics in COACH library in Vietnamese for all grades

The introduction of STEM education in high schools has been implemented through pilot projects and included in the secondary school curriculum. STEM education is emphasized in each subject, providing students with the opportunity to apply scientific and technological knowledge to solve practical problems in specific contexts. The orientation of implementing STEM education in high schools in Vietnam is to equip students with the necessary skills to meet the demands of the Fourth Industrial Revolution and globalization and to prepare them for future challenges. Along with the guidance on implementing STEM education in high schools, the Ministry of Education and Training of Vietnam has provided a minimum list of equipment and software necessary for teaching each subject and implementing STEM education. The use of software and digital resources is encouraged to actively engage students in learning, and the equipment list is specified for each level of education. Additionally, educational institutions may supplement or provide additional equipment while implementing STEM education topics. The integration of platforms like COACH software allows for the combination of traditional experimental equipment and sensors for data collection and automatic control, creating a more interactive and immersive learning experience for students. The software provides a comprehensive system of educational materials and equipment for STEM learning and is available in Vietnamese, making it accessible to all. The software is based on practical problems that help students explore and create solutions suitable for real-life situations. It also includes features such as data collection, data processing, video analysis, programming, and modeling. The COACH software has contributed to promoting innovation in Vietnamese education through STEM education as part of the digital transformation model to meet the demand for human resources in the age of Industry 4.0. Its development has been oriented toward exploiting modern educational methods, supporting diverse sample activities, and creating a community that connects textbook authors, teachers, and students to support learning and self-study activities.

The synchronization between traditional lessons and the COACH software helps students and teachers to take charge of the teaching and learning process, synchronize data in the COACH library, and have specific instruction templates that can be learned at home, in the laboratory, and in class with limited lesson time prescribed in the curriculum. Through the use of COACH software and a library of STEM topics, teachers can design lessons that follow the engineering design process or scientific inquiry process. The library includes topics that are easily accessible to teachers and students, with accompanying instructional videos and training activities. The STEM products that students create can be controlled or adjusted in the classroom, providing opportunities for hands-on learning. The synchronization between study sheets, publicly available video clips on YouTube, free software with libraries that teachers and students can access, and teaching devices prescribed in the education program has helped schools implement digital transformation more quickly and effectively. And so, this shift in education toward the integration of technology is expected to improve student engagement and learning outcomes, preparing them for the demands of the Fourth Industrial Revolution.

References

A. Agarwal, G.G. Gao, C. DesRoches, A.K. Jha, The digital transformation of healthcare: current status and the road ahead. Inform. Syst. Res. **21**(4), 796–809 (2010). https://doi.org/10.1287/isre.1100.0327

A. Alhubaishy, A. Aljuhani, The challenges of instructors' and students' attitudes in digital transformation: a case study of Saudi Universities. Educ. Inform. Technol. **26**(4), 4647–4662 (2021). https://doi.org/10.1007/s10639-021-10491-6

F. Brunetti, D.T. Matt, A. Bonfanti, A. De Longhi, G. Pedrini, G. Orzes, Digital transformation challenges: strategies emerging from a multi-stakeholder approach. TQM J. **32**(4), 697–724 (2020). https://doi.org/10.1108/TQM-12-2019-0309

B. Nguyen Van, H. Tuong Duy, D. Tran Minh, H. Nguyen Van, T. Chu Cam, T. Nguyen Anh, T. Doan, T. Tran Ba, *STEM Education in High Schools* (Vietnam Education Publishing House, 2019)

Politburo of the Communist Party of Vietnam, *Resolution No. 52-NQ/TW, Dated September 27th, 2019 on Some Principles and Policies for Proactively Participating in the Fourth Industrial Revolution* (2019)

A. Qayyum, *Handbook of Open, Distance and Digital Education* (Springer, New York, 2023), pp. 249–266. https://doi.org/10.1007/978-981-19-2080-6

P. Reddy, B. Sharma, K. Chaudhary, Digital literacy: a review of literature. Int. J. Technoeth. **11**(2), 65–94 (2020). https://doi.org/10.4018/IJT.20200701.oa1

I.H. Tolboom, The impact of digital transformation: a survey based research to explore the effects of digital transformation on organizations. *Delft University of Technology, Faculty of Technology, Policy and Management* (2016)

G. Vial, Understanding digital transformation: a review and a research agenda. J. Strat. Inform. Syst. **28**(2), 118–144 (2019). https://doi.org/10.1016/j.jsis.2019.01.003

Vietnam Ministry of Education and Training, *Circular No. 37/2021a/TT-BGDĐT Dated December 30, 2021 on Issuing a List of Minimum Teaching Equipment for Primary Schools* (2021a)

Vietnam Ministry of Education and Training, *Circular No. 38/2021a/TT-BGDĐT Dated December 30, 2021b on Issuing the Minimum List of Teaching Equipment for Secondary Schools* (2021b)

Vietnam Ministry of Education and Training, *Circular No. 39/2021b/TT-BGDĐT Dated December 30, 2021c, on the Issuance of the Minimum Teaching Equipment List for High Schools* (2021c)

Vietnam Ministry of Education and Training, *Document No. 909/BGDTĐT-GDTH Dated March 8th, 2023 Providing Guidelines for Organizing STEM Education Activities in Primary Education* (2023)

Vietnam Ministry of Education and Training, *Circular No. 32/2018/TT-BGDĐT Dated December 26, 2018 on the Issuance of the General Education Curriculum* (2018)

Vietnam Ministry of Education and Training, *Document No. 3089/BGDĐT-GDTrH Dated August 14, 2020a on the Implementation of STEM Education in High Schools* (2020)

Vietnam Ministry of Education and Training, *Decision No. 131/QD-TTg Dated January 25, 2022 on the Approval of the Project "Enhancing the Application of Information Technology and Digital Transformation in Education and Training in the Period of 2022–2025, Towards 2030 Direction"* (2022)

Vietnamese Government, *Decision No. 117/QD-TTg, Dated January 25th, 2017a Approving the Project Strengthening the Application of Information Technology in Management and Support of Teaching, Scientific Research Contributing to Improving the Quality of Education and Training In* (2017a)

Vietnamese Government, *Directive No. 16/CT-TTg, Dated May 4th, 2017b on Enhancing the Capacity to Access the Fourth Industrial Revolution* (2017b)
Vietnamese Government, *Decision No. 522/QD-TTg, Dated May 14th, 2028 Approving the Project "Career-Oriented Education and Guidance for Student Streamlining in Secondary Education in the Period of 2018–2025"* (2018)
Vietnamese Government, *Decision No. 749/QD-TTg Dated June 3rd, 2020b on Approving the "National Digital Transformation Program by 2025, with a Vision Towards 2030"* (2020)

Teaching and Learning Physics with Digital Technologies—What Digitalization-Related Competencies Are Needed?

Lars-Jochen Thoms, Sebastian Becker, and Erik Kremser

Abstract In the physics classroom, many phenomena are treated experimentally and theoretically that cannot be experienced completely, or not at all, with the human senses. Digital technologies can support teachers in the teaching of physics, and students in gaining knowledge, by bridging the gap between human perception and physical reality. However, physics teachers need appropriate digitalization-related competencies in order to use digital technologies in the classroom in a meaningful and didactically sound way. The DiKoLAN framework (Digital Competencies for Teaching in Science Education, orig. Digitale Kompetenzen für das Lehramt in den Naturwissenschaften) is a suitable tool for planning, structuring, implementing, and evaluating science-specific pre-service teacher training and is already being used at several German-speaking corresponding institutions. Appropriately trained teachers can thus improve their own teaching through the use of digital technologies and also promote the media competencies of their students. This chapter describes

L.-J. Thoms (✉) · E. Kremser
Chair of Science Education, University of Konstanz, Universitätsstraße 10, 78464 Konstanz, Germany
e-mail: lars.thoms@uni-konstanz.de

E. Kremser
e-mail: erik.kremser@physik.tu-darmstadt.de

L.-J. Thoms
Chair of Science Education, Thurgau University of Education, Unterer Schulweg 3, 8280 Kreuzlingen, Switzerland

Research and Knowledge Management, Thurgau University of Education, Unterer Schulweg 3, 8280 Kreuzlingen, Switzerland

S. Becker
Digital Education, Faculty of Mathematics and Natural Sciences, University of Cologne, Herbert-Lewin-Str. 10, 50931 Cologne, Germany
e-mail: sebastian.becker-genschow@uni-koeln.de

E. Kremser
Department of Physics, Technical University Darmstadt, Hochschulstraße 6, 64289 Darmstadt, Germany

M. Streit-Bianchi et al. (eds.), *New Challenges and Opportunities in Physics Education*, Challenges in Physics Education, https://doi.org/10.1007/978-3-031-37387-9_21

problem areas, technology-based approaches to solutions, demands on teacher education, the DiKoLAN framework as a structuring aid, and exemplary approaches to promoting digitalization-related competencies in physics teaching and physics teacher education.

Keywords Physics education · Technological pedagogical content knowledge (TPACK) · Digital competencies for teaching in science education · Digitale Kompetenzen für das Lehramt in den Naturwissenschaften (DiKoLAN) · Pre-service teacher education · Self-assessment · Digital literacy · Student project · Data acquisition · Video analysis

1 Physics is More Than What We Can Perceive with Our Senses

Many physical phenomena cannot be perceived with the human senses, can only be observed indirectly, or can only be grasped incompletely. In the teaching of such phenomena, the difficult accessibility results in limitations in the acquisition of knowledge and learning difficulties for the students. Selected examples are given below.

- In the physics classroom, many processes are studied at vastly different timescales. In a laser, for example, atoms are excited by absorbing energy, whereupon they remain in the excited state for about ten nanoseconds until they release the energy again in the form of laser radiation. It is well known that the orbital period of the earth around the sun is 1 year. Although we humans can develop a feeling for the duration of a year through the seasons and observations of processes in time spans of a few years and even decades are still conceivable, atomic as well as cosmological processes elude any sensory observation.
- It is also difficult to grasp the dimensions of the systems in which the physical processes under consideration take place. In the physics classroom, the approximate diameter of an atom is estimated with the simple oil film experiment (Franklin 1774), and it is determined that this lies in the range of 0.1 nm. The light of the sun, on the other hand, covers a distance of about 150 million kilometers on its way to the earth. We can hardly imagine either of these distances.
- In a narrower sense, the term 'light' refers to the part of the electromagnetic spectrum that can be perceived by the naked human eye. Many animals can also perceive electromagnetic radiation that goes beyond the visible light spectrum. For instance, bees can detect ultraviolet radiation, and some snakes can detect infrared thermal radiation with their pit organs.

- For ionizing radiation of radioactive substances or X-rays, humans have not developed any senses at all, so that they cannot be perceived in any way. Not only do associated physical processes remain hidden to learners without suitable technical aids, but ionizing radiation and X-rays also generally pose dangers to living organisms.
- Physics education, like its reference science physics, is concerned with the measurement of physical quantities. Without precise knowledge of the quantities to be measured and of the available measurement methods, the measurement tools used and their advantages, disadvantages and limitations, there is a risk that improper measurements will lead to incorrect results and ultimately to false assumptions about the physical phenomena. Especially in the fields of spectrometry and thermography, spectrometers and thermal imaging cameras are often used improperly, and the results obtained are not reflected critically enough.
- Many physical quantities cannot be measured directly but must be derived by measuring other quantities and then performing calculations.

2 Digital Technologies Can Support and Expand Our Perception

Digital technologies can support educators in their teaching in a variety of ways. The teaching and learning of physics in particular benefits enormously from the use of digital technologies if the discrepancy described above between the time scales, the orders of magnitude or the energy spectrum of the phenomena to be observed and the range of detection of the human sensory world can be bridged through such technologies:

- Simulations and animations can support the acquisition of knowledge and depict processes independently of their natural duration. With smartphones and tablets, students can easily represent even fast processes slowed down (slow motion) and slow processes accelerated (time lapse).
- Animations such as in Powers of Ten (Eames Office 1977) and other films inspired by it, as well as simulations, can at least provide orientation. With the help of (digital) media, small objects can be enlarged, or large objects can be made smaller.
- Special UV cameras can be used to see the world from a bee's point of view, and amazing effects can be achieved, e.g., with flowers that are highly reflective in the UV range, making them very easy for bees to see. Thermal imaging cameras record infrared radiation and display the temperature of the objects recorded in false color images.
- Technical aids can make phenomena accessible in physics lessons that would otherwise remain hidden from us. This is particularly relevant in the case of ionizing radiation from radioactive sources and X-rays, which can be truly dangerous for living organisms and for which we have no sensory organ. Thus, technical aids are relevant to safety when conducting experiments in physics

classes and make many experiments possible in the first place or greatly simplify the acquisition of knowledge.
- Computer-aided digital data acquisition allows, for instance, the automated recording of measured values at fixed points in time, even for processes that could not be recorded without digital data acquisition. In addition to computer-assisted digital data acquisition systems, students and teachers today have access to mobile mini-labs in the form of smartphones and tablets and the sensors built into them, which can be accessed at any time, even outside the school laboratory. The built-in camera opens up a wide range of applications, e.g., for documenting experiments, analyzing motion using video analysis tools and even measuring ionizing radiation. External sensors, which can be quickly and easily connected wirelessly to the mobile device, greatly expand the possible uses of smartphones and tablets for digital measurement acquisition (Kremser and Sekyra 2021).
- For multi-step calculations or calculations that must be repeated several times, widely used tools such as spreadsheet applications can be used to digitally process the recorded measurement data.

Due to the myriad possibilities for supporting physics teaching with digital technologies, physics education research has dealt with the use of computers in teaching from a comparatively early stage. Today, this development can be observed in the use of mobile devices with built-in sensors. Thus, learning with media in the sense of using digital technologies as a subject-specific means of gaining knowledge, as a learning tool for imparting knowledge and for acquiring competencies is just as much a part of physics education research and development as learning via media in the subject-related examination of digital technologies. Central questions are whether and how the teaching–learning process can be improved by the use of digital technologies.

2.1 Students Need to Acquire Digital Competencies in Physics Classes

In the examples described above, the learning of subject knowledge and the acquisition of subject-related competencies are primary goals of the use of media in physics teaching. However, for the German-speaking countries, the educational mandate has expanded significantly beyond these goals, as can be seen with the strategy of the Standing Conference of Ministers of Education and Cultural Affairs of the Länder "Bildung in der digitalen Welt [Education in the digital world]" for Germany (Kultusministerkonferenz 2016), the amendment of the Ordinance of the Federal Minister of Education and the Arts on the curricula of general secondary schools for Austria (Bundesministerium für Unterricht und Kunst 1984) or the Curriculum 21 for the German-speaking Cantons of Switzerland (Deutschschweizer Erziehungsdirektoren-Konferenz (D-EDK) 2014). The promotion of digitalization-related competencies of school-aged students has been declared another main goal of school education.

Digital literacy should be addressed in the context of each school subject, rather than as a separate subject for students' digital literacy. Thus, it is no longer a question of whether digital technologies should be used in physics education (and how they can facilitate physics learning). From the perspective of physics education research, this gives rise to the task of promoting digitalization-related competencies among students in the subject lessons, without this being at the expense of teaching subject knowledge and promoting subject-related competencies. Teachers in particular are called upon to do this, which requires physics teacher training that develops and promotes the relevant competencies among pre-service and in-service physics teachers.

2.2 *Physics Teachers Require Digital Competencies*

The digitalization-related competencies defined by the Standing Conference of the Ministers of Education and Cultural Affairs of the Länder in the Federal Republic of Germany must be an integral part of the subject curricula of all subjects (Kultusministerkonferenz 2016). Thus, each subject must make its own contribution to developing and promoting these competencies in students. To do so, teachers must first have these competencies themselves, but also profession-specific competencies, since each subject enables its own subject-specific approaches to integrating digital media in the classroom. Physics teachers therefore need extended subject-related competencies related to digitalization in order to be able to exploit the potentials of media use in physics lessons and to contribute to the competence development of their students. This implies a cumulative development of digitalization-related competencies across all phases of teacher education. In December 2021, the Standing Conference of the Ministers of Education and Cultural Affairs of the Länder in the Federal Republic of Germany therefore called for the development of state-specific competency frameworks in which the acquisition of competencies is to be divided among the subject sciences, the subject didactics, and the educational sciences in a binding manner (KMK 2021).

However, this step also highlights a current deficit in teacher education in Germany because the lack of such competency frameworks means that there are also no subject-specific standards for the education of teachers with regard to digitization-related competencies in particular. This deficit may be one reason why digitalization is only slowly making its way into the classroom, which is clearly shown by the International Computer and Information Literacy Study (ICILS) 2018, especially for science teaching (Eickelmann et al. 2018). For example, 57.4% of the eighth graders surveyed said they never use digital media to learn in class. When digital media are used, the most frequently cited form of use by the teachers surveyed was digital presentation in frontal instruction. This means that only analog forms of presentation such as overhead projectors or blackboards are substituted [cf. SAMR model (Puentedura 2009)]. Thus, teachers do not seem to have the necessary competencies to use the transformative potential of digital media for redesigning and reorganizing subject lessons. As a result, not only is the potential to didactically design subject lessons in a contemporary way using digital media unrealized, but the requirements for subject lessons to develop and promote digitization-related competencies in students are also not met.

Teacher training institutions are faced with the major task of making the development of digitalization-related competencies mandatory in subject-specific teacher training and are also obliged to do so by the state (Kultusministerkonferenz 2016; KMK 2021; Bundesamt für Kommunikation (BAKOM) 2020; State Secretariat for Education 2019).

3 DiKoLAN—A Framework for Digitalization-Related Science Teacher Education

While several proposals for general digitalization-related competencies have already been published by governmental (Deutschschweizer Erziehungsdirektoren-Konferenz (D-EDK) 2014; Kelentrić et al. 2017; National Institute of Educational Technologies and Teacher Training (INTEF) 2017) and supranational organizations (Redecker 2017; UNESCO 2018; UNESCO-IICBA 2012) as well as non-governmental stakeholders (Battelle for Kids 2019; International Society for Technology in Education (ISTE) 2017) and individuals (Falloon 2020), there is so far only one internationally received orientation framework specifically for the teaching of science subjects: DiKoLAN—Digitale Kompetenzen für das Lehramt der Naturwissenschaften [Digital Competencies for Teaching in Science Education] (Fig. 1) (Becker et al. 2020a; von Kotzebue et al. 2021; Thoms et al. 2021). DiKoLAN describes digital core competencies that prospective science teachers should have developed by the end of their pre-service teacher training.

The DiKoLAN orientation framework distinguishes seven central areas of competence (Fig. 1), which relate both to teaching–learning scenarios in schools and to courses at universities. In addition to four interdisciplinary areas, which include knowledge, skills, and abilities to implement digital communication, presentation, documentation, and information search, three areas specific to the natural sciences are explicitly highlighted. These include the handling and didactical embedding of data acquisition systems, the ability to process data digitally, and to perform and use computer-aided modeling and simulations. While the more general areas address competencies that could also be promoted in subjects other than science, the subject-specific areas address competencies that are mandatory for science education and are closely related to scientific ways of working (Thoms et al. 2021). The seven central competency areas are flanked by technical core competencies. These contain the necessary general basic skills and abilities to use hardware and software in general. The legal framework must also be taken into account.

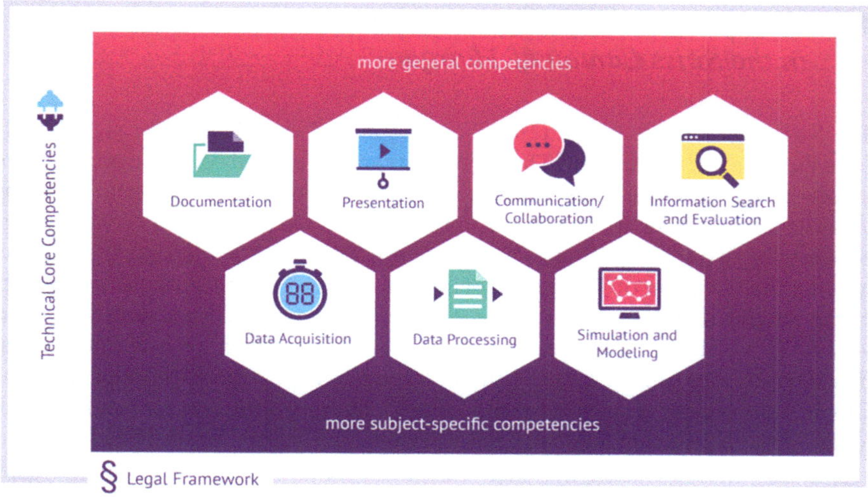

Fig. 1 Areas of competencies in the DiKoLAN framework (https:/dikolan.de) (von Kotzebue et al. 2021)

For each of the seven central competency areas, the competencies a prospective teacher should possess at the end of the university part of teacher education are defined. The competency expectations for an individual competency area are divided into three competency levels (*Name*, *Describe*, and *Use/Apply*) and, following the TPACK framework (Thyssen et al. 2020), into four areas of teacher action (*Teaching*, *Methods/Digitality*, *Content-specific context*, and *Special tools*).

The competency dimensions of the TPACK framework have already been investigated in numerous studies with respect to their particular expressions in different target groups as well as with respect to existing correlations between the dimensions and other person variables. Most of these studies used non-specific self-assessment instruments (e.g., Koehler et al. 2013, 2014; Schmidt et al. 2009).

4 Results of Implementing the DiKoLAN Framework

The DiKoLAN framework is already widely used in science education and science education research (Henne et al. 2022). Of particular interest, however, is the transfer of the framework to (a) other countries and (b) educational practice. Two selected examples are summarized below.

4.1 *Transfer and Adaptation to Teacher Education in the Swiss Canton of Thurgau*

The Swiss school system and teacher education in Switzerland differ significantly from the situation in Germany. Therefore, one can assume that country-specific adaptations have to be made when transferring the DiKoLAN framework from Germany to Switzerland. There are also not insignificant differences in the spoken language between the two countries. As an initial attempt, DiKoLAN was adapted for teacher training in the canton of Thurgau (Henne et al. 2022). For this purpose, the faculty of the natural science subjects at the Thurgau University of Education was interviewed individually about the fit of the DiKoLAN for their own teacher education. During these interviews, all the competency expectations formulated in the DiKoLAN from the areas of teacher action "Content-specific context" and "Special tools" were presented individually. Subsequently, the faculty members and the interviewer discussed the ability to adopt these competency expectations, the necessary adjustments in content and language, and need for deletions, if any. From the analysis of these interviews, it emerged that an adaptation is possible, but that this requires clear modifications, as well as linguistic reformulations. Moreover, the need for four different orientation frameworks specific to the four study programs (Kindergarten, Primary, Secondary 1, and Secondary 2) is clear.

4.2 *Design of a Seminar for the Targeted Promotion of Competencies According to DiKoLAN at the University of Konstanz*

Due to its connection to the TPACK framework, the DiKoLAN framework in particular, and the competency expectations, it contains an ideal basis for measuring corresponding competencies (von Kotzebue et al. 2021; Thoms et al. 2022a) and even for evaluating courses (Henne et al. 2022; Müller et al. 2022). At the Chair for Science Education at the University of Konstanz, a seminar is offered that explicitly addresses the promotion of the competencies described in the DiKoLAN orientation framework (Henne et al. 2022; Müller et al. 2022): "Fachdidaktik III—Digitale Kompetenzen für das Lehramt in den Naturwissenschaften [Subject Didactics III—Digital Competencies for Teaching in Science Education]". After an introductory session, the central competence areas of DiKoLAN are addressed successively on a weekly basis. For the evaluation of the seminar, the effectiveness of each seminar session is surveyed on the self-efficacy expectations in the associated competence area based on the competence expectations described in DiKoLAN. Accordingly, DiKoLAN serves not only as a template for an operationalized version of the seminar's learning objectives, but also as a starting point for the development of a congruent test instrument to survey subject- and digitalization-related self-efficacy expectations. From the perspective of professional practice, DiKoLAN can also aid in lesson planning when existing lessons or parts thereof are to be digitalized or when digitalization-related elements are to be integrated (Meier et al. 2021b).

5 Digital Competencies in Physics Teaching

It is not always possible to set up a completely new seminar that can explicitly and extensively promote the competencies described in DiKoLAN. In most cases, new approaches have to be integrated into existing curricula and established courses. In order for theoretically and empirically founded new teaching concepts to be applied in school practice, they must be enriched with examples of implementation for teachers and published in practical journals aimed at teachers (Girwidz et al. 2019; Girwidz and Thoms 2021). For these cases, examples of procedures that have proven to be successful are presented in the following.

5.1 Promoting Digitalization-Related Teaching Competencies

The pre-service teacher education project "Using digital media to promote experimental skills and train complex data analysis" was developed and has been implemented at the Chair of Physics Education at LMU Munich (Thoms et al. 2020a; Hoyer et al. 2020). The goal of the project is to provide prospective teachers with background knowledge on the targeted use of digital media in physics education. New content was integrated into existing seminars on learning and teaching physics and on school-related experimentation. In addition, an elective in-depth course was offered. The effectiveness of the implementations on attitudes toward and knowledge about the use of digital media in physics teaching was investigated by questionnaire surveys in a pre-post design. In addition, the implementation of what was learned in the learning materials created by the students was evaluated. The course "Teaching with digital media: multimedia and 3D printing in physics education" has a modular structure and covers the following contents:

- Didactic aspects of learning with multimedia (Mayer 2014),
- Digital media in physics teaching (Girwidz and Kohnle 2021; Dębowska et al. 2013),
- Digital data acquisition in on-site, remote and virtual labs (Thoms et al. 2018; Thoms and Girwidz 2017; Hoyer et al. 2019),
- Two- and three-dimensional representations of measured values (Hoyer and Girwidz 2018),
- 3D printing in physics education (Meier et al. 2022),
- Interactive learning and working material.

The learning content of the individual modules contributes overall to the mastery of the semester project. In this course, students design a digital self-learning course for students around an experiment in the Remote Lab, digital data acquisition, and data processing (Thoms et al. 2022b).

5.2 Promotion of Learner Competencies

Thoms et al. (2020b) describe a project-oriented teaching sequence to promote the digitalization-related competencies of students in physics classes, in which learners work in small groups to determine the spatial factor of gravitational acceleration using various digital measurement methods. The students are given a digital measurement method and are asked to (1) organize themselves within the project group and plan their project, (2) research background knowledge, plan the experimental procedure and carry out measurements themselves, (3) prepare a presentation of their project results, and (4) present their project results to their fellow students at a "student congress", selling "their" measurement method as the best of all measurement methods. In the described teaching sequence, several competencies from the strategy "Bildung in der digitalen Welt [Education in the Digital World]" are addressed (Frank and Thoms 2021).

Connecting physical learning content to authentic contexts has been shown in many studies to be effective for learning (Kuhn and Müller 2014). However, this context orientation often also entails an increase in complexity. Digital technologies can help here to reduce complexity by providing scaffolds. An example of this is learning physics in bionic contexts. In bionics, technical implementations are developed starting from biological observations or parallels between biology and technology are investigated. In both cases, physics is an important reference science. Thus, bionics is an excellent context for cross-curricular and interdisciplinary teaching in physics, biology, and technology. A thematic issue of the teacher-oriented journal "Naturwissenschaften im Unterricht Physik [Science in teaching physics]" is dedicated to this very interface and also explicitly addresses the use of media in physics lessons (Watzka 2022).

In digital video analysis, software is used to extract the time-dependent position coordinates of moving bodies from the video relative to a two-dimensional coordinate system (Brown et al. 2022; Freie Universität Berlin 2023). This information can then be used to determine the speed and acceleration of the bodies. The measurement data digitally recorded in this way can also be visualized in different forms of representation, for example in stroboscopic images or diagrams. Nowadays, technological advancements have progressed to such an extent that videos can be recorded in outstanding quality and with a high temporal resolution by the built-in digital cameras in smartphones or tablets. This also enables the analysis of movements that are too fast for the human eye. Using a video analysis app, the entire process of video analysis, from recording the video to analyzing it to visualizing the measurement data, can be performed on a single mobile device. By using mobile devices, students are given the opportunity to investigate movements from their everyday environment, e.g., to conduct freehand experiments in the schoolyard or playground.

On the one hand, the use of mobile devices in the classroom promotes the competence to digitally record and visualize measurement data, and on the other hand, it teaches how a device that is now commonplace for students, such as a smartphone or tablet, can be used as a digital measurement data acquisition system in physics lessons. The learning effectiveness of integrating this digital measurement methodology in the classroom has been empirically demonstrated in several studies (Becker et al. 2020b).

Links

- https://dikolan.de
- https://www.en.didaktik.physik.uni-muenchen.de/multimedia/index.html

References

Battelle for Kids, *Framework for 21st Century Learning* (2019). www.p21.org

S. Becker, T. Bruckermann, A. Finger, J. Huwer, E. Kremser, M. Meier, L.-J. Thoms, C. Thyssen, L. von Kotzebue, Orientierungsrahmen Digitale Kompetenzen für das Lehramt in den Naturwissenschaften: DiKoLAN, in *Digitale Basiskompetenzen, Orientierungshilfe und Praxisbeispiele für die universitäre Lehramtsausbildung in den Naturwissenschaften* eds. by S. Becker, J. Messinger-Koppelt, C. Thyssen (Joachim Herz Stiftung, Hamburg, 2020a), pp. 14–43

S. Becker, P. Klein, A. Gößling, J. Kuhn, Using mobile devices to enhance inquiry-based learning processes. Learn. Instr. **69**, 101350 (2020b). https://doi.org/10.1016/j.learninstruc.2020.101350

D. Brown, W. Christian, R.M. Hanson, Tracker: video analysis and modelling tool. Tracker (2022)

Bundesamt für Kommunikation (BAKOM), *Strategie Digitale Schweiz* (2020)

Bundesministerium für Unterricht und Kunst, *Verordnung des Bundesministers für Unterricht und Kunst vom 14. November 1984 über die Lehrpläne der allgemeinbildenden höheren Schulen; Bekanntmachung der Lehrpläne für den Religionsunterricht an diesen Schulen* (1984)

E. Dębowska, R. Girwidz, T. Greczyło, A. Kohnle, B. Mason, L. Mathelitsch, T. Melder, M. Michelini, I. Ruddock, J. Silva, Report and recommendations on multimedia materials for teaching and learning electricity and magnetism. Eur. J. Phys. **34**(3), L47 (2013). https://doi.org/10.1088/0143-0807/34/3/L47

Deutschschweizer Erziehungsdirektoren-Konferenz (D-EDK), *Lehrplan 21: Gesamtausgabe* (2014)

Eames Office, *The Powers of Ten* (1977)

B. Eickelmann, W. Bos, J. Gerick, F. Goldhammer, H. Schaumburg, K. Schwippert, M. Senkbeil, J. Vahrenhold (eds.), *ICILS 2018 #Deutschland. Computer- und informationsbezogene Kompetenzen von Schülerinnen und Schülern im zweiten internationalen Vergleich und Kompetenzen im Bereich Computational Thinking* (Münster, New York, 2019), p. 408. https://doi.org/10.25656/01:18166

G. Falloon, From digital literacy to digital competence: the teacher digital competency (TDC) framework. Educ. Tech Res. Dev. **68**, 2449–2472 (2020). https://doi.org/10.1007/s11423-020-09767-4

T. Frank, L.-J. Thoms, Digitale Kompetenzen beim Experimentieren fördern: Ortsfaktorbestimmung mit verschiedenen Sensoren im Physikunterricht. PhyDid B-Didaktik Der Physik-Beiträge Zur DPG-Frühjahrstagung **157**, 13–20 (2021)

B. Franklin, Of the stilling of waves by means of oil. Extracted from sundry letters between Benjamin Franklin, LL. D. F. R. S. William Brownrigg, M. D. F. R. S. and the Reverend Mr. Farish, in

Proceedings of the Royal Society of London. Philosophical Transactions of the Royal Society XLIV (1774), pp. 445–460
Freie Universität Berlin, *Viana* 2 (2023)
R. Girwidz, L.-J. Thoms, H. Pol, V. López, M. Michelini, A. Stefanel, T. Greczyło, A. Müller, B. Gregorcic, M. Hömöstrei, Physics teaching and learning with multimedia applications: a review of teacher-oriented literature in 34 local language journals from 2006 to 2015. Int. J. Sci. Educ. **41**(9), 1181–1206 (2019). https://doi.org/10.1080/09500693.2019.1597313
R. Girwidz, A. Kohnle, Multimedia and digital media in physics instruction, in *Physics Education*, eds. by H.E. Fischer, R. Girwidz (Springer International Publishing, Cham, 2021), pp. 297–336
R. Girwidz, L.J. Thoms, Physics education with multimedia applications in non-English teacher-oriented journals: an analysis of 491 articles about multimedia in physics education. J. Phys. Conf. Ser. **1929**, 012035 (2021) https://doi.org/10.1088/1742-6596/1929/1/012035
A. Henne, P. Möhrke, L.-J. Thoms, J. Huwer, Implementing digital competencies in university science education seminars following the DiKoLAN framework. Educ. Sci. **12**(5), 356 (2022). https://doi.org/10.3390/educsci12050356
C. Hoyer, R. Girwidz, A remote lab for measuring, visualizing and analysing the field of a cylindrical permanent magnet. Eur. J. Phys. **39**, 065808 (2018). https://doi.org/10.1088/1361-6404/aae35a
C. Hoyer, L.-J. Thoms, R. Girwidz, Field of a permanent magnet: remotely controlled measurement. J. Phys. Conf. Ser. **1233**, 012003 (2019). https://doi.org/10.1088/1742-6596/1223/1/012003
C. Hoyer, L.-J. Thoms, R. Girwidz, Lehren mit Multimedia, Fernlaboren und 3D-Druck im Physikunterricht: teaching with multimedia, remote labs and 3D printing in physics classes, in *Naturwissenschaftliche Kompetenzen in der Gesellschaft von morgen*. ed. by S. Habig (Universität Duisburg-Essen, Essen, 2020), pp.979–982
International Society for Technology in Education (ISTE), *ISTE Standards for Educators* (2017)
M. Kelentrić, K. Helland, A.-T. Arstorp, *Professional Digital Competence Framework for Teachers* (The Norwegian Centre for ICT in Education, 2017)
KMK, *Lehren und Lernen in der digitalen Welt: Die ergänzende Empfehlung zur Strategie "Bildung in der digitalen Welt"* (2021)
M.J. Koehler, P. Mishra, W. Cain, What is technological pedagogical content knowledge (TPACK)? J. Educ. **193**(3), 13–19 (2013). https://doi.org/10.1177/002205741319300303
M.J. Koehler, P. Mishra, K. Kereluik, T.S. Shin, C.R. Graham, The technological pedagogical content knowledge framework, in *Handbook of Research on Educational Communications and Technology*, eds. by J.M. Spector, M.D. Merrill, J. Elen, M.J. Bishop (Springer New York, New York, NY, 2014), pp. 101–111
E. Kremser, P. Sekyra, Eignung von wireless sensoren in experimental physik-Vorlesungen. PhyDid B-Didaktik Der Physik-DPG-Frühjahrstagung **14**, 461–468 (2021)
J. Kuhn, A. Müller, Context-based science education by newspaper story problems: a study on motivation and learning effects. Perspect. Sci. **2**(1–4), 5–21 (2014). https://doi.org/10.1016/j.pisc.2014.06.001
Kultusministerkonferenz, *Strategie der Kultusministerkonferenz "Bildung in der digitalen Welt"* (2016)
R.E. Mayer (ed.), *The Cambridge Handbook of Multimedia Learning* (Cambridge University Press, New York, 2014)
M. Meier, T. Schubatzky, M. Obczovsky, L.-J. Thoms, C. Thyssen, Fachdidaktische Perspektiven und Szenarien des 3D-Drucks im naturwissenschaftlichen Unterricht. MNU J. **75**(1), 79–84 (2022)
M. Meier, C. Thyssen, S. Becker, T. Bruckermann, A. Finger, E. Kremser, L.-J. Thoms, L. von Kotzebue, J. Huwer, Digitale Kompetenzen für das Lehramt, in *den Naturwissenschaften Beschreibung und Messung von Kompetenzzielen der Studienphase im Bereich Präsentation in Bildung in der digitalen Transformation* eds. by H.-W. Wollersheim, M. Karapanos, N. Pengel (Waxmann, 2021a), pp. 184–189
M. Meier, L.-J. Thoms, S. Becker, A. Finger, E. Kremser, J. Huwer, L. von Kotzebue, T. Bruckermann, C. Thyssen, Digitale Transformation von Unterrichtseinheiten: DiKoLAN als

Orientierungs- und Strukturierungshilfe am Beispiel Low-Cost-Photometrie mit dem Smartphone. Digital transformation of existing teaching units also for distance teaching: DiKoLAN as an orientation and structuring aid using the example of low-cost photometry with the smartphone, in *Digitalisation in Chemistry Education. Digitales Lehren und Lernen an Hochschule und Schule im Fach Chemie*, eds. by N. Graulich, J. Huwer, A. Banerji (Waxmann, 2021b), pp. 13–28. https://doi.org/10.31244/9783830994183

L. Müller, L.-J. Thoms, P. Möhrke, A. Henne, J. Huwer, Erprobung neuer Konzepte in der universitären Lehrerbildung für den Erwerb digitaler Kompetenzen nach DiKoLAN: Entwicklung und Untersuchung der Wirksamkeit eines Lehr-Lernmoduls im Bereich Simulation und Modellierung. Testing new concepts in university teacher training for the acquisition of digital competencies according to DiKoLAN: development and investigation of the effectiveness of a teaching-learning module in the field of simulation and modelling. Chemkon **29**(S1), 349–354 (2022). https://doi.org/10.1002/ckon.202200022

National Institute of Educational Technologies and Teacher Training (INTEF), *Common Digital Competence Framework for Teachers* (2017)

R.R. Puentedura, *As We May Teach: Educational Technology, From Theory Into Practice* (2009)

C. Redecker, *European Framework for the Digital Competence of Educators: DigCompEdu* (Publications Office of the European Union, Luxembourg, 2017)

D.A. Schmidt, E. Baran, A.D. Thompson, P. Mishra, M.J. Koehler, T.S. Shin, Technological pedagogical content knowledge (TPACK): the development and validation of an assessment instrument for preservice teachers. JRTE **42**, 123–149 (2009)

State Secretariat for Education, *Research and Innovation SERI, Orientierungsrahmen Grundkompetenzen in Informations- und Kommunikationstechnologien (IKT)* (2019)

L.-J. Thoms, R. Girwidz, Virtual and remote experiments for radiometric and photometric measurements. Eur. J. Phys. **38**(5), 55301–55324 (2017). https://doi.org/10.1088/1361-6404/aa754f

L.-J. Thoms, G. Colicchia, R. Girwidz, Audiometric test with a smartphone. Phys. Teach. **56**, 478 (2018). https://doi.org/10.1119/1.5055334

L.-J. Thoms, C. Colberg, P. Heiniger, J. Huwer, Digital competencies for science teaching: adapting the DiKoLAN framework to teacher education in Switzerland. Front. Educ. **7**, 802170 (2022a). https://doi.org/10.3389/feduc.2022.802170

L.-J. Thoms, A. Finger, C. Thyssen, T. Frank, Digitale Kompetenzen beim Experimentieren fördern: Schülerexperimente zur Messung der Periodendauer eines Fadenpendels und zur Bestimmung des Ortsfaktors. Naturwissenschaften Im Unterricht Physik **31**(179), 23–27 (2020b)

L.-J. Thoms, C. Hoyer, R. Girwidz, Mit digitalen Medien experimentelle Kompetenzen fördern und komplexe Datenauswertungen schulen, in *Digitale Basiskompetenzen*, eds. by S. Becker, J. Messinger-Koppelt, C. Thyssen (Joachim Herz Stiftung, Hamburg, 2020a), pp. 111–114

L.-J. Thoms, C. Hoyer, R. Girwidz, A teacher training course on using digital media for acquisition, visualisation and 3D printing of complex data, and for fostering pupils' experimental skills, in *Physics Teacher Education: What Matters?*, eds. by J. Borg Marks, P. Galea, S. Gatt, D. Sands (Springer International Publishing, Cham, 2022b), pp. 75–90. https://doi.org/10.1007/978-3-031-06193-6_6

L.-J. Thoms, M. Meier, J. Huwer, C. Thyssen, L. von Kotzebue, S. Becker, E. Kremser, A. Finger, T. Bruckermann, DiKoLAN: a framework to identify and classify digital competencies for teaching in science education and to restructure pre-service teacher training, in *Society for Information Technology and Teacher Education International Conference, Mar 29–Apr 2, 2021* (Association for the Advancement of Computing in Education (AACE), Waynesville, 2021), pp. 1652–1657

C. Thyssen, L.-J. Thoms, E. Kremser, A. Finger, J. Huwer, S. Becker, Digitale Basiskompetenzen in der Lehrerbildung unter besonderer Berücksichtigung der Naturwissenschaften, in *Digitale Innovationen und Kompetenzen in der Lehramtsausbildung*, eds. by M. Beißwenger, B. Bulizek, I. Gryl, F. Schacht (Universitätsverlag Rhein-Ruhr, 2020), pp. 77–98. https://doi.org/10.17185/duepublico/73330

UNESCO, *UNESCO ICT Competency Framework for Teachers* (2018)

UNESCO-IICBA, *ICT-Enhanced Teacher Standards for Africa (ICTeTSA)* (2012)
L. von Kotzebue, M. Meier, A. Finger, E. Kremser, J. Huwer, L.-J. Thoms, S. Becker, T. Bruckermann, C. Thyssen, The Framework DiKoLAN (digital competencies for teaching in science education) as basis for the self-assessment tool DiKoLAN-Grid. Educ. Sci. **11**(12), 775 (2021). https://doi.org/10.3390/educsci11120775
B. Watzka, Bionik, Naturwissenschaften im Unterricht Physik and Bionik im fächerübergreifenden Physikunterricht: Erkennen - Abstrahieren - Umsetzen. NiU Phys. **185**(32), 15227 (2022)

Distance Learning in Physics: Potential and Challenges

Michael Gregory and Steven Goldfarb

Abstract Lessons learned from the COVID-19 pandemic; while not without challenges, the pandemic lockdowns and the shift to distance learning also opened up many doorways for education and how both students and teachers can collaborate. Now that we have returned to the classroom, we can learn from our experience with distance education to facilitate international collaboration in education.

Keywords VSC · PPK · IPPOG · Perimeter Institute · CERN · Distance learning · Virtual camps · Virtual visits

1 Introduction

The global pandemic caused by COVID-19 made the years 2020 and 2021 challenging for all of us. By 2022, things were getting better, and we were emerging stronger, more experienced, and especially more connected than ever before. Nowhere is this truer than in education.

Teachers, students, and parents alike faced several challenges adapting suddenly to distance education. It wasn't always pretty, but the curious and motivated among us see challenges as opportunities for innovation. 2020–22 was a period of intense innovation in education. Innovation that arose out of necessity, but that taught us lessons that continue to open doors and provide more opportunities for unprecedented collaboration between teachers and students.

This chapter is written by a teacher and a particle physicist who were brought together by initiatives they took to inspire others with their love of science despite lockdown measures. It is divided into two sections, the first will focus on the initiatives of one eager teacher and his desperation to keep teaching, and the other will focus

M. Gregory (✉)
My Favourite Experiments, Paris, France
e-mail: myfavouriteexperiments@gmail.com

S. Goldfarb
School of Physics, University of Melbourne, Swanston Street and Tin Alley, Parkville VIC 3010, Australia

M. Streit-Bianchi et al. (eds.), *New Challenges and Opportunities in Physics Education*, Challenges in Physics Education, https://doi.org/10.1007/978-3-031-37387-9_22

on larger initiatives made by institutions. In both cases, the innovations which were made to connect students through distance education continue to be used to increase access to quality science education and outreach and make it easier for curious people around the world to fall in love with science.

2 Initiatives for Students and Teachers

2.1 Virtual Science Camps

"Virtual Science Camps" are online camps for students and teachers to perform experiments with simple materials from around the home. Started during the first weeks of COVID-19 lockdowns in France and around the world, they were designed to keep kids actively learning and doing hands-on science.

Founded by Michael Gregory in Paris, France, and featuring guest hosts from a dozen countries around the world, they were attended by hundreds of different kids and teachers from tens of countries. During their peak, virtual camps would take place daily, with schedules and lists of materials to gather for the experiments sent at the beginning of each week. This was a unique way for students from across the world to participate in the same virtual classroom with their kitchen as the laboratory. Many teachers attended the camps for inspiration on how to teach hands-on science through distance education. Some even invited administrators to show them what can be possible in an online classroom.

The international collaboration and innovative use of simple materials for hands-on science led to these camps being continued long after students returned to school in person. Among the most popular sessions were ones led by exciting guest hosts around the world, which in 2021 led to the creation of the specialised series of virtual camps "Particle Physics for Kids" (Fig. 1).

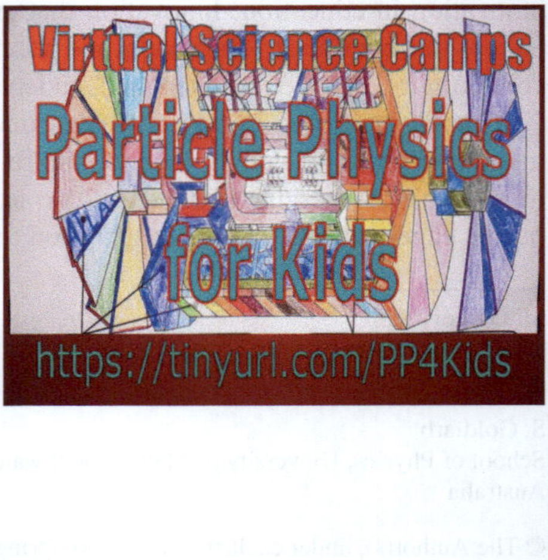

Fig. 1 Virtual science camps (@Michael Gregory)

2.2 *Particle Physics for Kids*

"Particle Physics for Kids" grew out of "Virtual Science Camps". The virtual camps brought together kids, parents, and teachers from dozens of countries through their passion for hands-on science, and this enthusiasm was worth maintaining even once distance education was finally being phased-out.

From the beginning, the virtual camps featured guest hosts from around the world, mainly friendly teachers and scientists who Michael had met on his travels. Some of the most exciting guests came from CERN in Switzerland and JINR in Russia. A dedicated "Particle Physics for Kids" series of virtual camps was a way to continue connecting excellent outreach with eager kids around the world.

This came at the same time that CERN was developing a new series of virtual visits to many facilities, and Particle Physics for Kids was the perfect place to pilot many of these new visits, as well as CERN's S'Cool Lab to test some of their new online science shows.

For more information on CERN virtual visits, see: https://visit.cern/virtual.

For information on ATLAS virtual visits, see: https://atlas.cern/Discover/Visit/Virtual-Visit.

As with his other virtual camps, many of the participants of Particle Physics for Kids were teachers, eager for new ideas to use in their teaching. After two successful runs in 2021 and 2022, Michael is taking a break from Particle Physics for Kids to work on plans for a new "Particle Physics for Teachers" series of camps, where the focus will be on sharing teachable ideas to incorporate particle physics topics into a general physics and general science classroom.

For more information, see: http://tinyurl.com/PP4Kids.

Fig. 2 YES INTERNATIONAL logo (@ Michael Gregory)

2.3 YES! International

YES! International is a gathering of school science clubs from around the world to share low-cost experiments. We hope to promote hands-on science in schools by sharing examples of interesting science that can be carried out in any country with minimal expensive equipment.

YES! for Young Enthusiastic Scientists is a science club founded in Paris, France in 2017, where older students develop hands-on experiments which they lead for younger students. This model fosters leadership skills and deeper thinking about scientific concepts in older students, while younger students benefit from seeing role models and exciting new experiments. At times YES! experiments have even been adopted by teachers to improve their practice.

In the summer of 2019, a handful of YES! students followed Michael Gregory to Ghana, where they worked with local teacher trainer Christoffer Akpeloo to run experiments in several schools across the Volta Region. School closures and travel restrictions made this impossible in 2020, so they began to travel virtually through the magic of Zoom, and YES! International was born!

YES! International started as a way for students in France to connect with students in Ghana. Each week a different school would volunteer for their club to present experiments which the others would repeat on the other side of the world. Following the success of this collaboration, YES! International meetings have also been attended by students in Italy, Albania, Greece, and Serbia.

New clubs are welcome to join. For more information, see the YES! International homepage: https://www.scienceonstage.fr/yes-international/.

Fig. 3 Experiment share a fruitful experience for teachers (©Michael Gregory)

2.4 *Experiment Share*

"Experiment Shares" are monthly meetings for teachers around the world to get together and share some of their favourite experiments (Fig. 3). Most meetings feature five to ten experiments from nearly as many countries and are attended by tens of teachers from up to twenty countries at once.

Piloted in August 2021, and modelled after YES! International meetings for students, inspiration came also from Perimeter Institute's Kitchen Wars, and Science on Stage. Run monthly since November 2021, over one hundred experiments have been shared so far with teachers in 25 or more countries. Late 2022 saw the addition of Spanish-language Experiment Share meetings. For more information, including recordings of past meetings and the schedule of upcoming meetings, see: https://www.scienceonstage.fr/experiment-share.

2.5 *Activities in Ghana*

Similar transformations and initiatives are taking place in the least developed countries (LCDs). Here are some remarks on our experience in Ghana, West Africa, where two schools are founding members of YES! International and a handful of teachers attend monthly experiment share meetings alongside teachers from Europe and around the world. Less available infrastructure, poor internet connections, and digital tools provide additional challenges for teachers here, but with innovation and initiative, it is possible to find solutions.

In Europe, access to computers and the Internet is often taken for granted. While in Europe, many places have started introducing one laptop per child policies, whereas many schools in Africa would be lucky if they could find one laptop per classroom or even per school. Innovative approaches have been seen, using one laptop shared amongst a whole class, including copying diagrams larger onto a blackboard and students taking turns consulting the screen and completing written tasks.

Internet access is another challenge, with no wired connections outside of larger cities and capitals, mobile hotspots over a metered connection to a 3G network are often the best connection available. This can be sufficient for videoconferencing provided that some attention is put into planning and execution, e.g. mapping a school site to find the areas with the strongest signal—at times, this can include hanging a phone from a tall tree or climbing onto the roof or out of a window can make the difference between a lagging video and a smooth connection. It is also worth having multiple phones with SIM cards from different providers to maximise the chances that at least one provider will have a strong network signal. And of course, remember to always top up with extra balance before any important meetings!

Despite the additional challenges faced, technology and the Internet are transforming education in Africa and allowing teachers and students to be more connected with the rest of the world than ever before (Fig. 4).

Fig. 4 HINTEGRAM Ambassador Domenic using one laptop per classroom (© 2021 HINTEGRAM)

3 Institutional Science Education Initiatives for Teachers

3.1 Science on Stage—Webinars and Resources

Science on Stage is a network of science teachers in 36 countries across Europe and beyond with a reach of roughly 100,000 teachers each year (Fig. 5). The network grew out of the Physics on Stage Festival, which started in 1999 but was changed to Science on Stage in 2004 to include all sciences. Science on Stage started as a

Fig. 5 Logo of Science On Stage Europe (reduce LOGO size to fit)

SCIENCE ON STAGE
EUROPE

Fig. 6 Logo SCIENTIX The community for science education in Europe (reduce logo size to fit)

festival, but through lasting connections between teachers, it has grown to become one of the largest teacher networks in the world. With the motto "From teachers for teachers", many excellent teaching resources are collaboratively created and spread throughout member states and beyond, including resource books, workshops, and webinars.

Science on Stage develops teaching resource booklets at both the national and the European levels. European resource booklets are collaboratively developed by teams of teachers from many different countries each contributing their expertise. Dozens of booklets have been written on topics varying from smartphones to science, sustainability goals, sport science, and cooking. They are all available to download for free on the Science on Stage Teaching Materials portal: https://www.science-on-stage.eu/teachingmaterials.

Many national delegations also create high-quality resources. Science on Stage Ireland creates books of demonstrations and teaching ideas. At festivals, each member of the Irish is responsible for gathering ideas for at least ten of their favourite experiments they have picked up from other teachers. They are compiled into books which are distributed in print form and online in.pdf form for free: http://www.scienceonstage.ie/resources/. Several experiments are also demonstrated in their video series, accessible on the same website.

In April 2020, in response to the pandemic, Science on Stage started offering webinars as a way to continue connecting teachers despite travel restrictions. Always from teachers for teachers, each webinar features one or more teachers sharing excellent experiments and teaching ideas. Some of the most popular ones have featured microchemistry, "What happens next?", and "It's not magic, it's science you don't see" series, now presenting part 5. For a list of upcoming webinars, see https://www.science-on-stage.eu/events.

3.2 *Scientix—Webinars, Blog, and MOOC*

Scientix is "the community for science education in Europe" (Fig. 6). It aims to promote and support a Europe-wide collaboration among STEM teachers, education

Fig. 7 Logo Perimeter Institute

researchers, policymakers, and other educational stakeholders to inspire students to pursue careers in STEM fields. Scientix was born in 2010 at the initiative of the European Commission, and European Schoolnet has coordinated the project since its first launch. This Brussels-based consortium of 34 Ministries of Education in Europe is an innovation driver in teaching and learning and fosters pan-European collaboration between schools and teachers.

Scientix Ambassadors are at the core of the teacher experience. Scientix has over 1000 ambassadors in 50 different countries. They are enthusiastic teachers who foster and support innovative education in STEM subjects in their countries and across Europe. They form a collaborative community that actively participates in several initiatives at the local, national, and European levels.

To find a Scientix ambassador near you, see: https://www.scientix.eu/in-your-country/scientix-4-teacher-panel.

Scientix regularly hosts webinars led by teachers and other stakeholders in education. During the pandemic, their frequency increased, and the focus shifted to supporting teachers through distance education. This further increased the presence and impact of Scientix across Europe and around the world. Scientix webinars are a great way for teachers to keep up to date on new ideas in science education, both from teachers, industry, and policymakers.

For lists of past and upcoming webinars, see: https://www.scientix.eu/live/scientix-webinars.

The Scientix Blog is a series of articles written by teachers, ambassadors, and other stakeholders in education. It is a place to share and keep up to date on teaching ideas, experiments, and projects. Topics range from particle physics masterclasses to homemade experiments, UN sustainability goals, educational travel, and more.

Check out the Scientix blog here: https://blog.scientix.eu/.

3.3 Perimeter Institute—Online Teacher Camp and Workshops

The Perimeter Institute for Theoretical Physics, in Waterloo, Canada (Fig. 7), is a leading centre for theoretical physics. Since its beginnings in 2000, outreach has been at the core of its vision, and it creates some of the world's best teaching resources for physics and the nature of science. With these resources, Perimeter Outreach delivers high-quality professional development for teachers locally and from around the world, notably through their annual "Einstein Plus" teaching programme.

Pandemic-induced distance education led Perimeter Outreach to significantly increase its online offer. The shift to online suddenly made Peritmeter's teacher programmes accessible to many more teachers around the globe. Following their success throughout the pandemic, many of these programmes continue to be offered online and are enjoyed by teachers around the world.

Resources: https://resources.perimeterinstitute.ca/.

Perimeter's Free Educational Resources for Teachers bring cutting-edge science into classrooms around the world. They are created over a 2-year development by Perimeter's outreach team in collaboration with researchers and the Perimeter Teacher Network. Organised by topic, they are flexible and modifiable, and teachers can use them for a single lesson or to teach whole curriculum units. All documents shared with teachers are available in editable versions so they can be adapted to regional/country differences and modified to best suit the classrooms where they are used. Teachers are encouraged to share any particularly successful modifications and use cases, and their feedback is used in Perimeter's iterative process to continually update and improve their resources.

"EinsteinPlus" is Perimeter's flagship teacher programme. It is like the Woodstock of physics teaching! Every year in early July, teachers from around the globe and across Canada converge in Waterloo for one week of activities, lectures, lab visits, and more. Teachers get to know Perimeter outreach staff and scientists, learn, and play around with resources, try them out for themselves, and provide feedback which is incorporated in the next iteration of development.

For more information, see: https://perimeterinstitute.ca/einsteinplus.

In 2021, EinsteinPlus was adapted into the Perimeter Institute Online Teacher Camp and became accessible to even more teachers around the world. Like Einstein-Plus, the Teacher Camp provided opportunities to interact and learn from outreach staff and scientists and to explore Perimeter's great teacher resources. The week included "Kitchen Wars", a competition between participants for innovative science demonstrations.

The enthusiasm and success have led to Kitchen Wars now being held annually: https://insidetheperimeter.ca/kitchen-wars-is-back/ (Fig. 8).

Launched in 2009, Perimeter Institute's Teacher Network now includes teachers around the world and has impacted millions of students worldwide. The goal is to provide energetic and accomplished teachers with the tools needed to reach additional educators in their region through professional development workshops. Teacher Network members deliver professional development workshops in their area, help other educators to integrate activities into their classroom, and advise and field test perimeter resources in development.

Perimeter's International Summer School for Young Physicists is like a student version of their teacher camp or EinsteinPlus. Originally run in person, it is now an exciting and challenging two-week online programme for Canadian and international high school students with a keen interest in theoretical physics. In the first 20 years, ISSYP has included over 900 students from 60 countries.

For more information on ISSYP: https://perimeterinstitute.ca/issyp.

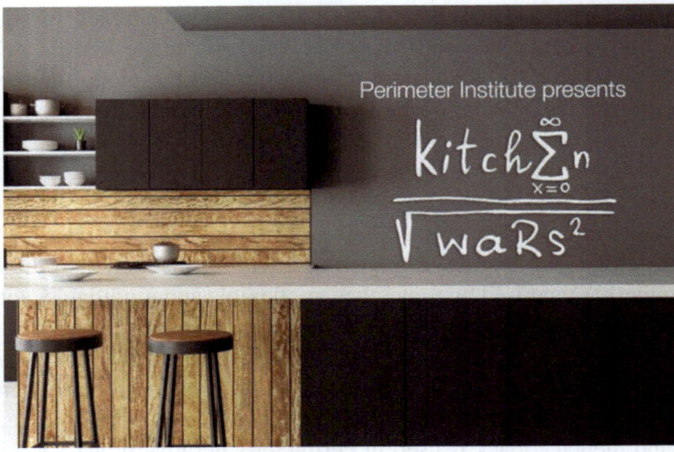

Fig. 8 Kitchen Wars (© Perimeter Institute 2022)

3.4 Virtual Visits to CERN Experiments

CERN is the largest international laboratory for particle physics in the world. It hosts a large variety of leading-edge experiments designed to explore the most fundamental nature of our universe, the elementary particles. Both the laboratory and the large international collaborations that build, run, maintain, and operate the experiments, host a variety of programmes targeting classrooms around the world. And tens of thousands of students and teachers visit the laboratory and its experimental sites, each year.

The experience gained by the visiting classes is an important one. Having the opportunity to interact with scientists involved in the latest research inspires young students and their teachers, allowing them to satisfy their curiosity about science and the latest methods used in the research. The lessons are invaluable, as they not only learn about the common, human questions being investigated and the technology being used to address those questions, but they gain an appreciation of the value of international collaboration and the need for diversity in science. Furthermore, it gives them a chance to meet role models and to seek answers to some of their very practical questions concerning careers in science.

Key to all of this is the human interactions between students, teachers, and scientists. As we learned during the pandemic, such experiences can be made available on a much more global scale through the usage of web-based video conferencing networks. In 2010, following the LHC start-up of Run 1, the ATLAS Experiment set-up equipment and developed a programme to reach out to classrooms, stakeholders, and public events around the world. Since then, virtual visits have grown in popularity and evolved to include all the LHC experiments, as well as other sites at CERN https://iopscience.iop.org/article/10.1088/1742-6596/898/10/102005.

At ATLAS alone, thousands of visitors, primarily students, and teachers have had the opportunity to connect, with many of them becoming regular clients.

4 Major National and International Educational Networks

4.1 Overview

The initiatives described above only touch on many educational programmes being made available through distance learning technology around the world. To coordinate these activities, streamline funding and improve visibility, several national and international networks now exist that can serve as access points for teachers and students. Among these are QuarkNet in the United States https://quarknet.org, Netzwerk Teilchenwelt in Germany https://www.teilchenwelt.de, Institut national de physique nucléaire et de physique des particules (IN2P3) in France https://www.in2p3.cnrs.fr/fr/formation-superieure-0, the public engagement components of the Science and Technology and Facilities Council (STFC) in the United Kingdom https://www.ukri.org/what-we-offer/public-engagement/public-engagement-stfc and Istituto Nazionale di Fisica Nucleare (INFN) in Italy https://home.infn.it/en. These networks foster the development of relations between local schools and researchers, offering common platforms and tools to reach classrooms across their countries.

Beyond the national initiatives, most of the large laboratories and international experiment collaborations have developed their educational programmes, going beyond local and remote visits. Material is continually developed to support efforts by collaboration members both at their sites and back at the home institutions. However, given the limited resources made available to these outreach teams and the similarity of programmes being developed in the various countries, labs, and experiments, it makes sense to coordinate efforts on a global scale.

Fig. 9 Logo of IPPOG The International Particle Physics Outreach Group

4.2 International Particle Physics Outreach Group (IPPOG)

The International Particle Physics Outreach Group (IPPOG) https://ippog.org (Fig. 9) is an international collaboration of scientists, science educators, and communication specialists working across the globe in informal science education and outreach for particle physics. It was founded in 1997 by CERN Director General Chris Llewellyn Smith to coordinate the efforts of these various independent entities to more effectively use the limited resources to reach classrooms and the public across the globe. Since that time, the group has expanded into an international collaboration which, at the time of writing, comprises 40 members, including 33 countries, 6 experiments, and 1 international laboratory (CERN). Two national laboratories, DESY and GSI, participate actively as associate members.

4.3 International Particle Physics Masterclasses

The largest and most well-known global activity of IPPOG is the International Particle Physics Masterclasses https://physicsmasterclasses.org. The programme pairs scientists involved in current research with students, typically hosted in the home institute or laboratory of the scientist or the classroom of the students, to engage in the excitement of fundamental research. Students learn about the physics goals and the functioning of an experiment, then get a chance to be physicists for the day by doing their analysis of data from the detector. At the end of the day, the students meet with other students via videoconference to discuss results, getting a first-hand experience of international collaboration. In the 2019 edition of the International Masterclasses (before the COVID-19 pandemic), 14,000 students from 54 counties and 225 institutions participated. The 2023 edition is on track to surpass these totals (Fig. 10).

While the focus of the Masterclasses is particle physics and much of the activity utilises data and tools from major experiments, the lessons are much broader in scope. Students' inherent interest in science is re-awoken through exposure to cutting-edge research activities and close engagement with the scientists involved in that research. They learn not only science but also the importance of scientific methodology and its relevance to the human understanding of nature. They also learn first-hand the excitement and value of international collaboration, and how it contributes to scientific advancement through human diversity and worldwide cooperation.

4.4 The Global Cosmics Portal

Another key programme supported by IPPOG is the Global Cosmics portal https://ippog.org/global-cosmic-rays-portal, which provides an organisational umbrella for a variety of experiments and data-sharing activities based on cosmic particle detectors

Fig. 10 Particle physics masterclass in Kathmandu, Nepal (Photo: S. Goldfarb/IPPOG © CERN 2022)

located in classrooms around the globe. Events like International Cosmic Day https://icd.desy.de organised by DESY, Netzwerk Teilchenwelt, IPPOG, QuarkNet, and Fermilab, and International Muon Week https://quarknet.org/content/international-muon-week organised by QuarkNet, expand the reach to locations where collider physicists might not have the opportunity to participate in Masterclasses.

It should be noted that, during the COVID-19 pandemic, attendance for Cosmic Ray programmes increased significantly, which is in stark contrast to what happened with Masterclasses. In many cases, detectors could be kept online, with students gaining access to data from home. Outside of the pandemic, such activities can also help to pair up schools with varying levels of resources. A well-equipped classroom, e.g. can share its data with another classroom lacking the ability to purchase a detector. This can occur across town, across the state, across the country, and borders.

4.5 The IPPOG Resource Database

One of the key challenges motivating the creation of IPPOG was the need for a forum for the exchange of both ideas and material between researchers and educators. IPPOG's twice-annual meetings provide an opportunity for scientists, communicators, and educators to get together to discuss the latest efforts, successes, failures, and ideas, to foster excellence in science education. The meetings include presentations, discussion sessions, working group meetings, and panel discussions to hear from experts in the field. Members leave the meetings informed and motivated to try new activities in their classrooms, institutes, or at public events.

To support their activities during and between meetings, a resource database has been developed as a part of the IPPOG website https://ippog.org/ippog-resource-database. The database provides access to a variety of material, including images, videos, games, text, posters, exhibits, etc., that can be used in the classroom, at home, or in public spaces. The material is searchable by topic, age group, target audience, language, media type, etc., and has been developed, procured, and tested by a group of scientists, educators, and students. Links to other databases and platforms are included, as well. The goal is to make this a central depository, easily accessed by teachers and students around the globe, and feedback is continually seeking to improve and add to the existing content.

5 Conclusions

Teachers and scientists alike, both individually and through networks and institutions are making great efforts to make science ever more accessible and engaging. We are entering an era of unprecedented collaboration and accessibility, in part thanks to lessons learned and innovations made during the COVID-19 pandemic.

In this chapter, we've highlighted several initiatives, from individual teachers, teacher networks, research institutions, and national, European and international collaborations. Key to all their success is their ability to connect people—whether it is connecting teachers to students, teachers to teachers, researchers to classrooms, etc., etc., the ability to make strong connections is key to getting research into the minds of curious people young and old. Necessity breeds innovation, and through COVID-19-induced isolation, we were able to innovate new ways to connect, and these innovations are helping us become closer than ever before. It is our sincere hope that you will find inspiration from this chapter to connect with one of these wonderful initiatives to help share our sense of curiosity and wonder.

References

ATLAS virtual visits. https://atlas.cern/Discover/Visit/Virtual-Visit

CERN virtual visits. https://visit.cern/virtual

S. Goldfarb et al., Sharing scientific discovery globally: toward a CERN virtual visit service. J. Phys., Conf. Ser. **898**, 102005 (2017). https://iopscience.iop.org/article/https://doi.org/10.1088/1742-6596/898/10/102005

M. Gregory, Particle physics for kids. http://tinyurl.com/PP4Kids

M. Gregory, Experiment share. https://www.scienceonstage.fr/experiment-share

M. Gregory, YES! International homepage. https://www.scienceonstage.fr/yes-international/

Institut national de physique nucléaire et de physique des particules (IN2P3) research network in France. https://www.in2p3.cnrs.fr/fr/formation-superieure-0

International Cosmic Day. https://icd.desy.de
International Muon Week. https://quarknet.org/content/international-muon-week
International Particle Physics Masterclasses. .https://physicsmasterclasses.org
Istituto Nazionale di Fisica Nucleare (INFN) in Italy. https://home.infn.it/en
Netzwerk Teilchenwelt particle physics education network in Germany. https://www.teilchenwelt.de
Perimeter Institute: Education Resources. https://resources.perimeterinstitute.ca/
Perimeter Institute: EinsteinPlus. https://perimeterinstitute.ca/einsteinplus
Perimeter Institute: International Summer School for Young Physicists. https://perimeterinstitute.ca/issyp
Perimeter Institute: Kitchen Wars. https://insidetheperimeter.ca/kitchen-wars-is-back/
Quarknet: U.S. national teacher and student training network supported by the National Science Foundation, Fermilab and the University of Notre Dame. https://quarknet.org
Science and Technology and Facilities Council (STFC) in the United Kingdom. https://www.ukri.org/what-we-offer/public-engagement/public-engagement-stfc
Science on stage Europe: events list. https://www.science-on-stage.eu/events
Science on Stage Europe: Teaching Materials portal. https://www.science-on-stage.eu/teachingmaterials
Science on Stage Ireland: teaching ideas selected by the Irish team. http://www.scienceonstage.ie/resources/
Scientix, European Schoolnet: Scientix Blog. https://blog.scientix.eu/
Scientix, European Schoolnet. https://www.scientix.eu/live/scientix-webinars
Scientix, European Schoolnet: Ambassadors by Country. https://www.scientix.eu/in-your-country/scientix-4-teacher-panel
The International Particle Physics Outreach Group (IPPOG). https://ippog.org
The IPPOG Global Cosmics Portal. https://ippog.org/global-cosmic-rays-portal
The IPPOG Resource Database. https://ippog.org/ippog-resource-database

Education and Interdisciplinarity: New Keys for Future Generations and the Challenge-Based Learning

Christine Thong, Aaron Down, and Anita Kocsis

Abstract Creativity, innovation and application are established concepts in exploring cross-disciplinary boundaries in science domains, however new possibilities for educational programmes are being enabled by advances in digital tools for communication, imagination and inspiration. These new developments allow educators to experiment with ways to enhance students' mindset and skills towards transformative thinking, sense of agency, questioning the status quo, cooperating to compete for enhanced outcomes across disciplines and to tackle uncertain and ambiguous futures. This chapter explores a tertiary education programme, challenge-based innovation A^3, that aims to do this by integrating different disciplinary concepts (e.g. systems thinking, behaviour change, Planet-Centric Design) for learners across different discipline backgrounds and cultures. Students innovate through creative problem solving to propose new societal applications using physics knowledge and related technology. Through imagining new products, systems, services and built environments in an experiential learning environment, students build capability towards being responsible, next generation innovators.

Keywords Interdisciplinary education · Challenge-based learning · Twenty-first century skills · Responsible innovation

C. Thong (✉) · A. Down · A. Kocsis
Swinburne Design Factory Melbourne, Swinburne University of Technology, John St, Hawthorn VIC, Melbourne 3122, Australia
e-mail: cthong@swin.edu.au

A. Down
e-mail: adown@swin.edu.au

A. Kocsis
e-mail: akocsis@swin.edu.au

M. Streit-Bianchi et al. (eds.), *New Challenges and Opportunities in Physics Education*, Challenges in Physics Education, https://doi.org/10.1007/978-3-031-37387-9_23

1 Introduction

In Rittel and Webbers (1973) acknowledgement of the challenges for planning in a pluralist society, the phrase "wicked problems"—ones that are difficult, impossible to solve, in which no definitive answers are possible—still holds relevance in science and interdisciplinary education. Compounding wicked societal challenges, for educators and employees, is the wicked problem of how to learn and thrive continuously given that entering the workforce today may result in 5 different careers and 17 employee changes over a life work span (Australians and AlphaBeta 2017).

Discussed in this chapter are the creative methods, tools and mindsets for future focused education that applies science; to immerse students from science and other disciplines with new tools, and explore transdisciplinary and pedagogic mindsets to define, locate and embrace the challenge of wicked problems (Rittel and Webber 1973, p. 159). The chapter explores Challenge-Based Innovation (CBI), specifically the CBI A^3 programme (tertiary education), which embeds learning interdisciplinary topics under a design methods framework, and transdisciplinary vocabulary of skills and tools to collaboratively address complex, societal challenges. Design-inspired thinking and practices are employed as they do not rely on exhaustive sets of rules and instructions. Design methods framework in CBI A^3 compliments the practice of "tame problems" such that analysing the structure of an organism or atom has set routines with a set mission (Rittel and Webber 1973, p. 160). Students learn future skills by applying technology and concepts from particle physics science to address United Nations Sustainable Development Goals (UN SDGs), and in the process become more comfortable with the discomfort and the associated skills to understand wicked, messy, and unexpected problems that will only increase into the future.

Groups of students in different global locations (e.g. Australia, Germany, the United States of America, Poland and Portugal) from different discipline backgrounds (including design, science, business, arts and engineering) are exposed to diverse methods, modes of thinking and expertise. Furthermore, realisation of one's own expertise becomes relative to another's in the pursuit of the challenge. The confluence of interdisciplinary practices, design and particle physics as demonstrated by the CBI A^3 programme exemplifies how different disciplinary inputs may stimulate different thinking and idea development. An added layer in the educational stimulus is the cultural diversity of the students in both real and virtual spaces. Digital and blended modes of interaction afford synchronous and asynchronous modes to allow for rich interactions with experts, innovation coaches and peers through temporal iterations made possible by design diagrams, formulae and other prototyping languages and artefacts.

This chapter enriches the toolkits of science to also consider and value different kinds of knowledge and explore interdisciplinary approaches to applications of science and creative reasoning. Interdisciplinary in its most simple form constitutes an appreciation of different discipline fields and the value of communicating across disciplinary boundaries to find ways to work together. Group decision making and encouragement to take risks in sharing ideas fosters confidence in learners to understand that the right answer may not be the most obvious route. Learning each other's disciplinary ropes pushes educators and students to negotiate other forms of scientific reasoning and toolkits (Redish and Cooke 2013) as explored in CBI A^3.

2 The Importance of Interdisciplinary Education

Learning experiences and skills developed through interdisciplinary, challenge-based learning to apply physics fundamentals to real-world solutions for society are explored through the lens of design thinking, systems thinking, planet-centric design, behaviour change, future thinking, technology translation methods. The interdisciplinary curriculum of challenge-based innovation empowers graduates with the ability to think, do and act towards positive societal transformation in a world, where deep technology embedded in particle physics will continue to develop exponentially. The methods and curricula require resources which are beyond the traditional scope of delivery. However, challenge-based curricula development is highly valued in order to deliver twenty-first century skills-based education that includes development of mindset to work with ambiguity, testing and failing with open-ended answers, encouraging creative methods, and collaborating across disciplines.

Drake and Reid (2018) propose integrated curriculum as an effective way to teach twenty-first century skills, providing a Know-Do-Be framework to guide K-12 teachers in developing curriculum. They categorise integrated curriculum into four different levels of integration:

(1) *Fusion*, where a single traditional subject is imbued with some other concepts;
(2) *Multidisciplinary*, where a theme is central to learning, but disciplinary boundaries remain;
(3) *Interdisciplinary*, where boundaries become less important and a specific skill drives learning; and
(4) *Transdisciplinary*, where discipline distinction falls away and a challenge or issue catalyses learning.

The two latter levels lend themselves towards simultaneously building many of the twenty-first century skills identified by the World Economic Forum (2017), where competencies (problem solving/critical thinking, creativity, collaboration and communication) and character qualities (curiosity, persistence/grit, initiative, adaptability, leadership and social and cultural awareness) are needed to build on Foundation Literacy skills (e.g. scientific literacy, ICT, etc.). Jia et al. (2021) support integrated curricula building interdisciplinary skills, with a study involving elementary school students blending the hands-on approach of Maker Education (that focuses on technology and creativity) with STEAM interdisciplinary principles (problem solving enhanced by interdisciplinary thinking) and an engineering design approach as an overarching learning framework. Similarly, CBI A^3 uses design methods and practices as the overarching framework to bring a combination of interdisciplinary and transdisciplinary concepts together to build the next generation of innovators who can adapt and have a sense of agency to use deep technology to propose applications for the future that will address societal challenges (Thong et al. 2021).

Transversal skills (discussed further in Chap 12 of Part 2) align with WEF (2016) twenty-first century competencies and character qualities are not bound by disciplines, and are applicable across industry sectors, professions and contexts. These skills are needed to address wicked, uncertain and evolving societal challenges, yet interdisciplinary curriculum is still not yet common practice in schooling (Moss and Godinho 2019) most likely due to the complex implementation requiring more educator collaboration, which takes a little extra time and/or resources. Many educators are overwhelmed and not properly equipped with the tools and capabilities to enable this complex implementation of integrative curriculum (Drake and Reid 2018), and sharing examples of *how* this can be done through CBI A^3 aims to lessen the overwhelm and breakdown topics involved in complexities related to challenge-based learning.

3 Digital Tools for Communication, Imagination and Inspiration

Listed as a twenty-first century *Foundation Literacy* (World Economic Forum (2017), Information and Communication Technologies (ICT) are something most populations in the western world take for granted as part of everyday life. Personal and portable, web-enabled devices mean modes of communication and access to all kinds of information is essentially instantaneous. Most K-12 and tertiary students use digital devices and software to discover, interact, explore, document, express and distribute their personal and professional ideas, values and emotions (think social media). This may be seen to democratise education by creating more accessible opportunities for more students to learn in global and cross-cultural communities of practice, access knowledge and insight from experts in different countries and remote locations and source-free online learning programmes like MOOCs. More recently, we can see how artificial intelligence (AI) is not only filtering but also synthesising and coherently documenting all possible information on a topic through open-access software like Chat GPT, posing not only the obvious challenges of plagiarism and written literacy for educators, but reducing the perceived need by students to have capacity to think critically and discern the information source and content presented to them (for further discussion on AI see the last chapter of the book). Appreciation for evidence-based decision making in science combined with critical and conceptual thinking has never been more important.

The positive possibilities and opportunities digital tools offer learners is vast. In this chapter, we categorise the learning opportunities into three areas (digital tools for communication, imagination and inspiration) as follows, and later give examples of how we leverage some of the tools in CBI A^3 to deliver the programme alongside the integration of different disciplinary concepts catalysed by the need to address wicked, complex societal challenges.

Digital ***communication*** enables remote learning (and working), connecting different organisations, institutions, learner groups and educators across communities, cities, countries and even into different parts of our solar system (satellites, space exploration). The speed of communication thanks to advances in data computing and the Internet allows for fast iterations of communication, where students may receive feedback, insight and responses synchronously and in short periods of time asynchronously. Through advances in telecommunication software, we can hear someone's tone of voice and see facial response, receiving greater tacit cues and information than ever before. Teams can share and store files with joint access and work simultaneously on these from different locations, using all sorts of file types—from written documents, images, videos, links and computer-aided design (CAD) that have been sourced or created. Access to different actors in learning, be they experts or other stakeholders related to a project challenge, gives students access to first-hand knowledge more readily than ever before.

Imagination is digitally enabled through software that allows students to explore, express, develop and collaborate on ideas in ways that require less hard skill development than in the past. For example, sketching or visualising an idea by hand or specific CAD software tools takes time and training. However, a collage of images or a montage of video can quickly create a conceptual sense for the aesthetic, functionality, context and emotion of an idea and can be put together with minimal skill using basic software packages. Creating and editing video can also be done. Image manipulation and programme layouts can be constructed with pre-set options to select from, often presented as suggestions. Online whiteboards also allow for collaborative brainstorming and idea development. There are, of course, still digital tools for communication which require training and skills. Coding can be played with to create different outputs and interactions. Virtual Reality can immerse others into a world or context, creating an experience from the imagination of the creator. Realistic renderings of products, the built environment and technical specification of how to materialise the ideas into production are other examples of this.

Inspiration-enabled digital functionalities closely relates to imagination, enabling learners access to others' work that benchmark and allows them to build on what is already done, challenge their existing thinking, see things that open their imagination in new ways and inform the pragmatic side of realising invention and ideas. The sources of inspiration include students' real-life contexts as experiential, and project-based learning featured in CBI A^3 is more than collecting information and returning it in some scholarly form. The current generation of digital natives is familiar with hybrids of both social and digital worlds and inspiration at the intersection between digital and the real influence how diverse digital methods may influence the breadth and depth of information acquisition. As argued in this chapter, students are positioned as creators with varying levels of access and literacy types and are hence encouraged to seek inspiration within their tool kit and also cater to a broad range of students' digital literacy skills (Kirschner and van Merriënboer 2013).

4 An Overview of Challenge-Based Innovation and the CBI A^3 Program

Challenge-based learning (Gallagher and Savage 2020) is commonplace in experiential, project-based and transdisciplinary education practices, and has inspired the *Challenge-Based Innovation (CBI)* as the name given to an initiative run by IdeaSquare, CERN. IdeaSquare is an Innovation Lab based at CERN, that facilitates collaboration between diverse actors, supporting radical innovation methods and mindset to generate solutions for the betterment of humankind (IdeaSquare CERN 2023). IdeaSquare host many different activities, events, educational programmes and Research and Development and Impact projects. The CBI initiative is part of IdeaSquare's educational programmes, aimed at developing the next generation of innovators and inventors (Leveratto 2021, p. 26). Under the CBI banner, different universities around the globe run different versions of CBI programmes, however, all are united by addressing UN SDGs using deep technology, typically from CERN. It is a complementary, co-creative partnership between IdeaSquare, CERN and Universities. Universities have the infrastructure and expertise to enrol, teach and deliver courses for students, whilst IdeaSquare create a rich learning environment based on open innovation practices that connects students with CERN experts and local NGO ecosystems, providing motivation, inspiration and engagement with world-class science.

CBI A^3 is a programme led by an Australian Higher Education Institution, Swinburne University of Technology, with current partner institutions from Germany (Hochschule Mannheim), the United States of America (Pace University and Pratt Institute) and Poland (Warsaw University of Technology). It is a 7-month programme, where students work as part of an interdisciplinary, global community of practice (Thong et al. 2021) aimed at developing skills and mindset empowering the next generation of globally responsible innovators. Students at the different institutions engage in the same curriculum framework, undertaking experiential, project-based learning to propose design solutions for 2030 that address an issue related to UN SDGs. Each year, CBI A^3 focuses on one specific UN SDG as inspiration, and student teams explore problem spaces that relate to their local ecosystems whilst understanding how these feed into a larger global issue. For example, for UN SDG 3 Good Health and Wellbeing, teams will explore challenges in Melbourne, Australia, Mannheim, Germany, etc., that connect to SDG 3. The UN's SDGs set the tone for the challenge and reinforce the point of interdisciplinarity. The challenges explored in the programme are less about the routines of science and methods of design per se and more about the purpose of managing and striving for solutions. Students experience the discomfort at various stages of the design process to eventually learn that potentially no solution may be found and/or at a certain point in the design journey there are logical and scientific inconsistencies to the problem that pivots the team to another yet-to-be conceived direction. You can learn more about CBI A^3 and explore past projects at cbi.dfm.org.au.

CBI A^3 has two distinct phases. Phase 1 explores a range of different problem and solutions spaces and involves mixing and matching to see where meaningful design solutions may lie so technology is not being pushed into markets for the sake of it. Phase 2 is developing the ideas and considering road mapping an implementation strategy to realise the idea in 2030. Deep technology typically is from the CERN knowledge transfer portfolio and has in recent years grown to include technology from the Australian Research Council Centre of Excellence for Dark Matter Particle Physics (Centre Dark Matter), Australia's Nuclear Science and Technology Organisation (ANSTO) and ATTRACT Academy (where ATTRACT has received funding from the European Union's Horizon 2020 research and innovation programme under grant agreement No. 101004462).

Design-inspired methods provide an overarching framework for integrating a range of different disciplinary concepts together to holistically and effectively address complex societal issues. These topics are discussed further in sections five to ten of the chapter. While small cohorts of students undertake the programme each year (10–16 students), survey data shows that after completing CBI A^3, students confidence to think radically, design for the future, work with deep technology and develop socially responsible innovations increases (Thong et al. 2021).

Digital literacy provides agency as *"learners* can contribute to the local and global knowledge ecosystem, learning through the act of producing and discussing rich media, applications, and objects" (Alexander et al. 2017, p. 2). CBI A^3 maximises digital and integrative learning technologies to support learning, educators' need to ensure they combine and integrate technology with the active learning engagements to produce the greatest yield (Brown et al. 2020, p. 14). Reflection, feedback and record of progress alongside the production of digital content both communicates content and is also a digital repository of information that students build and curate over time (Boud 2020, p. 11). Specific examples of digital tools for communication, imagination and inspiration used in CBI A^3 are discussed in the below section.

4.1 *Communication, Imagination and Inspiration Enabled Through Digital Technologies in CBI A^3*

Digital technologies enable us to ***communicate*** and collaborate across geographic locations and time zones within the CBI A^3 programme and are used in a variety of ways to support learning. Throughout the user-centred design process in CBI A^3, students use digital tools to access stakeholders' impact and understand the context better. They create short videos documenting interviews or site visits, generate surveys using software such as Survey Monkey, Microsoft Forms or Google Forms to collect user perspectives, and create short videos or digital presentations that explain an early stage idea. The digital format allows these ideas to be shared across the globe with various experts and collaborators. Emails, communication channels (such as Slack) and online blogs provide a way for students to connect, share work in progress and talk to external experts about their projects and thinking. As educators, we also leverage digital tools to bring expert lectures from various disciplines across the globe to students to enhance their learning experience.

CBI A^3 heavily embraces experiential learning, a non-theory or teacher-led endeavour that is holistic and embraces learning by and through doing (Dewey 1938), co-creation and interaction. We leverage Miro, a digital platform that enables co-creation and interaction as students experiment in the digital space. It is an industry-leading, cloud-based collaborative whiteboard platform designed to facilitate modern work for co-located, distributed and remote teams to communicate and collaborate without the constraints of the physical environment (Miro 2020). We use Miro as a technology tool and digital interface to support the collaboration activities in CBI A^3 that would normally take place in a physical studio classroom. At times, we also use this digital whiteboard as an interface for the delivery of learning content and interactive activities that can guide students through complex innovation processes and frameworks. As every action leaves a digital trace, it makes it easier to track process, the evolution of ideas and a team's thinking. The continuous documentation is a key element in being able to build on others' work and invite others' thinking. Miro provides a visual map and creates a shared understanding across members which improves remote communication. For example, we use online whiteboarding tools, like Miro, when introducing students to Systemic Thinking and Planet-Centric Design frameworks which allows students to populate templates guiding them to collaboratively map systems. These templates can easily be adjusted and iterated in a digital format in comparison to a printed physical template.

Imagination is fostered through the generation of new design ideas, where students can communicate mental models in teams and build on each other's ideas using collage, digital creation software like Canva and other digital programmes—which don't require significant drawing skills. Later in the programme, students with specific CAD skills (that take time to develop) have the ability to use these digital software packages to detail and develop visualisations that can capture imagination. By using digital tools that support building on others' ideas and allow for easy changes, alterations and tweaks, these tools help stimulate imagination with more freedoms than in a physical setting.

The CBI A^3 programme focuses on developing solutions for a 2030 future and requires students to think with imagination about the possible futures that may unfold. Students access the Internet for existing material that is relevant to the students' exploration. For example, we tap into online materials to access publicly available reports from the Future Today Institute, ARUP and CSIRO Futures around global trends and futures scenarios. These reports often depict future scenarios through visuals which promote imagination. More recently, student teams have used Open AI tools, such as Midjourney to create AI-generated images of futures to unlock their imagination.

Lastly, CBI A^3 uses digital technologies as ***inspiration*** with students. The aforementioned future scenarios and other technology trends that students access via digital tools provide inspiration for their work, prompting them to thinking radically and creativity about future possibilities in the light of this new digital stimulus. The various recorded videos and lectures from experts provide inspiration through the passion, tone and body language that is presented and conveyed in a digital video. Online access also provides a far greater pool of resources and different examples beyond that shared in the classroom, with global benchmarking and building on others work, further helping stimulate inspiration in a range of project development areas.

5 Beyond Design Thinking Models

Design thinking provides a general design process that is industry sector agnostic, applicable to any type of solution or outcome, be it a product, system, service, built environment or business model. The ubiquitous, double diamond design process (UK Design Council 2019) that guides the creation of new designs through stages of discover, define, develop, deliver, give a framework for understanding context, problems space and empathising with users in order to pinpoint the right scope of a project and generate, refine and execute a designed solution that can be implemented in the real world. In the process, design must synthesise a range of different considerations which may be described as the intersection of what is viable, desirable and feasible (Brown 2009). The synthesis of many different needs through creative problem solving, central to design practices, is what makes it a highly relevant overarching framework to bring interdisciplinary concepts, teams and learning together. However, traditional design thinking frameworks have limitations in their ability to deal with highly complex societal challenges, nor easily work when there is a fixed aspect of integrating deep technology. Design methods and approaches are evolving to respond to this, and CBI A^3 draws on Verganti's (2009) established notion that radical innovation occurs at the intersection of technology and design methods, as well as aligning with newer frameworks for responsible design practice offered by the UK Design Council (2021); systemic design that looks at different factors enabling design to catalyse change towards a net zero future. Design methods predominantly come from the field of design, with engineering design, business and ICT also drawing on design principles, design thinking and human-centred design methods. Outside of supporting challenge-based learning, design approaches may inspire physics learners and educators to use creative problem solving (e.g. lateral thinking) in expanding theoretical and experimental ideas, and enhance their ability to collaborate by empathising with professional stakeholders.

An activity to introduce people to generating new ideas via creative problem solving is the 30 circles challenge (Kelley and Kelley 2013). It can be done in a very short time period (10 min or less total), requires no previous knowledge of idea generation and demonstrates that there is no right or wrong answers, just many different options that can be explored. In CBI A^3, this is the warm up we typically run at the beginning of the very first idea generation session interdisciplinary student teams undertake. An A3 or A4 piece of paper with 30 blank circles on it is given to each student, and they are given three minutes to turn as many circles as possible into a recognisable object (e.g. flower, tennis ball, pizza). The speed forces students not to think about how well they have drawn something, or how good the object they've thought of is—they are going for volume of different ideas. It demystifies the creative process of idea generation. You can learn more about running the 30 circle challenge at https://hbr.org/2013/11/three-creativity-challenges-from-ideos-leaders.

6 Systems Thinking

Systems thinking offers a framework for breaking down and working with complexity, by seeing how all products, actions, interactions and services form part of a larger set of systems. The concepts of systems thinking ask students to consider things as an interconnected set of parts with relationships that influence one another—complex, loops, not linear or siloed. It self assists how different disciplinary perspectives, practices and professions can interconnect, but also provides specific tools in order to map and express, explore and develop and consider impact and unintended consequences thereof. CBI A^3 utilises a commonly used systems thinking framework, the iceberg model (Ecochallenge 2023) to illustrate these concepts and prompt students to think of various elements within a system. The iceberg model provides a tool for guiding systemic thinking as you move through the *Event*, *Patterns/ Trends*, *Underlying Structures* and the *Mental Model* levels of the model. There are many different types of systems that may be functional, organisational or ecological. Systems thinking concepts drawing on (but not limited to) engineering, business operations and management and environmental science fields. Outside of supporting challenge-based learning, systems thinking may inspire physics learners and educators to consider different systems in science, and the relationships between them, allowing for macro, meso and micro levels of detail to considered and influence thoughts on theory and experimentation design.

A practical activity used in CBI A^3 to introduce students to the topic is mapping the system components and relationships of a simple, everyday event, such as making toast. This is a commonly used exercise not specific to CBI A^3, with instructions for conducting this "Draw How to Make Toast" exercise freely available online (https://www.drawtoast.com/) as a simple and fun introduction to systems thinking. By choosing an object and actions that most people are familiar with, it makes the exercise accessible and demystifies thinking in systems. Students may draw, annotate and show the connections between different stages of turning bread to toast, as well as defining the boundaries of the system (e.g. does it start with taking a piece of bread from the pantry, the making of bread or the harvesting of wheat?). This exercise is enhanced by using a digital whiteboard, as it allows students to more easily merge their different interpretations to learn from each other. A follow-up activity CBI A^3 uses the iceberg model to guide students in deeper exploration of factors contributing to complexity in the challenge problem space they are trying to address or the potential solutions they are developing in response.

7 Planet-Centric Design

Planet-Centric Design is concerned with developing products and services without harming the planet. Planet-Centric Design expands on systems thinking, circular design and sustainable design methods to consider how design decisions may impact different planetary ecosystems, guided by a set of principles and toolkit (Planet-Centric Design 2022). CBI A^3 utilises the three principles, responsible, transparent and systemic (used in conjunction with design thinking features of desirable, feasible and viable), which are aimed at better navigating complexity and improved solutions for society that respect planetary boundaries. The Planet-Centric Design toolkit developed by Vincit does not require formal design profession training to utilise, so it is a great way to guide interdisciplinary teams with practical methods and frameworks to shape responsible thinking towards flora, fauna and humans. It relates to fields (but not limited to) engineering design (e.g. life cycle assessment), business (circular economies) and design (eco-design movements). Outside of supporting challenge-based learning, Planet-Centric Design may inspire physics learners and educators to consider a holistic, and responsible approach to how they problem-solve, particularly in experimental settings that may have planetary impacts.

A practical activity to introduce students to Planet-Centric Design, is a simple mapping of resource use developed by Carolina Faria, a former CBI A^3 educator and Planet-Centric Designer, specifically for CBI A^3 students. Student teams take a rough design idea they have conceptualised and consider the associated resources. To do this, they map on a linear scale of least important to most important pre-defined and broad resource categories, such as water, electricity, precious metals, trees and so forth. This exercise is enhanced by using a digital whiteboard, as it allows students to re-arrange hierarchies of resource importance more easily and quickly add extra resource categories relevant to their idea. The exercise provides a very quick way of prompting consideration of the planetary resource impact a potential solution may have. You can learn more about different Planet-Centric Design tools developed by Vincit, and download the toolkit at https://planetcentricdesign.com/method-tools/.

8 Behaviour Change

Behaviour change as a topic area enhances students interdisciplinary vocabulary by enabling them to consider leveraging techniques to bring about change. It provides students with considerations around how humans can be prompted to interact and use new solutions or inventions, ultimately assisting in their successful adoption. Behaviour change considers different factors influencing behaviour in order to intervene and transform the type of action/s being undertaken by a human/s. The range of factors shaping behaviour includes things like motivation, incentive, goals, emotions, prompts and effort. Understanding different techniques for each of these factors (e.g. fear, pleasure or desire) may be techniques to trigger an emotional response, that in turn motivates or incentivises different behaviours. There are different models and approaches relating to behaviour change, such as the Fogg Behaviour Change Model (Fogg 2009) and Flow Theory (Chen 2007). Concepts like gamification, where there is reward and gratification for achieving goals, can be another model. Conceptual understanding of behaviour change can offer student teams new ideas on how to effectively engage users in new and more socially and/or environmentally responsible behaviours. Behaviour change is deeply rooted in fields of psychology and neural sciences. Outside of supporting challenge-based learning, behaviour change may inspire physics learners and educators in collaborative settings, and instances of science communication and outreach that involve building people's attitudes towards science to be positive and appreciative.

An activity that makes exploring different practical techniques for behaviour change easily accessible is using the *Design for Intent*: *101 patterns for influencing behaviour through design* card deck developed by Lockton (2021). The card deck is freely available online http://designwithintent.co.uk/docs/designwithintent_cards_1.0_draft_rev_sm.pdf. As the title suggests, the cards offer 101 different potential ways to encourage behaviour change across six different "lenses" of *cognitive, visual, security, persuasive, error proofing* and *architectural*. CBI A^3 encourages student teams to use the card deck to generate new ideas from scratch, or further develop and evolve existing ideas they have for new products, services, systems or built environments. The broad exposure of these patterns increases their knowledge of techniques for behaviour change through design, regardless if they are integrated into the student teams work.

9 Futures Thinking and Scenarios

Futures thinking prompts students to consider multiple possibilities. The nature of the future being uncertain is a launch pad for engaging discourse about the many factors that impact our future and the actions we take to get there. Scenarios use imagination and creativity as a means to develop and explore alternative futures (Engeler 2017). These alternative futures may be optimistic or pessimistic, utopian or dystopian. By bringing the fields of foresight and futures into the classroom, it allows students to develop their anticipatory thinking skills and imagination. CBI A^3 leans on content from science fiction, design fiction and speculative design fields to draw inspiration and influence students thinking and framing of the future. The *Futures Cones* (Voros 2003) provides a simple framework to convey to students that wherever you are situated in time, there are multiple possible futures that could play out. The model illustrates that there are multiple possible futures which can be categorised with titles such as the plausible, the probable, the preferable and the preposterous. This provides students with a licence to think broadly and know no answer about the future is wrong. It also prompts a discussion and discourse around what is our *preferred* future? This allows students to unpack the changes that are required to achieve that future and the consequences of the decisions and innovations we make today.

Later in CBI A^3 projects, students are asked to generate *diegetic prototypes*. These prototypes can be described as "cinematic depictions of future technologies" (Kirby 2010) and allow students to create artefacts and solutions for concepts that may not yet be feasible. Outside of supporting challenge-based learning, futures thinking may inspire, physics learners and educators, new ways to consider different possible outcomes over different time horizons when designing big science experiments.

An activity aimed at exciting students to imagine futures, is *The Thing From The Future* by Situation Lab. *The Thing From The Future* is an imagination game that asks players to collaboratively and competitively describe objects from a range of alternative futures (Situation Lab n.d.). A series of physical cards are given to teams as prompts to create hypothetical objects from different near, medium and long-term futures. Players are given an *Arc* Card that describes the type of future (e.g. rapid growth, decline, disciplined, etc.), a *Terrain* card which describes contexts, places and topics (e.g. education, money, etc.), the *Object* card which describes the basic thing from the future (e.g. a book, building, device, etc.) and a *Mood* card that describes the emotions that might be evoked. This is a great activity to promote imagination and inspiration when designing objects for the future. In CBI A^3, we use this as a low threshold introduction to constructing a future scenario with a situated design solution. Being presented with boundaries for the conditions of a future scenario, it reduces the reliance on a learners current understanding of different societal, political or technological landscapes and trends. The Thing From The Future card decks can be purchased online, or a free version is available for download at https://situationlab.org/project/the-thing-from-the-future/.

10 Communicating Complex Scientific Information

One of the key challenges in any interdisciplinary endeavour is the ability to distil and share disciplinary concepts and knowledge with a broader audience outside that discipline effectively. When dealing with complex and technical specifications, characteristic to deep technologies, it is hard to communicate and share these with non-experts to enable collaboration and action. This type of knowledge translation is crucial for technology adoption into real-world applications. In CBI A^3, students are asked to understand features and functions of complex technologies in enough detail to be able to use them appropriately to address the challenge they are responding to. Translation of complex science knowledge to plain English is a skill derivative from technology management fields and science communication/outreach for a range of STEMM fields who have complex scientific and/or technical information to convey with different stakeholders and/or the general public. This area of study connects directly to physics science communication and outreach, and the activity developed by CBI A^3 below may inspire new perspectives on science communication and outreach practice.

An example of a practical activity developed by educators in CBI A^3, to support students in distilling complex technical and scientific information to plain English, is the preparation of "technology cards" as shown in Fig. 1. Further abstracting the type of information available found on technical specification sheets openly accessible from Knowledge Transfer offices, such as that at CERN, students take a specific piece of deep technology and aim to describe this as accurately as possible in their own words using plain English. The technology cards prompt the plain English descriptions with headings including name of technology, inventor, what does it do, how does it work (using metaphors where possible), unique characteristics, what can it do, what can it *not* do, looks like, powered by, materials used, cost, key images/diagrams and any questions related to the technology. Through the preparation of technology cards, students must increase their understanding of at least one piece of deep technology that will prepare them to talk to experts on that technology and ask more relevant questions. This develops skills in unpacking complex technical jargon for other deep technologies they may come across. Despite this benefit, it remains difficult to strike the right balance of information to assist conceptual understanding of the technology. For example, not being too surface level, too abstracted, or simply repeating technical jargon. The preparation of technology cards needs multiple iterations and feedback from people with higher levels of understanding of the specific deep technologies being worked with. The format of a card standardises the organisation and layout of information, making it easier for interdisciplinary student teams to understand cards prepared by other students, and create a larger, shared card deck that can serve a secondary purpose to be used in idea generation and development activities.

11 Conclusion

Challenge-based innovation is useful for developing transversal skills and interdisciplinary curricula. Both project-based and problem-based learning have synergies with CBI, however, what is distinct is the integration of the challenge component. The challenges as set out by SDG goals in CBI A^3 shifts the learning process from one of disciplinary expertise to an "all in" challenge that requires interdisciplinary expertise and diverse disciplinary thinking. As discussed in this chapter, and echoing IdeaSquare, CERN's CBI approaches, generating multiple perspectives, research and revision, iteration and testing against real world and future-facing situations in an enquiry cycle is a interdisciplinary and transdisciplinary task. To create and deliver CBI curriculum, resourcing and planning beyond the standard models of educational delivery is required, and justified by the learning value it creates, fostering a sense of agency to innovate using technology for societal benefit and future contexts.

Digital tools and approaches enable students across the globe to engage in this "challenge"—and provides an opportunity to learn in cross-cultural communities of practice and access international experts to support developing an interdisciplinary toolkit to navigate these complex challenges of the future.

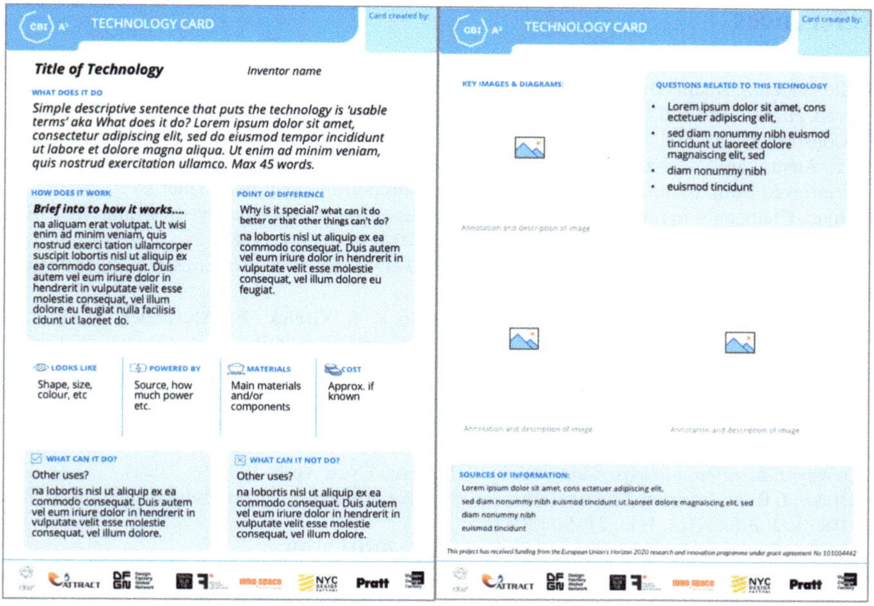

Fig. 1 CBI A^3 programme's technology card empty template, front and back (Credits CBI A^3 Australia)

As the debate moves from discussing what is required to develop transversal, twenty-first century skills to *how* these skills are developed, the examples in this chapter reveal digital communication, digital stimuli from wide sources for inspiration and imagination. The CBI A^3 programme is one piece of a much larger puzzle which acknowledges the creativity required and skills for challenges such as managing and shaping the megatrends in sectors and society consisting of several phenomena driving arcs or trends for change (Dufva 2020) forecast in nature, people, power, technology. CBI demonstrates that to be up for challenges requires that learners become well versed in translating, observing, and designing for responsible change through products, services, systems and the built environment.

Although the CBI A^3 example illustrated is at the moment applied at tertiary level, school teachers might find interest not only to be aware of their existence but also to adapt these concepts for younger levels of education. At the time of publication, development of a K-12 version of CBI A^3 was starting to be explored in an Australian context through Centre Dark Matter. The authors of the chapter welcome enquiries to provide, if necessary, further information on the possible implementation of CBI to apply deep technology to address societal challenges in their classroom.

References

B. Alexander, S. Adams Becker, M. Cummins, C. Hall Giesinger, *Digital Literacy in Higher Education, Part II: An NMC Horizon Project Strategic Brief*, vol. 3(4) (Austin, Texas. The New Media Consortium, 2017)

F.F.Y. Australians, AlphaBeta. *The new work smarts: thriving in the new work order* (2017). Retrieved from Melbourne http://www.fya.org.au/report/the-new-work-smarts/

D. Boud, Challenges in reforming higher education assessment: a perspective from afar. RELIEVE **26**(1) (2020). http://doi.org/https://doi.org/10.7203/relieve.26.1.17088

T. Brown, *Change by Design: How Design Thinking Transforms Organizations and Inspires Innovation* (Harper Collins, NY, 2009)

M. Brown, M. McCormack, J. Reeves, D.C. Brook, S. Grajek, B. Alexander, M. Bali, S. Bulger, S. Dark, N. Engelbert, K. Gannon, A. Gauthier, D. Gibson, R. Gibson, B. Lundin, G. Veletsianos, N. Weber, *2020 Educause Horizon Report Teaching and Learning Edition* (Louisville, CO: EDUCAUSE, 2020). https://library.educause.edu/resources/2020/3/2020-educause-horizon-report-teaching-and-learning-edition

J. Chen, Flow in games (and everything else). Commun. ACM. **50**(4) (2007)

J. Dewey, *Education and experience*, ed. by S. Schuster (New York, 1938)

S. Drake, J. Reid, Integrated curriculum as an effective way to teach 21st century capabilities. Asia Pacific J. Educ. Res. **1**(1), 31–50 (2018)

M. Dufva, Megatrendit 2020. Sitran Selvityksiä **162**, 2020 (2020)

Ecochallenge.org. A system thinking model: the iceberg. Ecochallenge (2023). https://ecochallenge.org/iceberg-model/

B. Engeler, Towards prospective design, in *The Design Journal: Proceedings of the 12th International Conference of the European-Academy-of-Design (EAD)*, eds. by L. Di Lucchio, L. Imbesi, P. Atkinson. Design for Next', Faculty of Architecture, vol. 20(1) (Sapienza University, Rome, 2017), pp. S4591–S4599

B.J. Fogg, A behavior model for persuasive design. in *Persuasive*, vol. 09, (Claremont, California, USA, 2009), pp. 26–29

S.E. Gallagher, T. Savage, Challenge-based learning in higher education: an exploratory literature review. Teaching High. Educ. 1–23 (2020)

IdeaSquare CERN, Homepage (2023). Retrieved 1 Mar 2023, https://ideasquare.cern/

Y. Jia, B. Zhou, X. Zheng, A curriculum integrating steam and maker education promotes pupils learning motivation, self-efficacy, and interdisciplinary knowledge acquisition. Front. Psychol. **12** (2021)

T. Kelley, D. Kelley, Three creativity challenges from IDEO's leaders. Harvard Bus. Rev. (2013). https://hbr.org/2013/11/three-creativity-challenges-from-ideos-leaders

D. Kirby, the future is now: diegetic prototypes and the role of popular films in generating real-world technological development. Soc. Stud. Sci. **40**, 41–70 (2010). https://doi.org/10.1177/0306312709338325

P. Kirschner, J. van Merriënboer, Do learners really know best? urban legends in education. Educat. Psychol. **48**(3), 169–183 (2013)

L. Leveratto, IdeaSquare progress report 2019–2020. CERN (2021)

D. Lockton, *Design with Intent* (2021). http://danlockton.com/design-with-intent/

Miro, Free digital whiteboard for online education (2020). https://miro.com/education-whiteboard/

J. Moss, S. Godinho, Enacting the Australian curriculum: primary and secondary teachers' approaches to integrating the curriculum. Aust. J. Teacher Educ. **44**, 24–41 (2019). https://doi.org/10.14221/ajte.2018v44n3.2

Planet Centric Design, Accelerating the transition (2022). Vincit. https://planetcentricdesign.com/

E. F. Redish, T.J. Cooke, Learning each other's ropes: negotiating interdisciplinary authenticity. CBE—Life Sci. Educ. **12**(2), 175–186 (2013)

H.W.J. Rittel, M.M. Webber, Dilemmas in a general theory of planning. Policy Sci. **4**(2), 155–169 (1973). https://doi.org/10.1007/BF01405730

Situation Lab, *The Thing From The Future. Situation Lab* (n.d.). https://situationlab.org/project/the-thing-from-the-future/
C. Thong, A. Cotoranu, A. Down, K. Kohler, C. Batista, Design innovation integrating deep technology, societal needs, radical innovation, and future thinking: a case study of the CBI A3 program. CERN IdeaSquare J. Exper. Innov. **5**(1) (2021)
UK Design Council, What is the framework for innovation? design council's evolved double diamond (2019). https://www.designcouncil.org.uk/news-opinion/what-framework-innovation-design-councils-evolved-double-diamond
UK Design Council, Beyond net zero: a systemic design approach (2021). https://www.designcouncil.org.uk/resources/guide/beyond-net-zero-systemic-design-approach
R. Verganti, *Design Driven Innovation: Changing the Rules of Competition by Radically Innovating What Things Mean* (Harvard Business Press, 2009)
J. Voros, A generic foresight process framework. Foresight **5**(3), 10–21 (2003)
World Economic Forum, Ten 21st-Century skills every student needs, in *Future of Work* (2017). https://www.weforum.org/agenda/2016/03/21st-century-skills-future-jobs-students/

The Quantum Bit Woman: Promoting Cultural Heritage with Quantum Games

Maria Luisa Chiofalo, Jorge Yago Malo, and Laura Gentini

Abstract Cultural fragility permeates large parts of the population, of all ages, and has been amplified in both its causes and effects by the pandemic. This has led to immediate and long-term damage to the economy, particularly in knowledge-based and art and science tourism, which sets up a vicious loop. The Quantum Bit Woman (WQuBit) intends to create a loophole by drawing on criticalities and transforming them into resources. The idea stems from three concepts. First, arts and science share creativity, experimental and formal literacies, which mutually fertilise each other. They represent an extraordinary, largely unexplored potential to develop ways of thinking about complex reality and to generate unprecedented reboots through innovation in content and language. Second, science-based technologies can be used to engage large population fractions in participatory processes, channelling creativity of the human mind via the design of games with a purpose. Third, quantum science implies enhanced imagination and creativity, so that quantum gameplay mechanics can boost purpose effectiveness. Moreover, science, and physics in particular, is part of Tuscan and Pisan cultural heritage. WQuBit braids these threads with art and science contamination: a quantum videogame is designed within participatory processes, with the purpose of boosting the cultural heritage fruition against cultural fragility, promoting its knowledge in engaging manners, and, within a citizen science approach, solving problems related to its accessibility.

Keywords Quantum physics literacy · Videogame · Cultural heritage · Science-based digital technology

M. L. Chiofalo (✉) · J. Yago Malo · L. Gentini
Department of Physics "Enrico Fermi", University of Pisa, Largo Bruno Pontecorvo, 3, 56127 Pisa, Italy
e-mail: marilu.chiofalo@unipi.it

J. Yago Malo
e-mail: jorge.yago@unipi.it

L. Gentini
e-mail: laura.gentini@df.unipi.it

M. Streit-Bianchi et al. (eds.), *New Challenges and Opportunities in Physics Education*, Challenges in Physics Education, https://doi.org/10.1007/978-3-031-37387-9_24

1 Introduction

Fragility and cultural poverty pervade the population at all levels of education, amplified by the exodus of young people from all possible social backgrounds to virtual reality, as reported by UNESCO (Unesco 2021). The pandemic has transported entire communities into virtual bubbles of space and time. The enormous damage caused by this shift, which has yet to be fully quantified, is manifesting itself through reduced autonomous problem-solving capacity, harm to intelligent and sustainable communities, widespread loss of cultural heritage enjoyment, and damage to the knowledge-based, art and science tourism economies. This evidently sets up a vicious loop.

In this paper, we propose an innovative research-based tool designed to create an effective loophole to address this cultural fragility condition. Our search for a solution is based on three seemingly unrelated concepts.

First, art and science are well-known to share creativity, experimental and formal literacies, which mutually fertilise each other. They represent a largely unexplored potential for thinking about complex realities and generating unprecedented reboots (Goorney et al. 2022).

Second, science-based technologies such as social networks, games and the Internet of Things can be utilised to engage large portions of the population in participatory processes, to channel the creativity of the human mind. These technologies can rapidly connect individuals across the globe, overturning the perception of space and time, and allowing the sharing and multiplication of resources by interconnecting devices. Additionally, part of the problem-solving intelligence can be transferred from individuals to devices. Now, the human mind is superior to computers in tasks requiring intuition and creativity. Therefore, the design of a human mind–machine interface that intertwines the best aspects of both could be an effective strategy in terms of solution quality and computing power. On the other hand, while all this can work to weld fragilities, abstraction from reality and reduced critical thinking will in general amplify them. Thus, despite the potential benefits of this paradigm shift, the question of how one can effectively design the human mind–machine interface to empower positive aspects while mitigating potential downsides remains.

In this framework, games can work as quite effective tools with well-suited traits (McGonigal 2011). In fact, games are characterised by objectives, that are centred on epic and emotionally engaging adventures; a set of rules that define constraints, but also provide opportunities to unleash creative thinking capable of overcoming skill limitations; a feedback system reinforcing motivation; and voluntary participation, allowing players to leave or keep trying without fear of the repercussions due to repeated failures. Games are demonstrated to be exceptional pedagogical tools (Montessori 1946), to enhance competencies (Winnicott 1971), and to foster community building. According to Bernard Suits, (Suits 2005) "playing a game is the voluntary attempt to overcome unnecessary obstacles". Gaming may be more difficult and gratifying than real life because players always have to push their skill limitations and use their abilities in the etymological sense of competition, which is working together to achieve a goal. This helps gamers to form stronger communities. In doing so, gamers are encouraged to push themselves beyond their comfort zones via skill challenges that are independent of their backgrounds and fields of study, in a positive feedback loop where motivations are enhanced, so that creative thinking is further enhanced, and works as an enduring resource for social development (McGonigal 2011; Suits 2005). In addition, games trigger emotions that are useful to improve the focus on what one is good at (without fear of failure), and frequently resemble epic adventures: these give players unrestricted access to the most complex scenarios, acting as a true playground where they can practise effective problem-solving techniques. Finally, games have a reward system based on non-perishable and immaterial products, generating feedback loop that amplifies motivation: in reality, effective resource management is one of the most desired characteristics of effective learning processes and, not by chance, of sustainable economies.

Thanks to all these similarities between gaming and problem-solving, as well as learning processes, many have begun to ponder if this large pool of gamers' intelligence may be used for activities that are beyond the capabilities of present computer systems. One solution is to create Games With Purpose (GWAPs). In fact, GWAPs-based citizen science approaches represent nowadays a powerful strategy in contexts of research, teaching and decision making (McGonigal 2011; Suits 2005; Ahn and Dabbish 2008). The design of associated educational materials that scaffold the development of the required competencies may therefore be done using games as a widely accessible and interesting entry point to complicated topics.

The effectiveness of this strategy is highlighted by a variety of factors (McGonigal 2011), especially in conjunction with so-called citizen science approaches. Here, individuals are given the opportunity to actively contribute to the resolution of specific research issues presented in an accessible, frequently gamified form, so that general participation and a feeling of purpose can be increased (Ahn and Dabbish 2008; Heck et al. 2018; Brown et al. 2013). This has motivated the use of serious games to undertake research in many scientific fields, beginning with the foundational project Foldit (Citizen science games—where science connects with games—FOLDIT game) for chemistry and biology, and afterwards extended to a number of physics domains. In fact, there are many successful examples of GWAPs both for education in classical and quantum physics, including those which tackle research problems (Heck et al. 2018; Kaur and Venegas-Gomez 2022; Cooper et al. 2010), and even for the promotion of cultural heritage (Seskir et al. 2022).

The third and last foundational concept inspiring our research-based solution is that quantum science implies enhanced imagination and creativity so that quantum gameplay mechanics can boost the purpose's effectiveness. In fact, the design of GWAPs has more recently been extended to the use of quantum versions of the underlying game mechanics. This may represent an additional advantage, for at least the following reasons. The process of scientific thinking hinges on phenomena observation, where conceptualisation is inferred via creative thinking, followed by model or theory building via formalisation in mathematical language, and closed by experimental validation and fact checking. Better than a closure, this last step re-opens to future progress: in experimental sciences, errors are part of the process and help to perfect it. In this cycle, creative, formal, and experimental literacies need to be structured. When dealing with classical physics, one can develop intuition after observing real-world occurrences, or can use some math tools that are available, at least to some extent, in a particular degree of education syllabus. When dealing with quantum mechanics instead, irrespective of it being a successful theory capable of effectively explaining microscopic phenomena, no tabletop experiments are usually available and the mathematical tools are too far advanced in conventional school contexts. As a result, the development of intuition with quantum physics problems becomes at the same time a hard challenge and a powerful opportunity (Pescarin and Videogames 2020).

Quoting the words of the Nobel laureate Tony Leggett, quantum mechanics "is much more than a theory: it is a new way of looking at the world, and it forces us out of our comfort zone, going beyond the conventional mind frames"; understanding quantum mechanics involves a paradigm shift that is arguably more extreme than any other in the history of human thought. Both in educational and outreach contexts, the inappropriate use of analogies to give quantum phenomena a familiar representation can lead to evanescent narratives or just historical storytelling, creating the risk of fostering misunderstandings. In addition, in undergraduate formal education courses, teachers run the risk of imparting knowledge that is mostly formal, scarcely useful to develop intuition, and in fact, a barrier in engaging the majority of students (Kohnle et al. 2014).

Despite the challenges faced in teaching and communicating quantum mechanics, some have suggested that quantum technologies, specifically quantum computing, could function as a powerful tool to inspire interest in the subject (Kohnle et al. 2015; Angara et al. 2020; Combarro et al. 2021; Salehi et al. 2022; Seskir and Aydinoglu 2021). This is not unexpected, given that the literature on quantum technologies has increased by almost three orders of magnitude over the past two decades (Salehi et al. 2022), and that quantum computing has been gaining more and more attention in light of the possibility of someday approaching the so-called quantum advantage (Bondani et al. 2022), that is the point at which a quantum computer will outperform the most powerful classical computer on a specific and useful task.

In fact and to begin with, quantum education has a practical scope: rapid developments in the second quantum revolution are leading to a shortage of a large enough and diverse quantum-technology workforce (Arute et al. Oct 2019). To solve this problem, many national governments (Kaur and Venegas-Gomez 2022) are creating dedicated outreach and education initiatives, like, for example, online courses, conferences, seminars, games, and community-focused networks (Arute et al. 2019).

The European Quantum Flagships, with its action "Quantum Technology Education—Coordination and Support Action" (QTEdu-CSA) (Dargan 2021) was born to foster both —the build-up of a competence ecosystem for a quantum-technologies workforce, and to boost outreach and education in quantum science. In QTEdu, this duple goal is reinforced by bridging the gap between academic and industrial quantum communities. In one and a half years across 2021–2022, QTEdu has been running 11 pilot projects in 25 European countries, to address educational needs and inform society about quantum technologies (Dargan 2021). Different QTEdu pilot projects have specialised in outreach and/or education for specific types of public: citizens of all ages, policymakers, teachers, students, media and industry representatives. Within the pilot projects, synergetic events have been organised such as joint live panels, conferences or quantum games jams (Education and (QTEdu) 2023) with the intent of engaging the public. New tools for effective science communication and education have been designed, some examples being new syllabi, videos and media products (Education and (QTEdu) 2023; Education and for everyone 2023), and formal educational courses for schools and universities have been introduced (QPlayLearn platform 2023; Education and open master 2023).

On top of this practical aim, quantum education and outreach could also encourage creativity, precisely because quantum phenomena elude common sense, and because experimental and mathematical literacies are more often limited. In fact, no direct sensory experience of quantum processes is available at the classical-to-quantum crossover's edges (Quantum teaching materials for schools 2023). It is significant to note that the latter circumstance affects the capacity of both non-expert citizens and quantum physics professionals, in developing intuition about nature's behaviours. Because of this limited capacity, the former develop limiting beliefs, and the latter struggle to push the borders of their fundamental and microscopic understanding. Both citizens and quantum scientists share the experience of the *periphery of knowledge* in what we call *culturo-scientific storytelling* (Goorney et al. 2022). As a result, creativity and imagination are vital abilities to be expanded by anyone who really needs to develop future scaffolding. In support of this imagination, for citizens, the design of alternative and complementary kinds of experimental and mathematical literacies is also especially helpful (Cooper et al. 2010). In summary, the storytelling of quantum science and technology with their underlying and fundamental ideas offers a rare chance to introduce citizens of all ages to a new way of thinking and to expand the use of senses other than those of ordinary perception.

The Quantum Bit Woman (WQuBit) is a research program lasting two years, started in July 2022, that builds on the above three foundational concepts. WQuBit braids these threads with art and science contamination: a quantum videogame is designed within participatory processes, with the purposes of boosting the cultural heritage fruition against cultural fragility, promoting its knowledge in engaging manners, and—within a citizen science approach—solving problems related to its accessibility. In doing so, WQuBit uses GWAPs to draw on social and cultural criticalities, and transform them into resources. In the spirit of scientific humanism, WQuBit strategically synthesises skills in art, quantum physics, neuroscience, computer science and gamification, to enjoy, understand and enhance artistic and cultural heritage. The produced videogame with purpose is intended to be playable online and in situ in places of arts and science in augmented reality, so that the cultural heritage promotion may happen at diverse levels of fruition. Within a gamified citizen science approach, the videogame is also designed to solve an open research problem of cultural heritage, while using quantum physics as a new, creativity-enhanced, paradigm. Moreover, science, and physics in particular, is part of Tuscany and Pisan cultural Heritage that the videogame is expected to promote. In the videogame, the compelling storytelling is animated by characters who crumble stereotypes (especially gender stereotypes) in epic adventures and perform tasks with the bizarre rules of quantum physics to enhance imagination. Players solve, in a collaborative citizen science approach, problems in the places they explore, with the problem–game interface appropriately designed for the best engagement and with inclusive technologies. As a multiplicative outcome, the videogame serves as an effective educational tool for quantum physics, aimed at the general public, and making plenty of use of the resources developed and available in the QPlayLearn platform (Education and for everyone 2023; Fein et al. 2019). The videogame design and methodology represent an example of physics outreach research (POR) and aim at creating a new paradigm that is exportable in contents and methods (Goorney et al. 2022).

The paper is organised as follows. In Sect. 2, we draw our research questions. In Sect. 3, we present the research methodology, meaning the core characteristics of our research, highlighting its unique features. In Sect. 3.2, we discuss preliminary results of our research and give a glimpse into the state-of-art of our videogame. Finally, Sect. 5 is dedicated to a discussion and to concluding remarks.

2 Research Questions

As discussed in the Introduction, the WQuBit research-based program aims at responding to a deeply general research question: how to create an exportable paradigm for content and methods, to design sustainable, effective, efficient and collaborative solutions to community problems. In our practical route to answer this general quest, four research questions (**RQ**) emerge as follows:

RQ1 How to use games to solve a problem of cultural heritage fruition. In particular, with special attention to both public:
RQ1.1 actually traveling to cultural heritage sites, and
RQ1.2 remotely accessing the cultural heritage sites, due to inaccessibility problems (e.g., pandemic-induced, or other).

Our investigation is contextualised within Tuscany cultural heritage, where the research team is based, and where funding has been obtained from the Tuscany Regional government, then complemented with funds originated from the proposal IQHuMinds (Integrating Quantum Machines and Minds for Quantum Technologies) submitted to Horizon 2020. Though localised in Tuscany, our paradigm and methodology can easily be generalised and exported.

RQ2 How to promote, through videogames, the so-called "hidden heritage". In fact, the definition of the perimeter about what should be considered cultural heritage, is *per-se* an open problem in humanistic studies (Foti et al. 2021).
RQ3 How to exploit citizens creativity and imagination to find innovative solutions to a given cultural-fruition problem. This RQ3 can be re-phrased after taking inspiration from the many examples of citizen science gamification experiments

designed to solve science problems, and then transferring that methodology to the fruition of cultural heritage.

RQ4 How can we foster players' creativity making use of quantum physics. As already discussed, quantum physics forces us to go beyond conventional mind frames, and to fall out of our comfort zone: we want to investigate how this process enhances creativity also in the general public.

3 Methodology

3.1 Methodology Traits

In order to answer our research questions in Sect. 2, we must create a multilevel intertwining that connects the diverse expertise of universities and companies, braids arts, science and technologies, and performs this within an international context. The WQuBit methodology is therefore imprinted with the following traits:

(a) **Interdisciplinarity**, to draw on different expertise in functional research and development: art and cultural heritage, gamification, physics and quantum technologies, and computer science. This creates an environment between art, science and technology that can engender a wide vision for creative processes, accelerate ideas and incubate intelligent solutions;
(b) **Internationality**, to include the best experiences and coherently with the international relevance of the heritage to be promoted;
(c) **Cross-sectorality** of universities and companies to guarantee diversified approaches to the problem, together with the public that have open, free access to the game;
(d) **Participatory setup** to channel the active contribution of very diverse actors, provide an expert environment for quality assessment, foster sustainability and create the conditions of replicability for newer initiatives. The setting of participatory processes makes use of suited methodologies which include quantum game jams and hackatons for co-design, testing, evaluation, task completion, dissemination and fruition;
(e) **Citizen science approach** through a videogame with a purpose that channels the cognitive potential of those who play. As already discussed, gameplay enhances imagination, emotional well-being, prosociality and competition, and focuses on the divergent resources of creativity and intuition;
(f) **Use of quantum physics** in defining the gameplay mechanics, to enhance creativity and imagination and to provide access to immersive environments for the players' learning.

In this context, the first step is the composition of the research team. The scientific working group of WQuBit reflects the traits (a)–(f) above. The scientific team is established within the Physics Department at the University of Pisa (DF-UniPi), and it is the coordinator of the research program activities, ensuring the coherence, effectiveness and efficiency of the research program, also facilitating the cooperation between other research program team members. The WQuBit scientific group is completed by the Department of Physics at the University of Helsinki (PD-UNIH), which provides expertise in the area of gamification of quantum physics. The WQuBit scientific working group is responsible for research questions **RQ3** and **RQ4** of Sect. 2 which need competencies in quantum physics, citizen science methods and gamification in order to be answered.

The cultural team leader of WQuBit is Fondazione Sistema Toscana (FST), an in-house body of the Tuscany Regional government, established to promote the regional territory and its identity with integrated digital communication tools: Web, multimedia productions and social media. FST also develops and supports activities in the film and audio-visual field, implementing regional sector policies in the field of education and training. FST, as the organiser of Internet Festival (IF, an international yearly event storytelling the world of Internet), is involved in the promotion and enhancement of the regional territory also through integrated digital communication tools. FST staff competencies are needed to answer cultural heritage-related research questions, as **RQ1** (both **RQ1.1** and **RQ1.2**) and **RQ2**. Moreover, FST contributes to the research program by designing and organising WQuBit-dedicated events at IF2022–IF2023, by evaluating videogame sessions, and by collaborating in the communication of the research program steps.

The WQuBit team also include the Institute of Informatics and Telematics of the National Research Council (IIT-CNR). The IIT-CNR brings its experience in the management of projects for the dissemination and diffusion of culture, like the design of educational activities mostly dedicated to students, in which (video) games and interactive laboratories are used to explain complex scientific topics. It can therefore act as an expert entity throughout the program and as an intermediary for dissemination, expanding the size of the research network.

The WQuBit team also includes two companies, Devitalia SRL and FABVISION SRL. Devitalia provides the research program with its own Cloud DevFarm solutions, while FABVISION provides tutoring and consultancy for technological solutions and digital graphics.

Finally, we highlight that part of WQuBit research methodology is involving the local community as an integrated part of the research group in a participatory setup: through dedicated events, as we will see in Sect. 3.2, the expert public was involved in the research program since the very beginning, generating social cohesion and meta-engagement, in a cyclic-economy approach of mutual fertilisation. In fact, in searching for the solution of a socio-cultural problem, we use a socio-cultural strategy.

The manner in which these traits are implemented in the methodology used is operatively presented in the following Sect. 3.2 and reported in more detail in Appendix A.

3.2 Operational Steps

The WQuBit research work and videogame design start from the identification of the founding elements of the videogame: target, purpose and genre. Together with the criteria for the choice of the setting, the identification of the locations related to art and science to be promoted, the storytelling and the characters, this constitutes the process of participatory co-design that involves all the research program actors, stakeholders, developers and gamers, giving also the latter opportunity to come into contact with researchers and developers.

In a spirit of collaborative design involving citizenship, the WQuBit team has designed and organised three different cultural events with the support of FST, dedicated to cultural workers, students and game designers: two of them are presented in detail here in the next section. Through the mutual fertilisation between these worlds, the WQuBit team formalised the story of the video game, as well as its quantum game mechanics. Thanks to the collaboration offered by Giovanni Timpano, a professional cartoonist and designer, and the Florence Comics & Art Academy "The Sign", the pre-production phase could also be completed in its artistic part (concept art and storyboard). Pre-production materials are expected to be presented in a dedicated event organised in collaboration with FST at IF 2023 with the presence of experts from gamification research, where a first step in validation and assessment can be performed.

Moreover, starting from the identification of targets, purposes and typology that emerged from the activities described above, it is in principle possible to identify the specific problem of fruition, valorisation and accessibility of cultural heritage one wants to solve, and embed the search for a solution as part of the outcome in the videogame. This process can channel the creativity and intuition of players through the videogame, from a cognitive and also a perceptive perception, through collaboration with visual neuroscience experts. Exploiting the cultural problem identification, the first testable and playable prototype is also expected to be presented at the IF 2023 event. There are two consequences of the existence of a playable prototype: first, it is a prerequisite for the realisation of the final version of the game, the beta-testing phase and the final release; and second, it gives the opportunity to start collecting data and analyse solutions to the cultural heritage accessibility problem, encoded in the game through gamification activities, bringing useful elements to the processes of valorisation and innovation of culture.

4 State-of-the-Art and Preliminary Results

The design of the video game began with a collaborative approach through the organisation of two dedicated events at the Internet Festival (IF) 2022. These events were tailored for two distinct audiences: professionals of the cultural and artistic heritage and game designers.

These events facilitated collaboration within the different communities, enabling individuals to share their skills and creativity, thus contributing to the research program's progress. The following sections describe the contents and methods of the events and provide a brief summary of the video game's plot that includes the feedback received during the events.

4.1 Educating to Beauty Through Playing

The "Educating to beauty through playing" workshop involved heritage professionals collaborating to develop the theme of "education to beauty" for the videogame, and to determine how this learning process can be represented in the game. During the workshop, the participants identified various elements of Tuscan cultural heritage, including tangible and intangible assets such as enogastronomy, literature and music. We also notice that participants included heritage elements which are often neglected by famous tourist routes. Therefore, we put significant effort into representing these hidden gems of cultural heritage in our game. Additionally, the participants contributed to designing narrative strategies to create an immersive and enjoyable game experience, guiding the player to discover their own interpretation of beauty (Tables 1 and 2).

Participants, divided into four working groups, tackled three 'Focus Points' (FP):

- FP 1—Cultural heritage and the case of Tuscany

The main aim of this FP is to define a standard paradigm of cultural heritage. Participants were also asked to identify three elements recognised by the group as cultural heritage, and associate them with three representative Tuscan examples. In addition, they were also asked to associate anti-examples. These latter 'anti-examples' are to be understood as opposed to the previous ones, i.e. not representative of heritage.

Table 1 A total of 29 participants attended the event and their different areas of expertise are summarised

Where	Online
When	17/11/2022
Moderators	Maria Luisa Chiofalo (UniPi), Jorge Yago Malo (UniPi), Sacha Alberti (FST), Daniele Cetrulo (FST)
Organisation and technical direction	Manuele Buono (IF)
Participants	29 (see also table below)
Conduction	Laura Gentini (UniPi)

Table 2 List of areas of expertise of participants

Museums curators	9
Architects (from Pisa's Architects order)	3
Public Administration workers (cultural heritage promotion offices or public libraries)	3
High school teachers	2
University students (humanistic, artistic and cultural majors)	3
Freelances (Journalists, artists)	6
Italian National research council staff	1
N.D.	2

The elements chosen by the groups, the examples and the anti-examples are expected to be used as the narrative "bonuses" and "maluses" components of the video game, respectively.

- FP 2—The values of culture as a narrative basis

The aim of this focus point is to identify which are the human and social values of cultural heritage. Identifying these values is necessary in order to understand how a society that does not possess them (such as the one in the game's incipit reported below) would appear; additionally, it elucidates the diverse goals or "bonuses" within the game in relation to the acquisition of human values necessary to accomplish the path towards the appreciation of heritage.

- FP 3—Scale of values and how to represent them in the video game

The aim of this focus point is to create a priority scale of the values identified in order to create the narrative path. Although each value is important, the WQuBit video game is a metaphoric journey symbolizing the progressive approach to cultural heritage, which as such must go step by step.

Participants found the workshop highly stimulating and useful for their work as well. In particular, two architects emphasised how the WQuBit video game could succeed in enhancing architectural heritage that is more complex to access, such as the natural landscape. A museum operator, on the other hand, dwelt on the potential of the video game to make a museum visit more vivid and interactive. In addition, all participants found the identification of progress within the video game with the progressive approach to cultural heritage particularly interesting, and generally liked the metaphor of the video game itself as a process of "education to the concept of beauty".

4.2 Quantum Game Mechanics—A Game-Design Workshop on Quantum Physics

The quantum game mechanics training workshop was proposed to the students of 'The Sign' game-design academy in Florence as part of their curricular activities, with the aim of introducing them to quantum mechanics and its gamification potential. In a spirit both collaborative and educational, the students first attended an interactive lesson on the principles of quantum physics, aimed at providing them with the basic tools, and then engaged in a group activity in which they used the concepts they had learnt to develop quantum game mechanics. The construction of the lecture and the general running of the workshop was taken care of by the WQuBit scientific team, with the support and collaboration of the game-design teacher Michele Lanzo.

Where	Online
When	17/01/2023
Participants	17 game-design students from "The Sign" academy, based in Florence
Moderators	Nicola Pranzini (Helsinki University), Jorge Yago Malo (UniPi), Vittoria Stanzione (UniPi), Dario Cafasso (UniPi) Paolo Braccia (UniPi), Michele Lanzo (Prof. "The Sign Academy"), Maria Luisa Chiofalo (UniPi)
Organisation and technical direction	Manuele Buono (IF)
Conduction	Laura Gentini (UniPi)

In the first part of the event, we decided to break down the basic concept of quantum mechanics for game-design students with an outreach approach, with particular attention to enlightening possible gamification pathways of quantum key concepts. To do so, we constructed and presented a minimal quantum dictionary that students used in the second part of the event and can possibly use in their future videogames. The minimal quantum dictionary includes:

- State of a system (Quantum and Classical);
- Quantum to classical crossover;
- Superposition;
- Quantum measurements;
- Tunnelling;
- Heisenberg's uncertainty principle;
- Wave-like behaviour in quantum mechanics;
- Entanglement;
- Teleportation protocol.

Students were then asked to choose some concepts from the dictionary and, working in small groups (three to four people in each group), design a game mechanic based on the chosen concepts. The ideas and concepts raised by the working groups are intellectual propriety of the participants and are not direct resources for the WQuBit game, nor presented here for the same reason. Nevertheless, we want to remark that participants were able to create visual analogies for the quantum concept they choose in order to create game mechanics. This constitutes excellent feedback for our outreach activity because it requires creativity on top of understanding. In fact, to gamify a quantum concept means being able to identify core characteristics of the physics phenomena underlying it, deeply understanding them, and constructing a new physical situation that, following the same logic, constitutes an analogy for the original quantum phenomena.

After the workshop, we asked students to give feedback about the idea of using quantum physics to design game mechanics, enlightening both strengths and weaknesses. Concerning strengths, students said (only a few examples of answers received are reported, translated from Italian):

- *"Quantum mechanics broadens the mind, forcing it to understand non-intuitive elements. This allows the designer to find new insights outside the usual logic circuits, leading to more innovative products that provide unique experiences for the player"*;
- *"With the different perspective it gives of reality, quantum mechanics offers unique creative opportunities. Being able to assume in a game that there are uncertainties where certainty is usually found, with the proper care, brings novelty to even the most classical of game mechanics. For instance, reinterpreting the movement of a character with statistical measures, or being able to manipulate the states of the environment through the superposition of states makes the game world truly unique."*;
- *"It helps to have a fresh look and insights into creating mechanics according to rules that are not part of common logic and knowledge, but which follow a pattern and defined laws"*.

On the other hand, students also said that using quantum physics to create games comes with some difficulties, especially related to the probabilistic nature of quantum measurement outcomes (only a few examples of answers received are reported, translated from Italian):

- *"The probabilistic nature and complexity of some concepts are typically difficult elements to handle in a game. Furthermore, it is necessary to develop the player's intuition for quantum concepts, as he does not inherit it from the real world."*;
- *"For the same reason they (the quantum inspired mechanics, N.d.A.) could make it (the game N.d.A.) more addictive, since not everyone could fully understand how it works, they make the game complicated and perhaps even frustrating for the player."*;
- *"It is difficult to give control of the uncertainty characteristic of quantum to the player, while still being able to provide him with fixed points where he can create*

strategies and dynamics. Making everything strongly quantum would not only be extremely alienating but would also become too influenced by factors beyond the player's control."

4.3 WQuBit Videogame Backstory

On a planet far from Earth, lives the species of 'Aliens'. Their society has developed around scientific progress, and the values of efficiency, pragmatism and rationality. Cooperation between individuals stops at the free circulation of scientific ideas, while individuality and private life are repressed as a source of imbalance. Their technology is advanced, and the exploitation of quantum mechanics and its principles is possible on a macroscopic level. The alien society is politically a dictatorship, or rather a technocracy, but no one, or hardly anyone, seems to care. Artistic sensibilities and 'non-conforming' behaviour are oppressed from childhood, making individuals perfect members of the scientific society, in which order reigns.

As a result, the Alien society did not develop any artistic sensitivity, cultural heritage or a concept of "beauty" and, because of that, Aliens live in a profound crisis.

The protagonist, V3NU5, works in the space exploration government organ. In the videogame prologue, she walks towards her workplace in City 97, a grey, standardised and ordered city. Her boss, AX184, informs her of the upcoming new mission. The government task force set up to solve society's problems has put up a plan to save the planet. In the "Happened things Report", it is written that in a distant galaxy, on the planet called 'Earth', there exists a race of "humans" who have devised a powerful technology capable of solving existing society's problems. Like other objects in the alien society, it may look like an ordinary object, in this case, a history book, but it is not. Emptied of its cultural value, it is merely a list, a practical tool designed for efficiency purposes.

The plan for saving the alien world is to go to Earth, take the blueprints of the mysterious technology and return with instructions to reproduce it on the home planet. V3NU5 was selected for the mission based on her commendable history, proficiency, dependability and devotion. However, she is also cautioned that a scout, 3R1S, was dispatched nearly a decade ago (in alien years), and his transmissions indicated encountering illogical beings, whose motives were not always pure. The communication abruptly ceased thereafter. V3NU5 is tasked to bring him back if she can, but this must not interfere with the main mission. At the end of the prologue, V3NU5 sets off on her mission.

V3NU5's expedition on Earth entails a voyage of discovering the essence of beauty, encompassing the comprehension of art, culture and their social significance. The protagonist and the player embark on both a metaphorical and physical journey towards cultural heritage, the enigmatic weapon that holds the potential to rescue the alien civilisation from impending ruin.

5 Discussion and Conclusions

After the first videogame prototype will be created, the final version playable online and in augmented reality will be addressed, aiming at a multilevel cultural heritage fruition, and at empowering curiosity through the immersive and collaborative mix of virtual and actual reality. First through the game prototype and then with its final version, WQuBit videogame is expected to enhance and promote (Tuscany's) heritages of art, culture and science, contributing to the economic development of Tuscany in this sector, providing ideas to optimise fruition problems and overcoming educational and cultural fragilities. In addition, WQuBit program is expected to create new skills for FST and the entirety of the research team, with an impact on the whole Tuscan system, given that FST is an in-house company of the Tuscany Regional government. These new skills and competencies are interdisciplinary for the involved researchers in a cross-sectoral context, enhancing and diversifying their occupability. From the science-outreach perspective, WQuBit is expected to increase curiosity and knowledge about quantum technologies and raise awareness about their growing everyday presence and impact in the era of the second quantum revolution. Overall, WQuBit intends to create a new exportable paradigm of sustainability in community problem solving, in which the creativity and skills of citizens, researchers, professionals and experts from universities and companies are integrated and channelled to answer research questions.

The WQuBit program can also have spin-offs on the regional system. In particular, operators can benefit first from ideas produced by the videogame design process and then from the analysis of the big data extracted from the game, containing players' proposed solutions to the cultural heritage accessibility problem. The regional system can also benefit from increased curiosity, interest and sensitivity to its services and products, both from citizens/users and from institutions and companies. The delivered outcomes of WQuBit, which are planned to be freely available and usable, can as well be the subject of territorial marketing, according to the core mission of the Tuscany Promotion Agency.

Finally, WQuBit is completely replicable in methodology and product creation concepts, changing the design details, the problem to be solved, or the context to be enhanced.

Overall, we feel that WQuBit can foster a wide-open access to unexplored trails of that periphery of knowledge (Goorney et al. 2022) where scientists and citizens fruitfully meet to expand their horizons, opening an original route to practise responsible research innovation (Vecco 2010; Stilgoe et al. 2013; Guston 2007; Schomberg 2012) within a scientific-humanism vision.

Acknowledgements The Authors thank all the WQuBit team members: Fondazione Sistema Toscana (FST), and in particular Adriana De Cesare, cultural research program manager for WQuBit; the scientific team from UniPi that includes Prof. Antonio Cisternino, Prof. Susanna Pelegatti (UniPi Computer science Department) and Prof. Michele Barsanti (UniPi Civil and Industrial Engineering department) as well as the authors themselves; The University of Helsinki Department of Physics(DP-UNIH), in particular Prof. Sabrina Maniscalco's group; the Institute of Informatics and Telematics of the National Research Council (IIT-CNR); the private companies Devitalia SRL and FABVISION SRL.

We thank also Prof. Augusto Smerzi (INO) and Giovanni Timpano, professional cartoonist.

This project has received funding from the European Union's Digital Europe Programme DIGIQ under grant agreement no. 101084035.

Appendix 1

The WQuBit research program has three main operative objectives (OO), each one of them associated with a list of related activities. For each operative objective, we report below a brief description of its scope, specific activities, expected results and assessment methods.

Operative Objective 1 (OO1)

Description

Identification of the video game's founding elements: target, purpose and genre

The realisation of the WQuBit videogame starts with the identification of the founding elements of the videogame, namely the target, purpose and genre, using participative methods. Together with the criteria for the choice of the setting, the identification of the art and science sites to be valorised, the storytelling, and the characters, this is a participatory co-design process involving all actors, from the research program team to stakeholders, developers and gamers. Gamers could also have the opportunity to come into contact with researchers and developers.

Activity 1.1: Ideation and organisation, within the framework of Internet Festival 2022, of an online co-creative co-design workshop, with the participation of leading gamification experts.
Activity 1.2: Online workshop realisation.
Activity 1.3: Formalisation of the idea emerged from the creative co-design activity 1.1, analysed with particular attention to its scientific value in terms of quantum physics outreach quality and efficiency.

Expected results and assessment:

- Definition of target, purpose and genre of video game;
- Engagement of stakeholders, that has to be verified through the modalities that already defined for Internet Festival events;
- Generation of interest in the WQuBit videogame from players and in the replicability of the research program from stakeholders and companies. The verification

takes place with an online survey distributed through social media, following a first communication action on the project activities and state-of-the-art.

Operative Objective 2 (OO2)

Description

Identification of the cultural problem to be solved with the video game

Starting from the identification of the target, purpose and genre that emerged from the activities of OO1, in OO2, the problem of the fruition of cultural heritage to be solved in a citizen science approach is identified. The codification of the problem in the video game is then realised, together with the narrative and graphic design.

Activity 2.1: Organisation of an Unconference that reunites the WQuBit team, together with the designer Giovanni Timpano, for three days of work on the technological and creative basis of the video game.
Activity 2.2: Realisation of the Unconference
Activity 2.3: Formalisation of the idea emerged from the Unconference.

Expected results and assessment:

- Definition of the research problem to be solved, its codification in the video game, the narrative and the graphic design of the video game;
- Growing engagement of stakeholders, that has to be verified through the modalities already identified in OO1;
- Growing interest in the WQuBit videogame from players and in the replicability of the research program from stakeholders and companies. The verification takes place with an online survey distributed through social media, following a second communication action on the project activities and state-of-the-art.

Operative Objective 3 (OO3)

Description

Videogame and game manual creation

Using the resources generated in OO1 and OO2, in OO3, the WQuBit team proceeds with the creation of the video game playable online and in augmented reality, generating curiosity through game mechanics that mix virtual and actual reality. Moreover, the team also creates the user's game manual. The process is highly participative through activities such as Quantum Game Jam (QGJ), co-design and hack and play workshops, which involves game developers, gamification experts and companies.

The video game must be fun and engaging for a general audience of different players. In order to achieve this goal, OO3 consists of three concatenated activities, each of which uses the results of the previous one and constitutes the input for the next one: activity (3.1) is the brainstorming, pre-production and prototype realisation; activity (3.2) is the testing phase; and finally activity (3.3) ends the program with the production, publication and distribution of the final version of the video game and user manual.

The characters and the story are conceived in activities (3.1–3.2) specifically to promote gender representation in STEAM (Science, Technology, Engineering, Arts and Maths), while activity (3.3) is focused on the cultural messages conveyed.

Although conventional, in WQuBit, all these activities are operated in an inter-disciplinary manner between cultural heritage, gamification, quantum physics, computer science and neuroscience.

Activity 3.1: It is dedicated to brainstorming, pre-production and prototyping, with special attention to the popularity factors of a video game: genre, interface and controls, storyboard, design quality, rendering and soundtrack. The brainstorming is in the form of an online quantum game jam (QGJ).
Activity 3.2: It is dedicated to testing the prototype, the user experience and the attractiveness of cultural content. The testing phase consists of two events: the first one is dedicated to students, and the second one is in the form of an online Hack and play, dedicated to a diverse audience of experts, invited to play with the prototype and provide feedback for activity 3.3.
Activity 3.3: It completes the production process of the video game and the user manual, their publication and distribution. Particular care is dedicated to activities of: (1) quality certification; (2) adaptation for multilingual and diversity-oriented use; (3) publication, dissemination and distribution for massive on-site and online use; (4) cultural messages conveyed on STEAM, heritage and edutainment. The launch of the video game is expected at an IF2023 event.

Expected results and assessment

- Creation of the video game and user manual. It is expected to test its functionality with students and stakeholders, as well as with online users, through dedicated and easily accessible online questionnaires distributed as a follow-on to the download of the video game;
- Growing engagement of stakeholders, to be verified through the methods already identified in OO1 and OO2;
- Growing interest in the WQuBit videogame from players and in the replicability of the research program from stakeholders and companies. The verification takes place through the analytics of the videogame downloads and online accesses.

References

L. Ahn, L. Dabbish, Designing games with a purpose. Commun. ACM **51**, 58–67 (2008)

P.P. Angara, U. Stege, A. MacLean, quantum computing for highschool students an experience report. in *2020 IEEE International Conference on Quantum Computing and Engineering (QCE)*, (2020), pp 323–329. https://doi.org/10.1109/QCE49297.2020.00047

F. Arute, K. Arya, R. Babbush et al., Quantum supremacy using a programmable superconducting processor. Nature **574**(7779), 505–510 (2019). https://doi.org/10.1038/s41586-019-1666-5

M. Bondani, M.L. Chiofalo, E. Ercolessi, C. Macchiavello, M. Malgieri, M. Michelini, O. Mishina, P. Onorato, F. Pallotta, S. Satanassi, A. Stefanel, C. Sutrini, I. Testa, G. Zuccarini, Introducing quantum technologies at secondary school level: challenges and potential impact of an online extracurricular course. Physics **4**(4), 1150–1167 (2022)

O. T. Brown, J. Truesdale, S. Louchart, S. McEndoo, S. Maniscalco, J. Robertson, T. Lim, S. Kilbride, Serious game for quantum research. in *Serious Games Development and Applications, 4th International Conference, SGDA 2013*, eds by M. Ma, M. Fradinho Oliveira, S. Petersen, J. Baalsrud Hauge. (Publisher Springer Berlin Heidelberg, Trondheim, Norway, 2013), pp 178–187. ISBN (electronic) 978-3-642-40790-1 and ISBN (print) 978-3-642-40789-5

Citizen science games—where science connects with games - FOLDIT game

E.F. Combarro, S. Vallecorsa, L.J. Rodríguez-Muñiz, Á. Aguilar-González, J. Ranilla, A. Di Meglio, A report on teaching a series of online lectures on quantum computing from CERN. J. Supercomput. **77**(12), 14405–14435 (2021)

S. Cooper, F. Khatib, A. Treuille et al., Predicting protein structures with a multiplayer online game. Nature **466**, 756–760 (2010). https://doi.org/10.1038/nature09304

J. Dargan, 15 leading quantum computing countries with national initiatives. *The Quantum Insider*, (Apr. 2021)

Quantum Technology Education (QTEdu), action of quantum flagship initiative. https: //qtedu.eu/. Available online, accessed on 01 Mar 2023

Quantum Technology Education for everyone, QTEdu pilot project. https://qtedu.eu/project/quantum-technologies-education-everyone. Available online, accessed on 01 Mar 2023

Quantum Technology Education open master, QTEdu pilot project. https://qtom.qtedu.eu/. Available online, accessed on 01 Mar 2023

Y.Y. Fein, P. Geyer, P. Zwick et al., Quantum superposition of molecules beyond 25 kDa. Nat. Phys. **15**(12), 1–4 (2019). https://doi.org/10.1038/s41567-019-0663-9

C. Foti, D. Anttila, S. Maniscalco, M.L. Chiofalo, Quantum physics literacy aimed at k12 and the general public. *Universe* **7**(4) (2021)

S. Goorney, C. Foti, L. Santi, J. Sherson, J.Y. Malo, M.L. Chiofalo, Culturo-Scientific Storytelling. Educ. Sci. **12**(7), 474 (2022)

D. Guston, Toward centres for responsible innovation in the commercialized university. in *Public Science in Liberal Democracy*, (University of Toronto Press, J Porter, P.W.B. Phillips, 2007), pp. 295–312. https://doi.org/10.3138/9781442684720

R. Heck et al., Remote optimization of an ultracold atoms experiment by experts and citizen scientists. Proc. Nat. Acad. Sci. **115**(48), E11231–E11237 (2018). https: //citizensciencegames.com/games/foldit/. Available online, Accessed on 01 Mar 2023

M. Kaur, A. Venegas-Gomez, Defining the quantum workforce landscape: a review of global quantum education initiatives. Opt. Eng. **61**(8), 081806 (2022)

A. Kohnle, C. Baily, A. Campbell, N. Korolkova, M.J. Paetkau, Enhancing student learning of two-level quantum systems with interactive simulations. Am. J. Phys. **83**(6), 560–566 (2015)

A. Kohnle, C. Baily, S. Ruby, Investigating the influence of visualization on student understanding of quantum superposition. in *2014 Physics Education Research Conference Proceedings*. American Association of Physics Teachers, (Minneapolis, MN: July 30–31, 2014) pp 139–142

J. McGonigal. *Reality Is Broken: Why Games Make Us Better and How They Can Change the World*. (Penguin Books, 1 edition 2011). ISBN-10 9780143120612 and ISBN-13 978-0143120612

M. Montessori, Work and play. Lecture 21–The 1946 London lectures. (Montessori-Pierson Publishing Company, 1 ed 1946)

S. Pescarin, Videogames, ricerca, patrimonio culturale. *Videogames, Ricerca, Patrimonio Culturale*, pp. 1–314, (2020). Franco Angeli Open access. ISBN: 9788835103004

QPlayLearn platform. https://qplaylearn.com/. Available online, Accessed on 01 Mar 2023

Quantum teaching materials for schools, QTEdu pilot project. https://qtedu.eu/project/quantum-teaching-materials-schools. Available online, Accessed on 01 Mar 2023

Ö. Salehi, Z. Seskir, I. Tepe, A computer science-oriented approach to introduce quantum computing to a new audience. IEEE Trans. Educ. **65**(1), 1–8 (2022)

R. Von Schomberg, Prospects for technology assessment in a framework of responsible research and innovation. in Technikfolgen abschätzen lehren: Bildungspotenziale transdisziplinärer Methoden, eds. by M. Dusseldorp, R. Beecroft VS Verlag für Sozialwissenschaften, Wies-baden (2012), pp. 39–61

Z.C. Seskir, A.U. Aydinoglu, The landscape of academic literature in quantum technologies. Int. J. Quantum Inf. **19**(02), 2150012 (March 2021)

Z.C. Seskir, P. Migdał, C. Weidner, A. Anupam, N. Case, N. Davis, C. Decaroli, İ. Ercan, C. Foti, P. Gora, K. Jankiewicz, B.R. La Cour, J.Y. Malo, S. Maniscalco, A. Naeemi, L. Nita, N. Parvin, F. Scafirimuto, J.F. Sherson, E. Surer, J. Wootton, L. Yeh, O. Zabello, M. Chiofalo, Quantum games and interactive tools for quantum technologies outreach and education. Opt. Eng. **61**(08), (July 2022)

J. Stilgoe, R. Owen, P. Macnaghten, Developing a framework for responsible innovation. Res. Policy **42**(9), 1568–1580 (2013)

B. Suits, *The Grasshopper: Games, Life and Utopia*. (Broadview Press, 2005). ISBN-10 155111772X and ISBN-13 978-1551117720

Unesco, Living Heritage in the face of covid-19 (2021). https://unesdoc.unesco.org/ark: /48223/pf0000377671. Available online, Accessed on 27 Feb 2023

M. Vecco, A definition of cultural heritage: from the tangible to the intangible. J. Cult. Herit. **11**(3), 321–324 (2010)

D.W. Winnicott, *Playing and reality*, (Tavistock Publications Ltd, 1 ed 1971)

Conceptual, Creative and Critical Thinking for Science, Philosophy, Business and Art

Anita Kocsis, Claudia Sciarma, Konstantinos Nikolopoulos, Christine Thong, and Grace McCarthy

Abstract The kaleidoscopic lens applied to this chapter explores how science is learned through different epistemological approaches by diverse cross-disciplinary content. Digitalisation in its myriad of forms affords democratisation of information, new modes of visualisation, interaction and fast, responsive modes of communication that both disrupt and bring opportunities for a vibrant learning ecosystem in science. The chapter canvases conceptual, creative and critical exemplars from design, art, business, philosophy and science reinforcing that agency for our students is paramount in building transversal skills to not only understand the concepts of what science is, also how it can be applied and what else it could be. The dualist debate of how learning the fundamentals is done has shifted from the student as a passive receiver of expert knowledge, to student-driven, authentic experiential learning. The provocations in this chapter include exemplars that underpin the philosophical implications and tensions via thought experiments like Schrödinger's cat; artistic expressions of particle behaviour in a cloud chamber experiment; coaching by business thinking to enhance creative competence; teamwork mindset; science outreach by art practices to provoke discussion of physics fact and the imaginary.

A. Kocsis (✉) · C. Thong
Swinburne Design Factory Melbourne, Swinburne University, John St, Hawthorn VIC, Melbourne 3122, Australia
e-mail: akocsis@swin.edu.au

C. Thong
e-mail: cthong@swin.edu.au

C. Sciarma
Internship European Southern Observatory, Karl-Schwarzschild-Strasse 2, D-85748 Garching Bei München, Germany
e-mail: sciarmaclaudia@gmail.com

K. Nikolopoulos
University of Birmingham, Edgbaston, Birmingham B15 2TT, UK
e-mail: k.nikolopoulos@bham.ac.uk

G. McCarthy
Faculty of Business and Law, University of Wollongong, Northfields Ave, Wollongong, NSW 2522, Australia
e-mail: gracemc@uow.edu.au

M. Streit-Bianchi et al. (eds.), *New Challenges and Opportunities in Physics Education*, Challenges in Physics Education, https://doi.org/10.1007/978-3-031-37387-9_25

Keywords Transversal skills · Interdisciplinarity · Vibrant learning ecosystem · Creative science · Conceptual thinking · Teamwork mindset

1 Introduction

This chapter inspired by the etymology of the kaleidoscope explores conceptual, creative and critical thinking to explore patterns and markers of transversal practices from science, business, philosophy, art and design that influence the broader science ecosystem. Despite the diverse discourse in this chapter, the dominant filter is the recognition that pressing problems that need to be addressed today requires even more conceptual, creative and critical thinking. The current century grand challenges need more than a single disciplinary domain to solve and rely on a collaborative approach hence multidisciplinary knowledge (Maxwell and Benneworth 2018). Transversal competencies such as the ability to work as a team, problem-solving, acceptance of different perspectives and critical analysis (Sá and Serpa 2018) are therefore key yet, in practice, challenging to achieve however needed in large-scale projects and infrastructures (Morton et al. 2015). Important to any generation are real-world factors that influence the meaningful learning of science thus "rest on the assumption that students' worldviews are commensurable with those of science as it is taught" (Roth and Roychoudhury 1994, pp. 6). Influencing students' worldviews in the 60's and today are multiliteracies; hence, the multiplicity of communication channels and increasing cultural and linguistic diversity (Cazden et al. 1996) today is compounded by the speed of visual, interactive digitalised access and exponential technological and computational advances (Friedman 2017).

Transversal practices, interdisciplinary skills and novel organisational and academic transdisciplinary approaches towards the goal of starting from school a vibrant life-learning ecosystem (Kuhn 2021) in science are presented. Creativity to explore across disciplines such as the sci-art, movements and hybridisation of science practices, as evidenced by complexity science, that is emergent and interdisciplinary at the frontier of science may trigger scientists from any discipline to consider their role at such frontiers. The chapters examples point to intersectoral pedagogies and philosophy suggesting that "interdisciplinary subjects do not link together the whole of one traditional discipline with another; particular subfields are joined together to make a new subject. The pattern is a varied one and constantly changing" (Pines 2018, pp. 1). We elaborate on constituents of Kuhns vibrancy factor in learning by exploring how novel concepts from one discipline are trialled and applied in another discipline to maximise knowledge. We offer up terms not necessarily commonplace with physics education and outreach to reinforce how the 'transfer' of concepts learned in one context and then used in different contexts (Kuhn 2021) an important objective of instruction, is made possible by coaching, design thinking and play once the purview of sport, design and business more recently discovered in science projects. What does it take to develop transversal capability and be equipped to adapt and potentially drop the tool you know, forge a new one, drop a method and unlearn and relearn a new one (McGowan and Shipley 2020, pp. 27)?

From the perspective of Murray Gell-Mann, Nobel laureate in physics in 1969, who coined the term synthesising mind, argued that "in the twenty-first century, the most important kind of mind will be the synthesising mind." (Gardner 2020, pp. 216). As the title of this chapter alludes, despite the different disciplinary epistemologies, the demand and urgency for solving pressing global challenges require both deep disciplinary expertise and willing collaborators who can synthesise ideas and methods (Gardner 2020, pp. 74).

1.1 Fostering Synthesising Minds to Adapt to Global Challenges

The resistance of a single frame, reinforcing heterogeneous skills, capabilities and critical thinking afforded by multiple intelligences (Gardner 2020; Gardner 2011a, b) has never been more apt. A shift of basic labour skills to higher-order thinking, production of services and intellectual capital based on knowledge-intensive activities has changed how we work, valuing "collective making" as such in this knowledge economy (Powell and Snellman 2004) and purports to accelerate technical advances and obsolescence. Similarly, cultural institutions, retail, education and digitalisation shaping the experience economy (Pine, Pine and Gilmore 1999) focus on sites of learning as an experience beyond the boundaries of physical walls acknowledging the visitor, customer, user and or participant as an active agent in the consumption and construction of knowledge.

Technology, talent and tolerance (Florida 2006) core to creative competence as seen by the increase of global science and technology innovation clusters, research and development laboratories, investment in start-ups and entrepreneurial initiatives has driven competition for talent resulting in a generation of a global creative class. This creative class, known as the "APP" generation, including generations *Y* and *Z* contribute to a dominant technologized internal world that is remote, distributed and at times anonymous live alongside external factors of pandemic, war, climate change and socio-economic effects. As technology and speed of technology drive most industry, learning and research, "quick and dirty methods and 'just in time knowledge' led by the dot.com industries is commonplace in the current generation (Leung and Bentley 2017). Speed and easy access to information coupled with diverse modes of Internet of things (IoT) consumption influence diverse paths to knowledge acquisition. Focusing on learning for this generation applied to real-world challenges transferable in employment is an increasing currency relevant to the university, industry and government triple helix consortium (Halilem 2010). University-industry collaboration recognising the need for skills beyond disciplinary acumen is demonstrated by a plethora of diverse disciplinary offerings in higher education curricula. Capability development in education informed by authentic learning methods derived from vocational situated experiences and creative capabilities in cross-organisational settings is argued to improve economic and innovation growth (Mavri et al. 2021) and transferable to the workplace.

Socially mediated learning affords learners and educators by legitimate peripheral participation to extend their expertise and practice to wider knowledge networks, enhancing knowledgeability (Wenger-Trayner and Wenger-Trayner 2015), extending one's own disciplinary expertise to complement another, across communities of practice to enhance potential novel learning and methods (Coddington et al. 2016). Co-design learning networks also reinforce how learning and working across domains and cultures (Björklund et al. 2019) by diverse disciplinary practices and voices encouraging design-inspired science alongside out-of-the-box thinking (Joore et al. 2022).

Given the above-mentioned challenges, how do we empower a generation of digital natives that have influenced digital media development and services to have the agency to conceive and shape the science ecosystem they want to work in? Are there clues for our young scientists with respect to the policies for very large scientific infrastructures in a mission-oriented approach driven by a political agenda, invention-oriented approach for conversions of scientific discovery as inventions potentially picked off by the market and system-oriented approaches to broaden and improve interaction amongst actors in the ecosystem (Wareham et al. 2019, pp. 12). How do the big to bigger initiatives driven with respect to innovation and productivity mandates balance the demand of the increasing appetite to move science off the bench and into the world given the pressure to lessen the gap between innovation and productivity?

As demonstrated by the forecasts of the global innovation index for increasing growth in science and technology clusters purported as two waves, the Digital Age, supercomputing, artificial intelligence and automation and the deep science innovation wave are to drive breakthroughs in biotechnologies, nanotechnologies, new materials and other sciences (Dutta et al. 2022) disrupting health, food, environment and mobility. Similar but different are the open movements connected to open source and open access (Burgos 2020). Such broad and deep skills derived from open science is claimed as a form of radical social innovation disrupting higher education (Stracke 2020, pp. 17) as interdisciplinary, intersectoral practices requiring synthesising skills and hybrid approaches to research are offered to stakeholders.

Notwithstanding the challenges of open science to innovation efforts in translating fundamental science to applied use (Chesbrough 2015, pp. 8), competition for skills reducing the gap between innovation and productivity (Dutta et al. 2022) includes transdisciplinary skills in addition to the work of fundamental science given finite and stretched financial resources. Extending on this dynamic is the example of the Open

Innovation in Science framework for practices across the scientific research process acknowledging academic scientists and "actors without formal scientific training, such as citizens, companies, or policymakers, as well as scientific actors working outside of academia" (Beck et al. 2020, pp. 5). Such co-production aims to address complex societal challenges that combine knowledge from different sciences with public and private sector stakeholders and citizens (OECD 2020).

The examples in the chapter do more than encouragement and inspiration as the forms and methods tap into a deeper acknowledgment to modes of learning. Kolb's concrete learning experiences manifest by doing (Kolb 1984, 1999, 2014) exemplify the 'particle dance' workshops and 'cloud chamber experiment discussed in the art sci chapters. Gardner's multiple intelligences theory reinforces the immediacy of thinking with things and experiential learning as illustrated by Taimina's physical prototypes of mathematical examples manifest as crochet to understand geometry concepts. To crochet, as in Fig. 1, demonstrates learning by doing to conceive curvature in maths used for differential geometry is experiential, aesthetic and potentially fun (Taimina 2018, 2020).

Similarly experiencing lightning strikes in a Faraday cage at the Deutsches Museum in Fig. 2 reiterates Dewey's position that learning is contingent on what we humans bring to the situation (Dewey 2005), and our embodied experience conceives the numbers of almost one million volts in two millionth of a second discharging up to 1000 amps. How do we capitalise in our learning the potential of said diverse intelligences; logical-mathematical intelligence, linguistic intelligence, spatial intelligence, musical intelligence, bodily-kinesthetic intelligence, the personal intelligences that include interpersonal feelings and intrapersonal intelligence and naturalist intelligence (Brualdi Timmins 1996; Gardner 1983; Gardner, 2020; Gardner, 2011a; Gardner, 2011b)?

Fig. 1 Image of hyperbolic crochet. (© Daina Taimina)[MS1]

Fig. 2 Deutsches museum lightening and Faraday cage exhibit[1]

2 Critical Thinking and Creativity Beyond Borders

2.1 Critical Thinking Across Philosophy, Psychology and Education

Critical thinking is considered by educators an important outcome of students' learning. The partnership for twenty-first century skills has identified critical thinking as one of the learning and innovations skills to prepare students for post-secondary school education. Indeed, many researchers affirmed that critical thinking skills can be taught (Lai 2011), and much of the educational literature concludes that "if students have to learn to think, they should be encouraged to ask critical questions" (Mason 2007). Critical thinking includes several components skills, such as: analysing arguments, judging or evaluating, making inference using inductive or deductive reasoning, making decisions and solving problems. Indeed, critical thinking entails both cognitive skills and dispositions, as flexibility, open-mindedness and fair mindedness, a propensity to seek reason and to be well-informed, and willingness and respect for diverse viewpoints. In addition to this, background knowledge of a given subject is considered a necessary condition for enabling critical thinking within the chosen subject (Lai 2011).

Despite researchers agreeing on the importance of critical thinking, there is a lack of consensus regarding its definition, and there are many definitions (Lai 2011; Hitchcock 2022). It is worth noting that literature on critical thinking ranges in three academic disciplines, i.e. philosophy, psychology and education, and each field has developed his own approach in defining critical thinking (Lai 2011). In order to underline how rich and multifaceted is defining critical thinking, we will mention three examples. In the philosophical tradition, definitions of critical thinking include self-directed thinking and reflective judging of what to believe or do. Instead, some cognitive psychologists include in the definition the use of critical and cognitive

[1] https://www.deutschesmuseum.de/museumsinsel/ausstellung/starkstromtechnik/hochspannungsanlage

skills which increase an outcome desirable for the thinker. Finally, in the educational approach, that is based on classroom experience, critical thinking is for example represented by analysis, synthesis and evaluation. In the critical thinking debate, there are some disagreements, for instance about the domain specificity of teaching critical thinking skills as well as the transferability to new contexts of the critical thinking skills learnt in a specific domain, and over the relationship of critical thinking with other type of thinking. For deepening this topic, one might want to consider "Critical Thinking", by David Hitchcock (2022), and "Critical thinking: A literature review" by Lai (2011). While critical thinking skills are often considered crosscutting and can be applied to many domains, it has also been highlighted the specificity of the criteria for critical thinking in different domains (Viennot and Decamp 2020). The transfer of critical thinking abilities to new contexts is likely to occur only if students have been specifically taught to practise their skills in a variety of areas.

The relationship of critical thinking to other types of thinking, such as problem solving, decision making and high-order thinking, is entangled in the definitions of these different types of thinking (Hitchcock 2022). For instance, historically "critical thinking" and "problem solving" were synonymous. Additionally, if we consider critical skills like tools for careful thinking about any topic for any purpose as broadly as possible, then decision making can be considered as a kind of critical thinking. Moreover, as we will see below, critical thinking overlaps creative thinking.

Critical thinking abilities relate to other important student learning outcomes, such as metacognition, motivation and creativity (Lai 2011). On the one hand, student motivation is a necessary precondition to learn critical thinking abilities; on the other hand, if a student is facing difficult and/or challenging tasks, which are particularly useful for improving thinking abilities, he or she may be more motivated than with easy tasks.

Metacognition can be simply defined as "thinking about thinking", and its relation with critical thinking is challenging. Indeed, some researchers maintained that critical thinking is a form of metacognition. Instead, others considered metacognition as being subsumed under critical thinking. Then, some researchers argued that self-regulation is the link between critical thinking and metacognition and others argued that they are distinct constructs (Lai 2011).

Many researchers have deepened the relation between critical thinking skills and creativity, and they believe they are the two sides of a coin. Indeed, intellectual products that a good thinker is able to produce are associated with creativity, and an individual needs critical thinking abilities to evaluate these intellectual products (Lai 2011). For details on studies of the methods of developing critical thinking skills and dispositions, see "Critical Thinking" (Hitchcock 2020) and "Developing critical thinking in physics" (Viennot and Decamp 2020).

2.2 Creativity as an Interdisciplinary Skill

Creativity is a very intriguing topic, and even if we are more familiar with "creative arts", creative thinking is also very relevant in science, and increasing students' scientific creativity is a good challenge in education. Creativity has been reported by Gaut (2010) as "the capacity to produce things that are original and valuable". But what does it mean precisely? The concept of creativity hides some issues, such as the relation of creativity to imagination, to knowledge and to tradition as well as whether creative processes are rational, can be explained, and are different in art and science (Gaut 2010).

There is a rich literature in psychology about creativity, and many significant theories with different approaches such as psychoanalysis, cognitive psychology, sociocultural and personality studies. Moreover, after the year 2000, there has been a revival of interest in creativity in philosophy (Gaut 2010; Gaut and Kieran 2018).

Great philosophers worked on creativity in the history of philosophy. For instance, according to Plato, inspiration is a kind of madness; instead, Immanuel Kant connected creativity to imagination. It is worth noting that according to Gaut (2010), an adequate treatment of the philosophical issues on creativity should pay attention to the psychological studies, and that in order to study the nature of creativity, it is necessary to consider not only philosophical studies on art, but also the philosophy of mind, science and epistemology. Indeed, one interesting contribution about creativity was provided by Henri Poincaré, who was a mathematician, theoretical physicist and a philosopher of science. Poincaré wrote about scientific creativity describing his own experience while he was stepping onto an omnibus (Heinzmann and Stump, 2022): "*At the moment when I put my foot on the step the idea came to me, without anything in my former thoughts seeming to have paved the way for it, that the transformation that I had used to define the Fuchsian functions were identical with those of non-Euclidean geometry*".

In other words, in the experience of Poincare, creativity is a swarm of ideas which combined randomly in his unconsciousness and among these combinations he selected the most promising with aesthetic criteria.

One of the topics discussed in literature is whether creative activity works in the same manner in all areas, including science and art. Some philosophers argued that both artistic and scientific creativity is a matter of problem solving. Instead, psychologists generally agreed that creativity operates in the same manner in both areas. However, some researchers believe that art is different from science since art usually does not deal with problem solving or as others put it only in art genuine creation occurs, since scientists are just discovering things (Gaut 2010).

Although creativity can be domain specific, sometimes it overcomes subjects' boundaries. This latter point is an important issue in modern life where science is embedded in multi-disciplinarity contexts; however, this will represent a challenge in teaching. Nonetheless, creative skills are likely to be valuable for graduates to face modern life challenges, and as we have already said, creativity is one of the essential competences to face the challenges of contemporary life. Indeed, based on the recent world undergoing technological evolution, we know that digital technologies will affect jobs. As a result, the occupational skills will be more and more dependent on creative actions and thought, and creativity will be essential in order to be able to adapt to a constantly changing work environment (Newton, Nolan and Rees 2022; Klieger and Sherman 2015). In this view, it is worth noting that some researchers consider fluency, flexibility, originality and elaboration as four aspects of creative thinking and when defining creativity as a cognitive activity to find new solutions to solve a problem, creativity easily applies to science (Rizal et al. 2020). Scientific processes have a creative nature, and physics is a truly creative endeavour as it uses imagination to make sense of what cannot be explained or directly seen (Newton, Nolan and Rees 2022). However, progress in physics and science in general is not usually considered creative. Indeed, people talks about "creative arts", but almost no-one talks about "creative sciences". Indeed, science tends to be associated with "problem solving", and not to "creativity". However, problem solving has a creative nature, because it can put together ideas from different domains in order to create something new (Newton et al. 2022).

The approach that science is far from creativity is partly due to the way science is taught and commonly widely presented. During their educational experience, students are used to memorising theories, formulae, facts and laws, and to finding the right answer, and they usually do not have the opportunity and the freedom to be scientifically creative and to learn from mistakes. Newton et al. (2022) use a good metaphor to explain this point: "*The shop front is neatly set out with highly polished theories, evidence, explanations and applications, while the back room has the real and messy story of people's creative thoughts, imaginative solutions and clever ideas (and those that turn out not to be solutions or quite as clever as they seemed at the time). But the shop window that presents physics as a static, finished subject tends to be what students see, and undergraduates' perceptions of various aspects of physics may hardly change over their time at university. This risk repelling those who want to invest something of themselves in a work in progress, a dynamic subject capable of change. However, it can also misrepresent the subject, even to those who study physics but never see its creative side, and it ignores a world in which creative competence will be increasingly important and a valuable asset*".

Once it is established that creativity is important, the question is: how can we teach it? There are no fixed procedures to teach creativity, and it is not easy to find a way to make students understand scientific creativity. In order to foster creative thinking, science content knowledge is a prerequisite. Besides, imagery and visualisation should have a central role in science teaching too. Science has a social nature, so it would be important to exercise scientific creative thinking not only individually, but also collaboratively.

2.3 Thought Experiments Beyond Science

Thought experiments are imaginary tests which involve something related to experience. They are used in different disciplines, such as philosophy and science, and they can be pedagogically effective (Klassen 2006). Thought experiments are basically devices of the imagination, employed for various purposes such as education, conceptual analysis, exploration, hypothesising, theory selection and implementation. They are used to investigate reality, conceived as widely as possible, include topics like electrons, people, universes, etc. (Brown and Fehige 2022). There is no conventional definition of thought experiments; however, thought experiments form a wide range of extremely diverse cognitive activities (Stuart et al. 2017). Simply put, we can state that a thought experiment is a hypothetical imaginary situation, thanks to which we can investigate the implications and outcomes of an idea or a hypothesis. Thought experiments are on one hand used in many different disciplines, including biology, economics, history, mathematics, philosophy and physics and are as well considered a research topic by researchers of different fields, such as philosophers, historians, cognitive scientists and psychologists.

To underline the diversity of mental activities that might be labelled as thought experiments, let us see some examples. In philosophy, there are, for instance, Plato's cave about reality (Stuart et al. 2017), Thomson's violinist about abortion (Brown and Fehige 2022) and Wittgenstein's beetle related to language. Instead, for science, we can mention Maxwell's demon in thermodynamics, Einstein's elevator in relativity and Schrödinger's cat in quantum mechanics (Stuart et al. 2017). However, the history of science has offered numerous thought experiments, which suggest that we can learn about the real world just by thinking about imagined scenarios. This epistemological challenge is the reason why thought experiments were primarily studied in philosophy of science and philosophy of philosophy. Besides, thought experiments' outcomes and results may be subjected to further real empirical testing according to Brown and Fehige (2022). We can also expand more our view on thought experiments. Indeed, many works of art can be fruitfully considered as thought experiments, such as the film "2001: A Space Odyssey" by Stanley Kubrick or the tragedy "Oedipus Rex" by Sophocles (Stuart et al. 2017).

Thought experiments are most often presented in a narrative form, and frequently with diagrams. According to Klassen (2006), the story-narrative form of a thought experiment is essential to encourage active student engagement and to support the creation of a significant student understanding. In a study conducted in Greece to investigate how two Einstein's thought experiments, called "Einstein's elevator" and "Einstein's train", can be used as didactical tools for teaching modern physics to upper secondary students, the narrative form and the minimum of mathematical formalism appear to be crucial in motivating students to follow the relevant steps of the experiments and to focus on the main concept of the physical theory (Velentzas and Halkia 2013). The narrative presentation triggers the creation of a mental model, and when this mental scenario is imagined, it may lead the thinker into further investigations of the involved concepts and this building of the mental process encloses the creative power of a good thought experiment.

Indeed, thought experiments have a creative function, for their capacity to generate, demonstrate and communicate new ideas, and a critical function, since they might help to problematize existing theories or hypotheses by showing their limitations and unforeseen issues or implications, and by questioning taken-for-granted assumptions (Kornberger and Mantere 2020). They are therefore pedagogically significant as can also enrich students' reflections on the physical world (Klassen 2006; Reiner and Gilbert 2000).

For instance, philosophy can be entirely introduced to students through thought experiments and for undergraduate teaching purposes one might want to consider "Doing Philosophy: An Introduction Through Thought Experiments" (Schick 2012).

When approaching a thought experiment, imagination is structured, goal-oriented and influenced by the previous knowledge on the topic. The thought experiment accesses knowledge which the person is not necessarily aware of and of which only a small part can be clearly explained. This implicit knowledge when coupled with the logical process of a thought experiment can produce new knowledge. When the students imagine the processes of a thought experiment, they can feel and manipulate forces, motions, etc., and they can deepen their understanding of physics. Because of their cognitive and creative nature, thought experiments are essentially an individual activity. However, activities with thought experiments can be also organised in a collaborative classroom set-up. Indeed, each student has his or her mental idea, and the accumulation of all the students' narratives lead to a common construction of meaning (Reiner and Gilbert 2000). In addition to this, the interactions between students and argumentation developed by some members of the group can help others who find processing the thought experiment difficult. In this perspective, thought experiments can be educationally more beneficial if performed in collaboration (Velentzas and Halkia 2013).

Additionally, exploring thought experiments in a classroom has a potential for deepening students' understanding of physics, including its philosophical consequences. To reach this goal, it is also important to present the actual purpose of the thought experiment and the historical context in order to help students to interpret the experiment comprehensively. For instance, the historical Schrödinger's cat thought experiment was originally set up in 1935 by Erwin Schrödinger to show a paradox of quantum mechanics, and shows what superposition means when applied to a macroscopic context. Today, it is possible to experimentally perform some aspects of quantum mechanics; however, the interpretations of quantum mechanics are still discussed, and these kinds of thought experiments can still be helpful. A Norwegian study on how upper secondary school students in Norway interpret Schrödinger's thought experiment shows that students interpret the experiment in a wide range of ways, some of which are consistent with interpretations by physicists today and during the development of quantum mechanics; however, some students miss the main points of the thought experiment. In light of these findings, the researchers believe that the student would need to be familiarised with the historical context in which quantum mechanics was originally formulated in order to deeply and fully understand Schrödinger's cat thought experiment (Myhrehagen and Bungum 2016).

3 Science|Arts Collaboration

Artistic practice, the making of art, is typically directly associated with creativity. This is a familiar connection that is rarely doubted. On the contrary, creativity in scientific research, and science more broadly, is less frequently recognised and acknowledged. In most cases, the process of generating new knowledge in the areas of natural sciences is seen as a dry, potentially algorithmic, procedure that lacks inspiration and creativity. This idea probably stems from the "strict" procedural protocols required by the scientific method, but also from the social conditioning, through educational systems around the world, on what constitutes creativity and the false dichotomy between the "interpretivists" and "positivists". It would be fair to say that in their majority the practitioners do not view themselves as creative persons in the course of their research. Moreover, they may feel uneasy with such a characterisation, due to the potential perceived lack of scientific rigour.

Regardless, it must be argued that scientists rely heavily on creativity, consciously or unconsciously. This realisation precipitated on one of the authors, a particle physicist, while collaborating with artists in their artistic practice and in the context of public engagement activities. In this process, it became evident that, despite obvious differences in tools and techniques, both specialisms are focusing on making the invisible, visible. In particle physics, the fundamental constituents of matter and their interactions, which are found at length scales well beyond our perception, are made accessible only by examining traces left in an enabling medium, the particle detector. This process is mirrored in artistic practice, where the artist expresses thoughts, emotions and insights through the making of marks and the manipulation of materials in visual arts or through movements and expressions in dance. Moreover, creating art and generating new knowledge share the same stages of development, which are well matched to the four stages proposed by Wallas (1926) to describe creativity: preparation, incubation, illumination and verification.

A brief account of the experience acquired through two science|arts collaborations will be discussed. In both cases, the collaboration between the artist and the scientist had a direct influence on the artistic practice and resulted in the making of art that aspired to express ideas of particle physics, but also explore their connections and consequences far beyond physics.

"The Sketchbook and the Collider" [https://www.thesketchbookandthecollider.com/] is an on-going collaboration between an artist (Ian Andrews) and a particle physicist (KN). Their fortuitous acquaintance in an artist community event led to initial discussions and exchange of ideas, and an artist residency at the University of Birmingham in 2018. The residency culminated in the exhibition "The sketchbook and the Collider". The exhibition "Collision Event" followed in 2019 at the Library of Birmingham, supported by Arts Council England, with subsequent exhibitions and events, most recently at the Laboratoire d'Annecy de Physique des Particules (LAPP) on the occasion of the 2022 Fête de la Science [https://cerncourier.com/a/the-sketchbook-and-the-collider/]. The work takes as its starting point the search for equivalents between the primary artistic language of drawing and the elementary particles and their interactions. It comprises three main elements: (a) the search for an intimate connection between pure visual language and elemental particle characteristics and interactions; (b) use of moving image work to explore actual movement and interaction in a drawing context equivalent to the interaction of particles; and (c) development of performative pieces that involve "live" drawing and the cooperation, participation and "interaction" of artists, scientists and members of the public.

The "Neutrino Passoire" [https://www.mairipardalaki.com/the-neutrino-passoire] is a collaboration between a dancer/choreographer (Mairi Pardalaki) and a particle physicist (KN). The starting point was the award of the 2015 Nobel Prize in physics "for the discovery of neutrino oscillations, which shows that neutrinos have mass" (Nobel 2015), which initiated a dialogue and exchange of ideas that culminated in a dance performance that follows the elusive solar neutrinos from their birthplace in the Sun, travelling through space and oscillating between flavours. In this journey, the neutrinos transverse matter, the Earth, our bodies, continuing unimpeded. The performance explores the idea that the human body is a colander (fr. passoire), letting neutrinos pass through without interaction, memory or trauma, follows-up to question notions prevalent in the public discourse, and leaves it to the audience to give their answers. The performance was presented for the first time at the University of Birmingham Arts and Science Festival 2016, and in several venues since (e.g. Brest, Paris, Orsay, Annecy, Bern).

These science|arts collaborations were founded in deep mutual respect between the representatives of the respective disciplines. The artists were interested to learn more about physics and understand the relevant concepts, at a level commensurate with their background knowledge (or lack thereof). Similarly, the physicist was invested to understand the creative process, despite their lack of experience in making art. This mutual and unprejudiced interest to understand the other was key to the success of the endeavour. As a matter of fact, these exchanges were initiated without a predefined agenda, and with no specific vision of what the outcome may look like, if there is an outcome at all. This led to natural connections to be formed, and analogies between the creative processes in the respective disciplines to be drawn. A continuous process of exchanges of ideas followed that has been on-going for several years, and continues. This allows for the artistic outcome to be refined, evolve, and take different directions as the understanding increases and new ideas emerge.

From the scientist point of view, given the limited prior exposure in the artistic creative process, it was fascinating to experience the creation of art. It was realised that artists, like scientists, begin with what we would call in scientific terms "literature review", they gather information and understand the subject of interest, as well as the possible techniques and media that can be employed. Subsequently, they proceed with the "design" phase, where they explore how to best convey their message—whatever that may be, and involves the bulk of the creative process. This is followed by the "implementation" phase, where the art work is created. Finally, the outcome of the process is "disseminated" in various forms (exhibitions, performances, etc.), and it is discussed with other artists and the audience. The choices need to be motivated and substantiated, and the produced work is potentially refined or evolves based on the received feedback. This process is a one-to-one mapping to the process of scientific research.

These successful collaborations led to the idea of the development of science|arts workshops for school children that mirror the collaborative process between artist and scientist. In the following, these creative pedagogy workshops will be briefly described, along with some reflections. The design and delivery of these one-day workshops was enabled and supported by the EU-funded CREATIONS project [https://cordis.europa.eu/project/id/665917], aspiring to create an engaging science classroom. The project's focus was in creative STEAM (Science, Technology, Engineering, Arts and Maths) education in late primary and secondary science education. These efforts are part of a much wider debate regarding creativity and creative pedagogies, which has received much attention in the literature. STEAM approaches enable connections: Learners link knowledge and environment, becoming more creatively engaged and responsive to their communities, and teachers make connections with peers and external partners (Colucci-Gray et al. 2017). The CREATIONS Literature Review (Chappell et al. 2015) found that STEAM pedagogies have the potential to be generative and exciting, enabling discussion and access to abstract scientific ideas, and facilitating understanding of pupils' thinking. A prerequisite is the presence of skilled teachers, who are effectively and sustainably trained and supported. For completeness, the CREATIONS definition of creativity in science education is given: "Purposive and imaginative activity generating outcomes that are original and valuable in relation to the learner. This occurs through critical reasoning using the available evidence to generate ideas, explanations and strategies as an individual or community, whilst acknowledging the role of risk and emotions in interdisciplinary contexts".

A main motivation for these science|art workshops was to creatively introduce scientific ideas, generate an interest in science, as an inherently creative subject, and stimulate a "creative curiosity" about the world to school children. Further motivations were to make particle physics more accessible, support students in developing self-confidence in relation to science and research and introduce them to teamwork. The aspiration was for these effects to persist for a long time after the workshop. Beyond these science-inspired motivations, it was crucial from the beginning that work produced and methods involved are artistically valid and that the experience would prove beneficial also for students studying art. This could mean for example providing potential coursework with only limited additions to the work produced during the day. Thus, genuine connections between the separate subjects and innovative curriculum design were encouraged, promoting interdisciplinary collaborations.

It was immediately clear to the collaborators and becomes obvious from the stated motivations, that science and arts were two equally important elements in the process. The relationship between the two disciplines is that of two pillars that reinforce the student experience. This means that a bridge is constructed between art and science subjects that are currently disconnected in the school curriculum. The design was further informed by the established CREATIONS pedagogical features (Chappell et al. 2015).

The "particle physics and visual arts" workshop experience comprises a range of methods and materials used in current fine art education at first year degree level (Andrews, 2018). An initial range of drawing exercises, gradually developing from a traditional base to increasingly experimental, is followed by sculpture, photography and performance. The students engage with basic concepts regarding the structure of the microcosm and on approaches used to generate knowledge. For example, the interaction of a particle in the detector sensitive material, which is used to construct images that are interpreted as physics processes, finds analogies in artistic techniques. As the workshop progresses, the outcomes are placed onto a single A1-sized "ideas" sheet. Thus, the pupils assume ownership of the product, which is further enhanced by scientific explanations added as annotations. Conveying concepts briefly in their own words offers another learning opportunity. Eventually, the "ideas" sheet becomes the focus of a series of group critiques during the workshop, as shown in the figure below, a common practice within the arts which is the equivalent of the scientific seminar.

The "particle dance" workshop shifts particle physics learning from the traditional classroom to the dance studio (Nikolopoulos 2020). This enables pupils to approach the subject matter through an embodied approach. The structure of the workshop encourages informal learning, avoids potential student preconceptions regarding the science classroom and activates interdisciplinary connections. The students realise that dance is not only a means to express feelings and emotions, but also to convey ideas, even on topics that are not usually perceived to be connected with dance. The

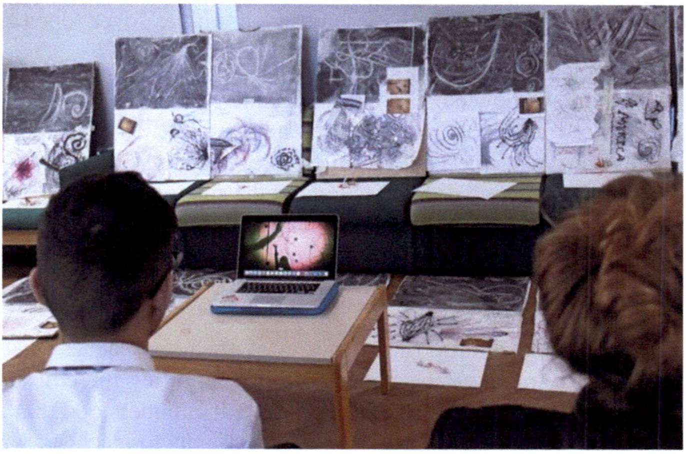

Fig. 3 Group critique during the "particle physics and visual arts" workshop. (Credits K. Nikolopoulos)

workshop consists of two parts: Initially, the students learn about the structure of the microcosm, and they offer connections of the particles to dance through short choreographic movements, inspired by the particle characteristics. Subsequently, they learn about particle interactions, and, working in teams, produce a choreography of an interaction. They assume complete responsibility both for the development of the choreography and the choice of music. The agency of the students to decide on the music had a huge impact on their empowerment and in them taking ownership of the process. It was realised that this was the first time the students were not asked to "dance to the music", but they were put in charge of creating a complete dance performance. This workshop was included as a case study in an investigation of the role and manifestation of dialogue and materiality in science|arts creative pedagogy (Chappell et al. 2019).

Reflecting on this process a few things became apparent:

- Obtaining a complete outcome, either an "ideas" sheet or a complete dance performance was providing agency to the students and enabled them to communicate their work with others. This was crucial for their empowerment with respect to the learning process.
- The science and art pillars support each other to the mutual benefit of both. For example, it was observed that students that were more inclined towards the arts, would find interest in the scientific elements of the workshop, and vice versa. This was particularly encouraging, in terms of developing self-confidence in areas that they were not familiar with.
- In order for the outcome of the learning process to be valid both artistically and scientifically, particular effort was devoted to ensuring that misconceptions were challenged and understanding was achieved.
- During the workshops, students were introduced to the idea that our understanding is incomplete and ever evolving. This came as a shock to many students and reflects the traditional education approach presenting "certainties" over "open questions".

Furthermore, it is important to keep a balance between "creative teaching" and "teaching of essential knowledge, skills and understanding". These two aspects should not be placed in opposition (Cremin and Barnes 2018), but rather work in synergy. Dry teaching of science, developing skills and understanding may not appeal to the students, while creative pedagogy without the support of essential knowledge and techniques may generate a misconception about the tools and processes of the scientific method. In this effort teachers are key, and in particular, the collaboration between art and science educators can result in a fruitful collaboration beneficial to both disciplines and the pupils.

Regardless, of the potential positive impact of such approaches in science|arts education, it is key to reiterate the importance of supporting the educators in this effort, both through training, but also through time. It is understood that the school curriculum in most cases places far too many constraints to teachers, with limited available time for innovative approaches, let alone for regular inquiry-based science education or creative pedagogies. Teachers need to be supported in developing confidence in their ability to implement such pedagogies, and time/space to engage in these sustainably against the pressure of performative education culture.

4 Art in Science Uniting Diverse Backgrounds

In efforts to make science engaging and easily understood by many different types of learners, practices in science communication and outreach often use plain English (removal of technical jargon), metaphors, different visualisation techniques, physical models and simplified hands-on experiments to demonstrate concepts and theories in science for general public audiences. While the purpose of science communication and outreach may be distinct, for example, the International Particle Physics Outreach Group (IPPOG, see Part 2 chap 9 and for more on IPPOG see https://ippog.org/goals) aims to foster appreciation of fundamental science methods and knowledge, encourage evidence-based decision making and build trust with global and diverse communities; to achieve these goals, many of the techniques utilised resonate conceptually and pedagogically with other discipline practices like art and design.

Art and design practices are learnt via a studio environment; where hands-on making experiments with materiality and form, creating artefacts by manipulating materials or digital mediums, and by default learning through doing. In design practices, abstraction of technical knowledge to form that can be easily understood is common practices. For example, a smart phone has a relatively simple external casing and digital user interface, with no need for consumers to understand the complex electronics and components functioning inside. In art practices, an exhibition, for example, of an artist work of glass sculptures when viewed by a patron does not require the patron to understand the craft of glass manipulation techniques needed to create the sculptures in order to appreciate or form an opinion of the artwork. Further, creative practices in art and design also use different visualisations techniques and physical models to explore new artworks or designs, as well as being the outcome of art and design, accessible to the general public through exhibition, products, services, systems and built environments. As established earlier in the chapter, creative practices can aid in learning and collaboration for interdisciplinary and transdisciplinary project exploration.

A practical example where visualisation practices are used to aid learning of interdisciplinary students about physics concepts is the cloud chamber experiment run by the former S'Cool LAB at CERN (https://scoollab.web.cern.ch/scool-lab-cloud-chambers). Aimed at secondary school students 14 years and above, the cloud chamber experiment at S'Cool LAB is a basic particle detector that can make behaviours of cosmic particles, such as alpha and beta particles visible to the human eye.

This is done through ionising particles, based on the device invented by C.T.R. Wilson in 1895 (Chaloner 1997). In order to make natural radiation and cosmic particles visible, dry ice, isopropanol and UV light torches are used to construct cloud chambers in a darkened room, as depicted in Fig. 4. Through observing what they see and describing the characteristics of the particles they have illuminated, discussion of how particles such as muons, electrons and positrons behave is catalysed. To aid discussion, students are asked to re-create via drawing on whiteboards, visualisations of their observations, making explicit their individual mental models of observed particle behaviour. The whole activity lasts between 75 and 90 min. Based on a group of 16 tertiary students from different countries and backgrounds undertaking the S'Cool LAB cloud chamber experiment in 2019, we can explore the use of visualisation to provoke conceptual thinking across an interdisciplinary and intercultural cohort. Even though the drawings they produced took only 5 min approximately and were produced on a whiteboard, the visualisations are a collection of artistic expressions that forced the students to reflect on how they might best describe and communicate what they personally observed.

Figure 5 depicts the visualisations produced from this group of tertiary students who were from different disciplines including computer science, philosophy, film studies, design, strategy and biomedical fields and from universities in Australia, Germany and North America, with Asian cultural heritage also in the mix. You can see a range of different interpretations of observed particle behaviour, some very similar, some radically different, and everything in between. There were many different shapes with dotted techniques and also similar shapes, but with different "fillings", e.g. shape may be a similar teardrop, but depicted through dots, or scratchy lines, or solid wave lines to create the outline and content of the teardrop. The different drawing techniques are all aimed at describing the fall off effect of alpha particles as electrons are knocked off in the ionising process, creating a moving line that lasts only moments as a point of visual reference. Facilitated by a trained science

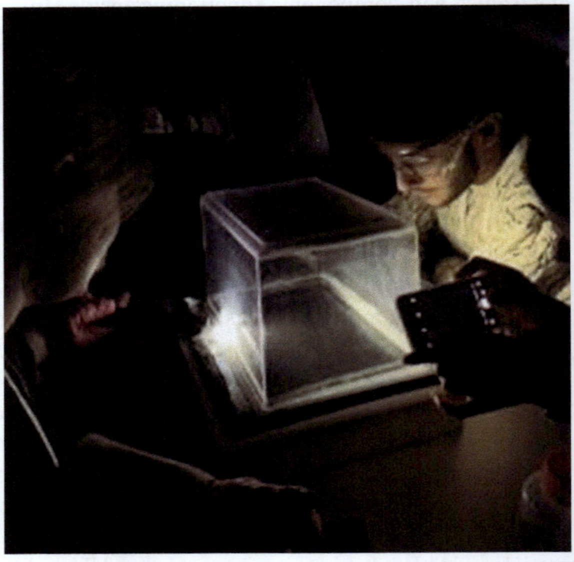

Fig. 4 Cloud chamber experiment (Credits CERN)

Fig. 5 Drawings from tertiary students from different countries and different disciplines backgrounds, conceptualising observations from a cloud chamber experiment they undertook at S'Cool LAB CERN in 2019. (Credits S.Cool Lab CERN)

instructor, students discussed what visualisations they agreed with or not and why, many identified other visualisations they thought were more accurate than their own, and this enabled discussion that looped back to model cosmic particle behaviour.

Through the act of individually creating a visual interpretation of observed particle behaviour, and then reconciling individual interpretation with others a dual layer of reflection enhances the engagement with conceptual thinking for science in particle physics. Creative practices of visualisation provoke conceptual thinking, moving beyond the procedural thinking used to conduct the cloud chamber experiment initially (Tall et al. 2001) to enhance appreciation and understanding of science across disciplines and cultures. The provocation of conceptual thinking through visualisation could be experimented with further, engaging digital visualisation tools where 2D, 3D or 4D visual representations could be manipulated live during discussion and continue to enhance deeper learning in transdisciplinary cooperation and pedagogies for conceptual thinking in science (Etkind and Shafrir 2013).

5 Navigating the Science Ecosystem of Connectivity and Careers

As explored in the above example's creativity in science and science in creativity has undeniable benefits yet in the practice from one discipline influencing the other, how do we also demonstrate the requirement for the integrity and the demands of the discipline to not set up false expectations for learners? Traversing learning physics and careers in physics are challenging as in any discipline; however, opportunities for learners drawing from complimentary methods may prove useful in science careers. In considering how science views itself in respect to the evolution of the science ecosystem how do young and mature learners mitigate the ecosystem of competition, connectivity and careers? (Baumberg 2018, pp. 225) How can physics learners thrive and what are the clues in fields such as business and entrepreneurship for example that reinforce not just the fundamental questions of physics that may be investigated and the purpose for the problem, routine or challenge to be solved? The implications for those working in pure science as simplifiers and those working in applied science as constructors (Baumberg 2018, pp. 10) may consider other practice, action or purpose-driven careers likened to "constructors' whose creative processes emerge not from one individual or a specific research group but often from a large network of cooperating and competing research teams" (Baumberg 2018, pp. 19). How do physics learners for example understand their data, findings within the broader array and or network for example? This snapshot of science work and the people within a science ecosystem may then call on skills, capability and teamwork mindset that includes future generators, synthesisers, imaginators and disruptors poking, inspiring and driving science. Coaching and business methodologies suggest not only the direct professional benefits of extending ones learning repertoire and fundamentals of transferability. How can critical thinking for example in business be applied to the problem in physics and vice versa?

6 What Can Physics Teachers and Students Learn from business?

The vast amount of data available to businesses today threatens to overwhelm our human capacity for conceptual, critical and creative thinking. How do we collaborate to tackle the global challenges the world is facing? And how can we expect the unexpected, identify the black swans of our time and prepare for the next pandemic? Over centuries, each discipline and profession has honed ways of thinking and knowing peculiar to its own context. Now is the time to look beyond our own borders and see what we might learn from other disciplines. For example, what can physics' teachers and students learn from business?

One of the features of conceptual thinking in business is that before starting a project, the project team first seeks to understand what the project is really about, why it is important and what each stakeholder needs; in other words, there is some thinking before the doing. An apocryphal example tells the story of three medieval stonemasons who were asked what they were doing. One said he was building a wall, one that he was working to feed his family, and the third that he was building a great cathedral to honour God. The latter was reputed to be the best builder of the three. Understanding the relevance and purpose of a project makes it more likely that the project will succeed and that stakeholders will be satisfied with the outcomes. Such an approach is also relevant in the sciences where increasingly the needs of different stakeholders need to be considered. Furthermore, as our knowledge increases, and many people contribute a small piece of the puzzle, it is important to know how such small projects contribute to the bigger picture.

Stakeholder analysis is also a key part of strategy development in business. Understanding the perspectives of different stakeholders can help generate new opportunities. In science projects, the questions team members or reviewers ask can help strengthen a proposal or report by bringing in different perspectives. Businesses have long been wary of the danger of paralysis through analysis, in other words of spending so much time analysing data that we do not take appropriate action. Fortunately, we can now make use of technology to help harness the vast amount of data generated (Ducange et al. 2019). Artificial intelligence combined with big data analytics offers the ability to identify patterns in the data and ask meaningful questions. Our ability to understand what the data means requires critical thinking. Whether in business or in physics, the ability to make sense of data distinguishes the human from the machine. Technology can also help us develop innovative business models through the combination of artificial intelligence, Blockchain, Cloud Computing and Data Analytics (Akter et al. 2022), thus providing opportunity for creative thinking.

While some people in business are natural conceptual thinkers, many people benefit from a coach who acts as a sounding board, to help them tease out the nuances of an idea. As Nancy Kline (Passmore and Sinclair 2020) says, the quality of our thinking improves when someone else listens to us. Time spent thinking about a problem before we attempt to solve it is time well spent as it helps ensure we spend time solving the right problem. A combination of critical and creative thinking is invaluable in generating innovative solutions.

Conceptual thinking includes the ability to conceptualise, to think in the abstract, to think about non-tangibles and to think about possibilities. While critical thinking is rational, it may foster self-limiting beliefs. Creative thinking can help unblock our thinking by reframing a challenge when we think "we can't do this because to free up creative thinking, a business coach might ask: "Let's suppose we could do x, what would that look like?"—and of course then we can use creative thinking approaches to figure out how to get there. Similarly, creative thinking in the physical sciences enables us to visualise what we cannot currently see and then conduct experiments to confirm our hypotheses. There are many ways we can foster our creative thinking abilities which we can then apply in combination with our critical thinking to achieve the best outcomes. For example, we can do a quick brainstorming activity at the start of a class where students are challenged to come up with alternative uses for a disused offshore oil platform, going around the classroom multiple times until there are no more ideas. Or students can be asked to come up with ideas for a new variety of sandwich, but each idea must start with a different letter of the alphabet. Students might be asked to think about an existing product or solution and then use SCAMPER (Eberle 1971) where they think about what to substitute, combine, adapt, modify, put to another use, eliminate a particular part or reverse or rearrange. Students get used to such prompts and soon generate large numbers of ideas which then be improved on or deleted.

Another approach to creativity is design thinking, which is often associated with Stanford University who developed an approach to teaching engineering students to think creatively about product design, emphasising three main stages: starting by identifying a customer need or problem, brainstorming potential solutions and prototyping the selected solution (Auenammer and Roth 2021). The emphasis on empathy and the human need for a solution is a salient feature of design thinking. This approach has been adopted in companies, universities and start-up incubators around the world. An Indonesian study by Pratomo et al. (2021) showed that students who were taught design thinking showed higher levels of creativity than students who had not. Creativity activities like SCAMPER (Eberle 1971) can be used in conjunction with design thinking. Recently, augmented and virtual reality has been used to help students deepen their understanding of the potential customer and the problem they need solved, and for example to engage with "wicked problems" such as sustainability (Earle and Leyva-de la Hiz 2021).

Tackling the huge problems of our time such as climate change and the UN Sustainable Development Goals requires businesspeople and scientists to have sufficient understanding of each other's worlds to know what questions to ask, to understand each other's worldview and how to work effectively together. We have a responsibility to understand enough to ask questions, not simply to accept what we are told but to challenge what is being said. This is increasingly important these days, so that we are not swayed by greenwashing (de Freitas Netto et al. 2020) and fake news (Machete and Turpin 2020). Dialogue between discipline and professions will help enhance our thinking skills and ultimately our ability to make meaningful contributions to society from the perspective of each of our disciplines.

7 Conclusion

In this chapter, we presented less of a how to and more of a how might we by embracing the multiple ways of having access to, being inspired by and ultimately wanting to pursue physics with a purpose for science learning and career development. The kaleidoscopic lens explores how science is learned through different epistemological approaches framed by diverse cross-disciplinary content.

Our future generations shape the science ecosystem, and it is our responsibility not only within the domain of physics to offer robust learning in which solutions are learnt and potentially applied to new contexts founded on a vibrant learning ecosystem that includes thinking with things, create conducive learning environments, educate for action and take teaching and learning seriously (Kuhn 2021).

It is demonstrated that thought experiments can be pedagogically effective and employed in the classroom as a useful tool to teach physics with a broad vision. Indeed, the story-narrative form of thought experiments is essential to encourage active student engagement and to support the creation of a significant student understanding (Klassen 2006), by problematizing existing theories or hypotheses showing their limitations and unforeseen issues or implications, and by questioning taken-for-granted assumptions (Kronenberger and Mantere 2020).

We have also shown how creativity is relevant in both art and science, as explored in the art and science activities and collaborations in this chapter. Science and art projects are increasingly used in science communication too, for their potential to encourage critical reflection on science and to engage audiences (Fleerackers et al. 2022). Indeed, some of the main motivations for the science|art projects were to introduce scientific ideas in a creative manner, generate an interest in science as a creative subject, and stimulate students' "creative curiosity" about the world. Demonstrated by business practices are examples of the value of creativity as a type of critical thinking ambidexterity to clarify and communicate beyond the purview of physics. Coaching, design thinking, prototyping and think out loud methods assist students to clarify not only what they are learning and help to consider drawing on alternate methods that may not be commonly used in physics education to enhance mindset.

To sum up, drawing on conceptual, creative and critical thinking across intersecting disciplines in science, philosophy, business, art and design, we elucidated the imperative to consider alternate regimes of learning for professional, transversal skills for the sake of communicating and working with others. Physics, and science in general, is not an isolated pursuit, as part of a broader network and is often the results of international teamwork; in other words, scientists live in the here and the now of society and they are probably influenced by different happenings. Ideally, this should be reflected in teaching practice and approaching physics teaching in an interdisciplinary way can be empowering in two respects. On the one hand, it can give students a more realistic idea of scientific processes including its creative nature and other fields' influences (Viennot and Decamp 2020). For instance, when students study Albert Einstein's theories, explaining to them that Einstein considered David Hume's and Ernst Mach's philosophical writings useful (Northon 2010) can be stimulating for students, and it gives them a more multifaceted idea of Einstein's work. On the other hand, it can help students to improve their conceptual, critical and creative skills. Nevertheless, the challenges of transferability are still open in finding a way to transfer critical, conceptual and creative skills learnt in each school subject to everyday life issues including political and social issues (Hitchcock 2022).

Universal and translatable modes of communication such as coaching and design thinking for example shift the focus to solving the right problem in the problem-solving process. It might be argued that dialogue between disciplines, professions and epistemologies may be considered interference which could risk diluting disciplinary expertise. However, the chapter reiterates that a focus on points in the curriculum and/or workplace that is both disciplinary and culturally diverse can lead to meaningful contributions that we have not yet anticipated from our respective disciplines. We conclude that our intention is not to dissolve disciplines nor detract from the tenets of physics, but rather to focus on the power of curiosity, creativity and critical thinking at a time when getting information in an era of digitalisation and AI synthesis in a keystroke is easy, yet knowing what questions to ask and how to communicate both questions and solutions is core to a vibrant learning ecosystem.

The examples provided elevate the importance of conceptual, critical and creative thinking to improve the capacity to understand the nature of science and enriching physics learning. Appreciating the differences and similarities across art, science, business and design can improve the enrichment of the learning and innovations skills by the partnership for the twenty-first century to prepare students for post-secondary school education and as such an invaluable suite to face and shape the challenges of future employment.

References

S. Akter, K. Michael, M.R. Uddin, G. McCarthy, Transforming business using digital innovations: The application of AI, blockchain, cloud and data analytics. *Annals of Operations Research*, pp. 1–33 (2022)

I. Andrews, K. Nikolopoulos, Introducing particle physics concepts through visual art. Phys. Educ. **53**, 054001 (2018). https://iopscience.iop.org/article/https://doi.org/10.1088/1361-6552/aad276 and arXiv:1806.11490.

J. Auenammer, B. Roth, The origin and evolution of Stanford University's design thinking: From product design to design thinking in innovation management. *Journal of Product Innovation management, 38*(6), 623–644 (2021)

J.J. Baumberg, The secret life of science: how it really works and why it matters, (Princeton University Press, 2018)

S. Beck, C. Bergenholtz, M. Bogers, T.-M. Brasseur, M.L. Conradsen, D. Di Marco, A. Effert, The open innovation in science research field: a collaborative conceptualisation approach. Indus. Innov. 1–50 (2020)

T.A. Björklund, T. Keipi, S. Celik, K. Ekman, Learning across silos: design factories as hubs for co-creation. Eur. J. Educ. **54**(4), 552–565 (2019)

J.R. Brown, Peeking into Plato's heaven. Philosophy Sci. **71**(5), 1126–1138 (2004)

J.R. Brown, Y. Fehige, Thought experiments. in *The Stanford Encyclopedia of Philosophy*, eds. by E.N. Zalta, U. Nodelman Winter 2022 edn (Metaphysics Research Lab, Stanford University, 2022)

A.C. Brualdi Timmins, Multiple intelligences: gardner's theory. Pract. Assess. Res. Eval. **5**(1), 10 (1996)

D. Burgos, Radical solutions and open science: an open approach to boost higher education, (Springer Nature, 2020)

C. Cazden, B. Cope, N. Fairclough, J. Gee, M. Kalantzis, G. Kress, M. Nakata, A pedagogy of multiliteracies: designing social futures. Harv. Educ. Rev. **66**(1), 60–92 (1996)

CERN, S'Cool LAB Cloud Chamber Experiments. https://scoollab.web.cern.ch/scool-lab-cloud-chambers

C. Chaloner, The most wonderful experiment in the world: a history of the cloud chamber. British J. History Sci. **30**(3), 357–374 (1997). https://doi.org/10.1017/S0007087497003130

K.A. Chappell et al., Dialogue and materiality/embodiment in sciencelarts creative pedagogy: their role and manifestation. Thinking Skills Creativ. **31**, 296 (2019)

K. Chappell, L. Hetherington, H. Ruck Keene, C. Slade, M. Cukorova, D2.1 The features of inquiry learning: theory, research and practice (2015), pp. 106

H. Chesbrough, From open science to open innovation. (Institute for Innovation and Knowledge Management, ESADE, 2015)

A. Coddington, C. Giang, A. Graham, A. Prince, P. Mattila, C. Thong, A. Kocsis, Knowledge-abity culture: Co-creation in practice. DRS2016 RESEARCH PAPER (2016). https://doi.org/10.21606/drs.2016.134

L. Colucci-Gray, J. Trowsdale, C.F. Cooke, R. Davies, P. Burnard, D.S. Gray, Reviewing the potential and challenges of developing STEM education through creative pedagogies for 21st learning: how can school curricula be broadened towards a more responsive, dynamic, and inclusive form of education?. (British Educational Research Association, London, 2017), p. 105

CREATIONS - developing an engaging science classroom. http://creations-project.eu/, CORDIS Project In formation: https://doi.org/10.3030/665917

T. Cremin, J. Barnes, Creativity and creative teaching and learning. in *Learning to Teach in the Primary School* (Routledge, 2018) pp. 428-442

J. Dewey, *Art as Experience*, Paperback edition 2005 (The Berkley Publishing Group, Penguin, 1934). ISBN: 0-399-53197-1

J. Dewey, *How We Think, Experience and Education.* 2nd Edition, People's Education Press Beijing 23

P. Ducange, M. Fazzolari, M. Petrocchi, M. Vecchio, An effective decision support system for social media listening based on cross-source sentiment analysis models. Eng. Appl. Artificial Intell. **78**, 71–85 (2019)

S. Dutta, B. Lanvin, S. Wunsch-Vincent, L.R. León, *Global Innovation Index 2022*: *What is the Future of Innovation-Driven Growth*? 15th edn. (WIPO, 2022). ISBN: 978-92-805-3432-0 (print) ISBN: 978-92-805-3433-7 (online)

A.G. Earle, D.I. Leyva-de la Hiz, The wicked problem of teaching about wicked problems: Design thinking and emerging technologies in sustanability education. *Management Learning 52*(5), 581–603 (2021)

R.F. Eberle, Scamper: games for imagination development. BuffaloN. YDOK (1971)

M. Etkind, U. Shafrir, Teaching and learning in the digital age with pedagogy for conceptual thinking and peer cooperation. in *INTED2013 Proceedings, Conference 4th–6th March 2013*, (Valencia, Spain, 2013), pp 5342–5352. https://www.researchgate.net/profile/UriShafrir/publication/265783984_Etkind_Shafrir_Pedagogy_INTED-2013/links/

A. Fleerackers, P. Brown Jarreau, J. Krolik, Why create SciArt? an investigation into science artists' goals and professional journeys. JCOM 21 (06), A05 (2022). https://doi.org/10.22323/2.21060205

R. Florida, The flight of the creative class: the new global competition for talent. Liberal Educ. **92**(3), 22–29 (2006). ISSN 0024-1822

S.V. de Freitas Netto, M.F.F. Sobral, A.R.B. Ribeiro, G.R.D.L. Soares, Concepts and forms of greenwashing: a systematic review. Environ. Sci. Euro. **32**(1), 1–12 (2020). https://enveurope.springeropen.com/articles/https://doi.org/10.1186/s12302-020-0300-3

T.L. Friedman, Thank you for being late: an optimist's guide to thriving in the age of accelerations (Version 2.0, With a New Afterword). (Picador USA 496), (2017)

H. Gardner, Frames of mind: a theory of multiple intelligences. New York: Basic Books (1983)

H.E. Gardner, Frames of mind: the theory of multiple intelligences. Paperback. Basic books p 528 (2011a). ISBN-10 0465024335

H.E. Gardner, The unschooled mind: How children think and how schools should teach: basic books. Twentieth Anniversary Edn 2nd (2011b). ISBN-13 9780465024391

H.E. Gardner, A synthesizing mind: a memoir from the creator of multiple intelligences theory. (MIT Press, 2020), p. 304. ISBN: 9780262542838

B. Gaut and M. Kieran (Eds), *Creativity and Philosophy*. (Routledge 2018)

B. Gaut, The philosophy of creativity. Philos Compass **5**(12), 1034–1046 (2010). https://doi.org/10.1111/j.1747-9991.2010.00351.x

N. Halilem, Inside the triple helix: an integrative conceptual framework of the academic researcher's activities, a systematic review. J. Res. Admin. **41**(3), 23–50 (2010)

G. Heinzmann, D. Stump, Henri Poincaré. in *The Stanford Encyclopedia of Philosophy*, E.N. Zalta (ed.) (Summer 2022 Edition). https://plato.stanford.edu/archives/sum2022/entries/poincare/

D. Hitchcock, Critical thinking. in: *The Stanford Encyclopedia of Philosophy*, E.N. Zalta, U. Nodelman (eds.) (Winter 2022 Edition). https://plato.stanford.edu/archives/win2022/entries/critical-thinking/

D. Hitchcock, Seven Philosophical Conceptions of Critical Thinking: Themes, Variations, Implications. In *Critical Thinking and Reasoning*. (Brill 2020) pp. 9–30

IPPOG, About; Goals. Accessed 17 Feb 2023. https://ippog.org/goals

P. Joore, T. Björklund, C. Thong, E. Zancul, Co-creating the future through design-based education in innovation hubs. CERN IdeaSquare J. Exper. Innov. **6**(2) (2022). https://doi.org/10.23726/cij.2022.1413

S. Klassen, The science thought experiment: How might it be used profitably in the classroom? Interchange **37**, 77–96 (2006). https://doi.org/10.1007/s10780-006-8401-5

A. Klieger, G. Sherman, Physics textbooks: do they promote or inhibit students' creative thinking. Phys. Educ. **50**(3), 305 (2015). https://doi.org/10.1088/0031-9120/50/3/305

D.A. Kolb, Experience as the source of learning and development. Upper Sadle River: Prentice Hall (1984). ISBN: 0132952610

D.A. Kolb, (1999) Learning style inventory. (McBer and Company Boston, MA)

D.A. Kolb, Experiential learning: experience as the source of learning and development, 2nd edn (FT press. 2014) ISBN-10: 0-13389240-9 and ISBN-13: 978-0-13–389240-6

M. Kornberger, S. Mantere, Thought experiments and philosophy in organizational research. Organiz. Theory **1**(3), 2631787720942524 (2020)

S. Kuhn, Transforming learning through tangible instruction: the case for thinking with things. (Routledge, 2021), p. 222. Available on Taylor & Franci eBooks https://doi.org/10.4324/9781003129073

E.R. Lai, Critical thinking: a literature review. Pearson's Res. Reports **6**(1), 40–41 (2011)

L. Leung, N. Bentley, Producing leisured laborers: developing higher education courses for the digital creative industries. J. Arts Manag. Law Soc. **47**(2), 148–160 (2017). https://doi.org/10.1080/10632921.2016.1259133

P. Machete, M. Turpin, The use of critical thinking to identify fake news: a systematic literature review. in *Conference on e-Business, e-Services and e-Society*. (Springer Cham, 2020), pp 235–246

R.F. Malina, Leonardo: our fourth decade. Leonardo **30**(1), 1–2 (1997)

R.F. Malina, Alt. Art-Sci: we need new ways of linking arts and sciences. Leonardo **44**(1), 2 (2011)

M. Mason, Critical thinking and learning. Educ. Philos. Theory **39**(4), 339–349 (2007). https://doi.org/10.1111/j.1469-5812.2007.00343.x

A. Mavri, A. Ioannou, F. Loizides, Cross-organisational communities of practice: enhancing creativity and epistemic cognition in higher education. Internet Higher Educ. **49**, 100792 (2021). https://doi.org/10.1016/j.iheduc.2021.100792

K. Maxwell, P. Benneworth, The construction of new scientific norms for solving grand challenges. Palgrave Commun. **4**(1), 1–11 (2018). https://doi.org/10.1057/s41599-018-0105-9

H.E. McGowan, C. Shipley, The adaptation advantage: let go, learn fast, and thrive in the future of work, (John Wiley & Sons, 2020), p 272. ISBN-10 1119653096

L.W. Morton, S.D. Eigenbrode, T.A. Martin, Architectures of adaptive integration in large collaborative projects. Ecol. Soc. **20**(4) (2015) art. 5. https://doi.org/10.5751/ES-07788-200405

H.V. Myhrehagen, B. Bungum, From the cat's point of view': upper secondary physics students' reflections on Schrödinger's thought experiment. Phys. Educ. **51**(5), 055009 (2016). https://doi.org/10.1088/0031-9120/51/5/055009

D.P. Newton, S. Nolan, S. Rees, Creative thinking in university physics education. (IOP Publishing, 2022), p. 128. ISBN-10: 0750340266 and ISBN-13: 978-0750340267

K. Nikolopoulos, M. Pardalaki, Particle dance: particle physics in the dance studio. Phys. Educ. **55**, 025018 (2020). https://doi.org/10.1088/1361-6552/ab6952 arXiv:1912.11128

J.D. Norton, How Hume and Mach helped Einstein find special relativity. Discourse on a new method. in *Reinvigorating the Marriage of History and Philosophy of Science, Open Court*. eds. by M. Friedman, M. Domski, M. Dickson, pp. 359–387

OECD Addressing societal challenges using transdisciplinary research. (2020). https://doi.org/10.1787/0ca0ca45-en

Nobel Prize Outreach. The Nobel Prize in Physics, (2015). https://www.nobelprize.org/prizes/physics/2015/summary

J. Passmore, T. Sinclair, Humanistic approach and the time to think model. in *Becoming a Coach*, (Springer, Cham, 2020), pp. 119–126. https://doi.org/10.1007/978-3-030-53161-4

B.J. Pine, J.H. Gilmore, The experience economy: work is theatre and every business a stage. (Harvard Business School Press, Boston Massachusetts, 1999). ISBN 0-87584-819-2

D. Pines, Emerging syntheses in science. (CRC Press, 2018). ISBN 13 978-0367320454

L.C. Pratomo, D.K. Wardani, The Effectiveness of Design Thinking in Improving Student Creativity Skills and Entrepreneurial Alertness. *International Journal of Instruction 14*(4), 695–712 (2021)

W.W. Powell, K. Snellman, The knowledge economy. Ann. Rev. Sociol. **30**(1), 199–220 (2004). https://doi.org/10.1146/annurev.soc.29.010202.100037

M. Reiner, J. Gilbert, Epistemological resources for thought experimentation in science learning. Int. J. Sci. Educ. **22**(5), 489–506 (2000). https://doi.org/10.1080/095006900289741

R. Rizal, D. Rusdiana, W. Setiawan, P. Siahaan, Creative thinking skills of prospective physics teacher. J. Phys.: Conf. Ser. **1521**, 022012. (IOP Publishing 2020). https:/doi.org/https://doi.org/10.1088/1742-6596/1521/2/022012

W.M. Roth, A. Roychoudhury, Physics students' epistemologies and views about knowing and learning. J. Res. Sci. Teach. **31**(1), 5–30 (1994). https://doi.org/10.1002/tea.3660310104

M.J. Sá, S. Serpa, Transversal competences: their importance and learning processes by higher education students. Educ. Sci. **8**(3), 126 (2018). https://doi.org/10.3390/educsci8030126

E.B.-N. Sanders, P.J. Stappers, Co-creation and the new landscapes of design. Co-Design **4**(1), 5–18 (2008). https://doi.org/10.1080/15710880701875068

H. Schick, *Entwicklungspsychologie der Kindheit und Jugend. A textbook for teachers training and school practice.* 1st Edition W. Kohlhammer. 2012 ISBN: 3-17-020879-9

C.M. Stracke, Open science and radical solutions for diversity, equity and quality in research: a literature review of different research schools, philosophies and frameworks and their potential impact on science and education. in eds D. Burgos, Radical solutions and open science lectures notes in educational technology. (Springer, Singapore, 2020), pp. 17–37. https://doi.org/10.1007/978-981-15-4276–3_2

M.T. Stuart, Y. Fehige, J.R. Brown, Thought experiments: state of the art. in *The Routledge companion to thought experiments* (2017). ISBN 9780415735087

D. Taimina, Crocheting adventures with hyperbolic planes: tactile mathematics, art and craft for all to explore. (CRC Press, 2018), p. 200. ISBN 9780367375072

D. Taimina, Crocheting adventures with hyperbolic planes. in *Proceedings of Bridges 2020: Mathematics, Art, Music, Architecture, Education, Culture*, eds by C. Yackel, E. Torrence, K. Fenyvesi, R. Bosch, C.S. Kaplan, pp. 17–18. ISBN 978-1-938664-36-6. http://archive.bridgesmathart.org/2020/bridges2020-17.html

D. Tall, E.M. Gray, M.B. Ali, L. Crowley, P. DeMarois, M. McGowen, D. Pitta-Pantazi, M. Pinto, M. Thomas, Y.M. Yusof, Symbols and the bifurcation between procedural and conceptual thinking. Can. J. Sci. Math. Technol. Educ. **1**(1), 81–104 (2001). https://doi.org/10.1080/14926150109556452

A. Velentzas, K. Halkia, The use of thought experiments in teaching physics to upper secondary-level students: two examples from the theory of relativity. Int. J. Sci. Educ. **35**(18), 3026–3049 (2013). https://doi.org/10.1080/09500693.2012.682182

L. Viennot, D. Decamp, Developing critical thinking in physics. vol. 7 (Springer, 2020) ISBN: 978-3-030-43772-5

G. Wallas, The art of thought. in *Internet Archives*, (J. Cape, Ltd London, 1926), p. 319. https://archive.org/details/theartofthought

J.D. Wareham, L. Pujol Priego, M. Nordberg, P. Garcia Tello (2019) Systematising serendipity for big science infrastructures: the ATTRACT project. ESADE Bus. Sch. Res. Paper **271**. https://doi.org/10.1016/j.technovation.2021.102374

E. Wenger, Communities of practice: learning, meaning, and identity. (Cambridge University Press, 1999), p. 336. ISBN: 9780521663632 https://doi.org/10.1017/CBO9780511803932

E. Wenger-Trayner, B. Wenger-Trayner, Learning in landscapes of practice. in *Learning in Land-Scapes of Practice. Boundaries, Identity, and Knowledgeability in Practice-Based Learning*, eds. by E. Wegner-Trayner, M. Fenton-O'Creevy, S. Hutchinson, C. Kubiak, B. Wegner-Trayner (Published by Routledge, 2015), pp. 13–30. ISBN: 9781138022195

CERN Science Gateway: Example of Informal Contemporary Physics Education in an Authentic Research Environment

Daria Dvorzhitskaia, Patricia Verheyden, Julia Woithe, and Annabella Zamora

Abstract Contemporary physics education has an immense potential to positively influence students not just in their course and career aspirations but also in their fundamental understanding of the Nature of Science. However, teachers might face a number of challenges when integrating modern physics topics into their teaching. In this chapter, we will discuss the following three challenges and opportunities in contemporary physics education: working with invisible entities, aiming for authentic experiences and interactions and evaluating ideas and designs with the target audiences. We showcase how we navigated these aspects in the development of CERN Science Gateway, a new education and outreach facility at CERN, Geneva, Switzerland. Moreover, we describe the advantages of digital tools in these contexts and derive recommendations for teachers.

Keywords CERN · CERN Science Gateway · Informal science education · Particle physics · Authenticity · Evaluation

D. Dvorzhitskaia · P. Verheyden · J. Woithe (✉)
CERN, Esplanade des Particules 1, 1211 Geneva, Switzerland
e-mail: julia.woithe@cern.ch

D. Dvorzhitskaia
e-mail: daria.dvorzhitskaia@cern.ch

P. Verheyden
e-mail: patricia.verheyden@cern.ch

D. Dvorzhitskaia
University of Zurich, Zurich, Switzerland

A. Zamora
University of Lausanne, 1015 Lausanne, Switzerland
e-mail: annabella.zamora@unil.ch

M. Streit-Bianchi et al. (eds.), *New Challenges and Opportunities in Physics Education*, Challenges in Physics Education, https://doi.org/10.1007/978-3-031-37387-9_26

1 Introduction

In this chapter, we discuss the following three challenges and opportunities in informal contemporary physics education: teaching about invisible entities, aiming for authentic experiences and interactions and evaluating ideas and designs with the target audiences. We showcase how we navigated these aspects in the development of CERN Science Gateway, a new education and outreach facility at CERN, Geneva, Switzerland. Moreover, we describe the advantages of digital tools in these contexts and derive recommendations for teachers.

We would like to start this chapter by thanking a large team of highly motivated and skilled colleagues, volunteers, external partners and advisors who contributed to CERN Science Gateway with their time, effort, passion and patience, allowing us to write this chapter reporting on their behalf.

1.1 Challenges and Opportunities of Contemporary Physics Education

Contemporary physics education has an immense potential to positively influence students not just in their course and career aspirations but also in their fundamental understanding of the Nature of Science. In particular, the ROSE study found that high-school students are intrigued by unanswered questions and phenomena that scientists cannot yet explain (Sjøberg and Schreiner 2012). Examples of modern physics topics that spark particularly high levels of interest are the universe and its history (Levrini et al. 2016; OECD 2016) and space science (DeWitt and Bultitude 2020).

Moreover, contemporary science education allows teaching about science in the making, showcasing modern scientific practices and aspects of the nature of scientific knowledge (Woithe et al. 2022a, b). This provides excellent opportunities to address common misconceptions about science and scientists such as (old, white, male) physicists working alone in their laboratory, and introduce learners to today's reality of scientific research including, for example, huge collaborations across the world that require a significant amount of work time spent on communication.

However, teachers might face a number of challenges when integrating modern physics topics into their teaching. First, it often deals with invisible entities and phenomena that we cannot observe directly with our eyes but that we can only measure indirectly. Second, authentic research equipment and experiments are usually not accessible to schools because of their complexity, cost and safety constraints leaving limited opportunities for enquiry-based learning. Third, there is limited research about students' conceptions and suitable learning progressions.

Finally, in particle physics in particular, there are only few explicit links to high-school curricula (Kranjc Horvat et al. 2022b), and teachers might require professional development programmes to update their limited content knowledge before feeling comfortable teaching particle physics (Kranjc Horvat et al. 2022a).

1.2 Informal Science Learning

For all these reasons, teachers often use informal learning offers at out-of-school learning places when introducing modern science topics (e.g. visiting research institutes, laboratories or science centres). For example, the International Particle Physics Outreach Group (IPPOG, https://ippog.org) coordinates a variety of educational activities and public engagement events for particle physics across the globe. CERN offers a broad range of programmes and activities for high-school teachers and students (education.cern) as well as guided tours, exhibitions and online resources (visit.cern).

Learning at informal learning places can have positive effects on affective and cognitive outcomes such as interest, curiosity or understanding even in the case of short-term interventions (Chi et al. 2015; National Research Council 2009). There are several characteristics that make science learning in out-of-school settings special such as the absence of grade pressure and curriculum pressure, free-choice exploration driven by interest and curiosity, or the contact with scientists.

1.3 CERN Science Gateway

CERN Science Gateway (sciencegateway.cern) is a new education and outreach facility at CERN, Geneva, Switzerland. Its construction started in December 2020, and still ongoing by the time we were writing this chapter (Fig. 1). CERN Science Gateway was officially inaugurated on the 7th of October 2023 and is open to the public since the 8th of October 2023. It includes (a) exhibitions with interactive hands-on exhibits, real scientific objects, immersive environments to discover CERN, our universe and its unresolved mysteries as well as the quantum world at the very smallest of scales, (b) education laboratories with hands-on experiments and enquiry-based learning activities to let visitors discover their inner scientist, and (c) science shows with awe-inspiring experiments in a theatre-style setting to let visitors explore the exciting stories and scientific methods behind major advances in science and technology. CERN Science Gateway will welcome diverse audiences of all ages with the goal to inspire, to foster curiosity in science, to help visitors make sense of the science that shapes their lives and allow young visitors in particular to explore careers in science and technology. The design of its educational offer was based on a long iterative process building onto evidence from previous studies, front-end evaluations and multiple iterations to optimise its educational value.

Fig. 1 CERN Science Gateway during its construction phase in November 2022 (© CERN 2022, Maximilien Brice)

2 Teaching About Invisible Entities

2.1 Scientific Reasoning and Modelling of Invisible Entities

Science often deals with invisible entities and phenomena that cannot be observed directly but allow only indirect measurements. Particle physics may even be the field with the least amount of tangible research objects because particles themselves are best described as abstract excitations of quantum fields that do not possess tangible properties such as size or colour. This is not a problem for particle physicists who developed an astonishing theory and can calculate interactions or predict the outcome of experiments with remarkable accuracy. But this level of abstraction is a problem for people outside the field, such as school students and their teachers, who might not have the advanced mathematics knowledge required to access quantum field theory.

Furthermore, research in learning psychology suggests that the development of complex scientific reasoning skills is a multi-stage process linked to brain development (Lawson 2004). It is assumed that the basic patterns of scientific reasoning are already present at birth. However, very young children require physical objects to manipulate, and they struggle to mentally manipulate hypothetical things. Only around the age of 12, do learners start demonstrating hypothetical reasoning skills.

However, understanding complex particle physics analyses requires hypothetico-deductive reasoning with invisible entities that can only be measured indirectly. According to Lawson (2004), only learners in their late adolescence in the so-called post-formal stage of intellectual development are equipped for this.

The preference for concrete physical objects might be reflected in the way students struggle with understanding the concept of scientific models. Even though already 12-year-old students can understand the basic principles of particle physics, they seem to struggle with the model character (Wiener et al. 2015). Indeed, many students hold naive views of scientific models, for example, by being unaware of their tentative nature or by mistaking them for replicas of real objects (Treagust et al. 2002).

2.2 Implementation at CERN Science Gateway

To address these challenges, CERN Science Gateway offers a variety of opportunities to make the invisible world of elementary particles visible, for example, by using digital tools. In addition, the exhibitions house a cloud chamber and a spark chamber, which are particle detectors, able to make the tracks of ionising particles visible.

- In cloud chambers, ionising particles cause localised phase transitions in a super-cooled, supersaturated alcohol vapour leading to a track of droplets that can be illuminated and then observed with the naked eye. In addition to a continuous diffusion cloud chamber in the exhibitions, the education laboratories at CERN Science Gateway offer a workshop that includes assembling DIY cloud chambers using dry ice, an aluminium plate and a fish tank (Woithe 2016) (Fig. 2).
- When electrically charged high-energy particles traverse a spark chamber filled with a helium–neon gas, they ionise the gas atoms generating free electrons. If a high voltage is applied at the right moment, these electrons are accelerated towards

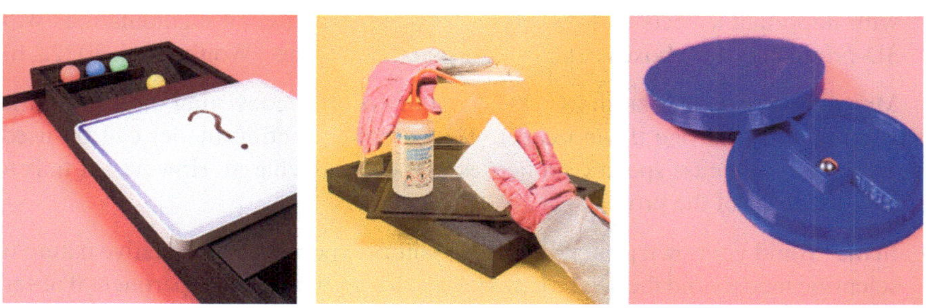

Fig. 2 Three hands-on activities designed for the education laboratories at CERN Science Gateway to make invisible entities visible and teach about the Nature of Science. From left to right: "Seeing the invisible" workshop, DIY cloud chamber, mystery box for the "Higgs in a box" workshop (© CERN 2022, Guillaume Durey and Patrick Thill)

the nearest electrode causing an avalanche of secondary electrons and finally a spark that can be registered even with our eyes.

CERN Science Gateway makes use of different digital tools to create even more advanced visualisations of invisible particles and their properties.

- The education laboratories use pixel detectors together with visualisation software to record tracks of ionising particles and generate coloured images in real time. These images allow students to investigate the different amounts of energy deposited in the detector material and compare different types of particle tracks with the tracks observed in cloud or spark chambers.
- The digital exhibit "Build your own experiment" in the *Discover CERN* exhibition has been designed to simulate the interaction of particles with matter. Visitors can simulate and visualise the passage of different types of particles through different types of materials. This exhibit uses the Geant4 software package, a powerful toolkit that is used at CERN, for example, for detector simulations.
- The exhibition *Quantum World* aims to visualise and have visitors experience quantum phenomena through immersive scenography and hands-on exhibits. For example, visitors will be able to play a digital round of uncertainty tennis while finding out more about the uncertainty principle.

Moreover, CERN Science Gateway offers opportunities to reflect on the Nature of Science in general and the nature of scientific models in particular. For example, the education laboratories offer hands-on activities involving different types of mystery boxes, which are an effective tool to enable students to experience elements of scientific discovery and explicitly reflect on NoS (Fig. 2).

- In the "Seeing the invisible" workshop, participants roll ping pong balls into a hidden obstacle inside a box, observe the scattering pattern and develop a model of the invisible structure.
- In the "Higgs in a box" workshop, participants compare milestones in the discovery of the Higgs boson by the ATLAS and CMS collaborations in 2012 with important aspects of NoS while conducting a series of manipulations with 3D-printed mystery boxes containing a spherical magnet (Woithe et al. 2022a, b).

Many physics textbooks show a model of a proton as consisting of three quarks (two up-quarks and one down-quark). This simple scientific model can be used, for example, to discuss specific properties of the strong charge. However, it fails to explain the mass of the proton.

- In the "Build your own proton" exhibit, visitors experience the tentative nature of scientific models and the need for models to change when new experimental observations contradict an existing model. In particular, visitors are asked to combine elementary particles to form particle systems, for example, protons. Subsequently, visitors use a balance scale to compare the masses of the selected elementary particles with the total mass of a proton. They discover that the proton is much more than a combination of just three quarks. Only once they add additional particles to the three quarks (in particular, gluons, the interaction particles of the strong interaction), the scale shows an equilibrium. At the end, an animation shows a visualisation of a more complex model of a proton.

2.3 *Recommendations for Teachers*

Invisible entities that can only be measured indirectly are not just a challenge in particle physics and CERN Science Gateway but an inherent feature of many physics' phenomena. We advise teachers to use this as an opportunity to think about suitable and multiple visualisations as well as explicit instruction about the nature of scientific models, in particular the tentative nature of models and their limitations. Here, simplified visualisations in textbooks such as atomic models (Wiener 2020) or the quark model of the proton can serve as fruitful learning opportunities. Teachers can even use simple hands-on activities such as a 3D-printable quark puzzle (McGinness et al. 2019a, b) to enhance their explicit teaching of the Nature of Science. In addition, students can be asked to develop their own models, for example of the inside of mystery boxes (Horvat et al. 2022; Woithe et al. 2022a, b), to practise their modelling skills.

Particle detectors are fascinating research tools that allow us insights into the world of invisible entities. Even though advanced particle detectors are usually not affordable for schools, previous education projects at CERN developed low-cost DIY versions of cloud chambers (Woithe 2016) and pixel detectors (Keller et al. 2019) as well as many other activities intended for the use in physics classrooms (https://cern.ch/scoollab/classroom-activities).

One way to make quantum field theory accessible to high-school students and their teachers is working with Feynman diagrams, which are pictorial representations of the underlying mathematical expressions describing particle interactions (Dahlkemper et al. 2022; Woithe et al. 2017).

3 Aiming for Authentic Experiences and Interactions

3.1 *Authenticity in Science Education and Outreach*

Authenticity in Science, Technology, Engineering and Mathematics (STEM) education and outreach has been the subject of numerous studies (Anker-Hansen and Andrée 2019; Beerends and Aydin 2021; Braund and Reiss 2006; Clark et al. 2014; Hampp and Schwan 2015; Hutchison 2008; Saffran et al. 2020; Schriebl et al. 2022). They concluded that authenticity in STEM education and outreach is desirable and perceived as something to aim for. We will show how authenticity takes shape in STEM education and outreach at CERN Science Gateway and how it could be pursued by teachers.

Authenticity is a complex concept relying on social interactions and is nevertheless a widely used term. From a philosophical perspective, it can be described as something of "*undisputed origin or authorship*", something "*faithful to an original*" or a "*reliable, accurate representation*" (Varga and Guignon 2020). Based on an extensive review, Schriebl et al. (2022) identified 22 aspects of authenticity and demonstrated its multiple dimensions such as locations, equipment and practices. Authenticity can be analysed as an outcome of social interactions. It is a process in which "people build a shared agreement over what is to be considered authentic" (Vannini and Williams 2009, p. 12). This process is an interactive negotiation of meanings and representations (Beerends and Aydin 2021).

Another aspect of authenticity in STEM is the opportunity for public groups, and especially school students, to interact with a variety of people working in the field of science and technology. Coming in contact with "real" scientists and engineers, even for a brief amount of time, can help address the still-prevalent stereotype of scientists (Stamer et al. 2020) in case of both students and teachers (Houseal et al. 2014; Woods-Townsend et al. 2016) and can have an impact on students' aspiration of pursuing a scientific career (Fadigan and Hammrich 2004; Hochberg and Kuhn 2019). Moreover, our research also showed that positive interaction with scientists at CERN is a strong predictor of students' interest in physics in the context of hands-on education laboratories (Woithe et al. 2022a, b).

3.2 Implementation at CERN Science Gateway

As a public laboratory with many member states, CERN has a long tradition of communicating its work to different audiences (CERN 1956). Education and outreach activities, such as education laboratories, guided tours in the facilities and exhibitions, are an inherent part of the laboratory's mission. In our experience with teachers, students and other visitors, we found that seeing real places where science is done is important. In addition, whatever the discipline they have been teaching, teachers were emphasising their need of finding accessible and relatable content for them and their students (Zamora 2020).

With this mission, CERN Science Gateway content has been designed with authenticity and relatability in mind. The building is located right next to the laboratory's site and is therefore part of the CERN campus. Thanks to this proximity, visitors encounter CERN scientists and engineers in several different ways.

First of all, members of the CERN community are present in the exhibition spaces and education laboratories at all times, acting as guides, helping visitors navigate the exhibits, facilitating hands-on workshops and providing further physics explanations.

In addition to people dedicating their time to outreach and education activities in the exhibitions, members of the CERN community are also present digitally on several interactive multimedia screens, sharing authentic insights about their work and their personal and professional journeys. The diversity of the selected CERN members (in terms of gender, age, ethnicity, country of origin, career stage, personal interests, etc.) provides all visitors with an opportunity to identify with a person working in STEM. This feature of CERN Science Gateway exhibitions also highlights the diversity of professions and activities contributing to particle physics research, making visitors aware about a variety of possible career paths in the field of science and technology and a wide applicability of STEM skill sets.

Alongside the screens described above, the exhibitions encompass scenographic elements, hands-on and multimedia exhibits and real objects (for example, a piece of a particle detector). The scenographics and hands-on exhibits have been designed in close collaboration with CERN scientists and engineers across different departments of the laboratory. This collaboration aims to ensure the relevance and diversity of the selected topics and to demonstrate the benefits of interdisciplinarity in science and technology. For example, the exhibit "Tunnel building" combines civil engineering and geological knowledge. Each exhibit also provides visitors with an opportunity to discover more about a given topic—for example, relevant scientific and technological applications in society. Furthermore, many authentic objects on display have been designed and developed at CERN and serve as a medium through which many interesting stories about research happening at the laboratory are told. For example, a real particle accelerator accessible to visitors in the *Discover CERN* exhibition has the potential of blurring an imaginary border between a scientific research space and a public space (Meyer 2011) and transmitting a message about "science-in-the-making".

In relation to the previous points, CERN Science Gateway also aims at emphasising the idea that modern particle physics research can only be carried out by large collaborations (rather than lonely individuals), sometimes drawing upon the expertise and effort of thousands of people across the world (Birnholtz 2007; Chompalov et al. 2002). This message, for example, has driven the design of some interactive hands-on exhibits that require visitors to collaborate in order to solve a specific challenge. One of such exhibits is "Find the Higgs", which consists of three large multi-touch screens installed next to each other on a big table, enabling several visitors to work on a task all together at the same time. Even though in a simplified way and on a much smaller scale, this exhibit reflects the collaborative and multidisciplinary nature of CERN's outstanding achievements (El-Kebir et al. 2019; Hoffmann 2012).

One example of an enquiry-based learning activity we designed with a high intended level of authenticity for the education laboratories of CERN Science Gateway is the "Power of Air" workshop. During the workshop, students use 3D-printed components and toy balloons to systematically investigate balloon hovercrafts. This activity is designed to raise awareness of how engineers at CERN exploit the power of air to move 1000+ tonnes detector slices by means of an air pad system. To further increase the authenticity of the activity, volunteers from CERN's scientific community will accompany students in their learning process and discuss their findings (Thill et al. 2022).

Finally, visits to CERN Science Gateway can be combined with guided tours of CERN's research facilities—adding another layer of authenticity to the visitor experience.

3.3 Recommendations for Teachers

The review of science education literature of Anker-Hansen and Andreé (2019) argued that "*designing authentic school activities where students make authentic contributions implies a complete reframing of education*" (p. 61). However, this implies a considerable challenge. According to Schriebl et al. (2022), authenticity can have benefits in science education, if students themselves recognise it as such. However, they noticed a gap between how teachers and students perceive authenticity and connection to reality. Therefore, teachers and students could together start this process by building an agreement on what authenticity is.

Moreover, a common advice given to teachers is to go out of the classroom to authentic contexts of science and research, like CERN Science Gateway. Science museums, zoos and botanical gardens are described as more exciting places to learn about science (Braund and Reiss 2006). However, from our research with primary school teachers in the area local to CERN, we learnt that field trips are difficult to organise and may not be feasible for all schools (Zamora 2020).

But there are other solutions when aiming for authenticity in classrooms:

- If going to a laboratory with a school group is practically challenging, inviting a scientist, engineer or technician in the classroom to talk about their work may be much easier from a practical point of view. Moreover, virtual visits, such as CERN's, are also a new trend which enables live interaction with scientists and engineers while not requiring a class of students or a scientist to travel physically.
- The learning task can be designed considering complexity, collaboration, equipment and data close to scientific practices (Schriebl et al. 2022). Driver et al. (1996) argue that it is important for students to engage in actual scientific work and experience the process of obtaining scientific knowledge. There are many digital resources available online to access material and data from laboratories and research institutes. For example, using open data from the CMS experiment was appreciated by teachers in an outreach project (Veteli and Lassila-Perini 2021). Other examples are a 3D-printable model of the ATLAS detector (Woithe et al. 2020) or a model of particle traps (McGinness et al. 2019a, b).
- Teaching STEM starting from contexts, applications and links between sciences, technologies and societies (STS) can result in improving students' attitudes and understanding of science (Bennett et al. 2007). This approach can be a fruitful opportunity to explore collaborations between STEM teachers and teachers of other subjects, such as History or Geology, bringing the learning experience closer to the reality of interdisciplinary scientific collaborations.

4 Evaluating Our Ideas and Designs with the Target Audiences

4.1 Evaluation in Informal Science Education and Outreach

Another aspect that characterises our work at CERN Science Gateway is evaluation research, which has accompanied the development of the exhibitions and education laboratories from the start in 2019.

Evaluation (also called audience research, visitor studies, etc.) comes in a variety of forms depending on the evaluation goals, scope of the project and the field of expertise. Nevertheless, the idea that evaluation is crucial for enhancing the quality and impact of science communication and science education initiatives is widely accepted in both research and practitioners' communities (Falk and Needham 2017; Guba and Lincoln 1989; Jensen and Lister 2017; Spicer 2017; Ziegler et al. 2021).

Drawing upon our expertise in science education and science communication research, here we describe our experience with front-end and formative evaluation of CERN Science Gateway educational offers. These types of evaluation aim to empirically determine aspects of an evaluated entity (this could be an interactive exhibit, a hands-on activity, a text paragraph, etc.) that should be changed in order for it to be more suitable with respect to its purpose, as well as exploring practical and conceptual ways in which such improvement could be done. We believe that it can provide inspiration and guidance for physics teachers who share our values of evidence-based education.

4.2 Implementation at CERN Science Gateway

The development of learning activities for the education laboratories at CERN Science Gateway has been guided by the model of educational reconstruction, which provides a framework for improving science teaching and learning not just in schools or universities but also in out-of-school settings (Duit et al. 2012). In particular, the model emphasises the need to study and analyse the science subject matter (e.g. important key concepts and principles) but also the importance of investigating student and teacher perspectives (e.g. research on interests, pre-instructional conceptions and attitudes) before designing new learning activities. Consequently, the design and evaluation of teaching and learning environments is a complex iterative process. The model of educational reconstruction has been used successfully to develop learning activities in modern physics topics such as general relativity (Kersting et al. 2018), special relativity (Kamphorst et al. 2023), dark matter (Woithe and Kersting 2021) and particle physics (Woithe 2020).

For example, before designing the "Power of Air" activity around balloon hovercrafts described above, we studied the CERN context of air pads used to move heavy detector slices, but we were also considering developing an activity about the vacuum in CERN's particle accelerator beam pipes. However, previous studies demonstrated that young learners struggle with the concept of air in static situations and with the concept of vacuum in particular but are able to reason in situations where air moves (Driver et al. 2014). Consequently, we decided to drop the idea of a vacuum workshop for young learners.

For the exhibitions at CERN Science Gateway, we employed a variety of small-scale evaluation activities at different stages of the project. In the very beginning, we carried out an empirical study on potential visitors' perceptions (both cognitive and affective) of physics concepts that are key to the exhibitions at CERN Science Gateway: matter, atom, particle, Big Bang, quantum, etc. The results of this study equipped us with an evidence-based understanding of preconceptions that could potentially hinder visitor learning (Dvorzhitskaia 2020). For example, the outcome of the study made us realise that some of the participants believed that the Large Hadron Collider was used to accelerate atoms, neutrons or "light". This misconception made us pay extra attention to the framing of the *Discover CERN* exhibition, which, among other messages, communicates the idea that only electrically charged particles (in particular, protons and ions) can be accelerated in the Large Hadron Collider.

As soon as ideas for the exhibition design and content began to take shape, the focus of our evaluation narrowed down to scenographic elements and exhibition texts. For example, the design of the *Back to the Big Bang* exhibition relies upon visitors realising that they are mentally travelling further in space and back in time as they physically move through the exhibition. Discussing the preliminary design renderings with a school class allowed to identify parts of the storyline that did not come through in the current version of the exhibition clearly enough (Dvorzhitskaia 2021). For example, we learnt that students who participated in our small-scale evaluation study were not at all familiar with a visual representation of cosmic microwave background (CMB) radiation, with some students assuming it was a visual of the Sun, a star or the Big Bang. Moreover, the way students perceived the CMB visual seemed to have played a role in their interpretation of the overarching scenographic narrative (for example, the direction of time).

At a later stage, we also started to prototype and evaluate hands-on and multimedia exhibits. In line with the equity, diversity and inclusion values underpinning CERN Science Gateway, tests of several interactive exhibits were carried out with blind and visually impaired visitor groups. Overall, taking into consideration the fact that a majority of the exhibits at CERN Science Gateway have been conceptualised in-house, it has been of utmost importance for us to ensure that these exhibits are cognitively accessible and interesting for our target audiences.

Most of the individual exhibit testing activities have therefore been realised in close collaboration with school teachers, who kindly agreed to dedicate an hour or two of their class time for it. Typically, a school class was divided into groups of 5–6 students, who then took turns to interact with an exhibit prototype (physical or multimedia) as a team, and afterwards completed a short anonymous questionnaire

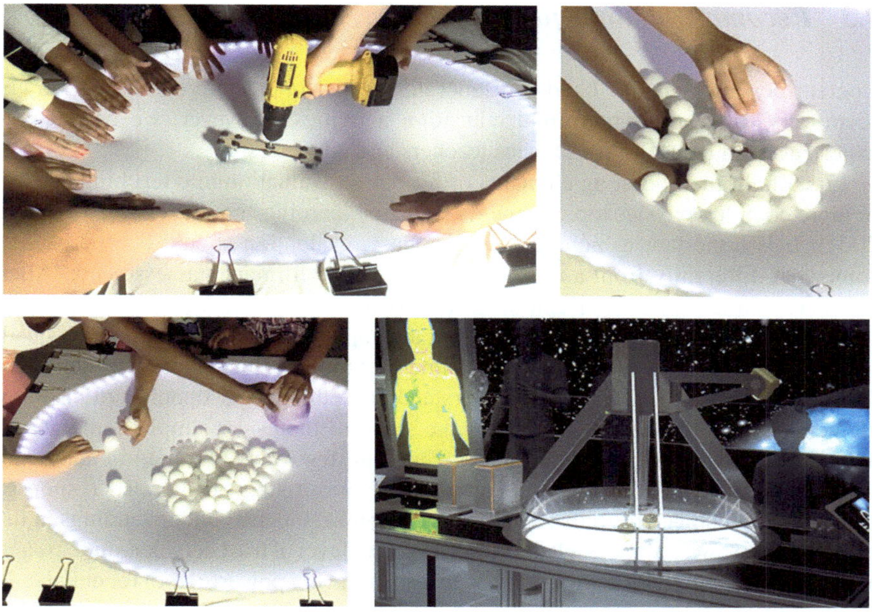

Fig. 3 Testing the prototype of an exhibit on gravitational waves with primary school students in a French school in 2021 (© CERN 2022). Bottom right: rendering of the final exhibit design (© CERN 2022, design by Tinker Imagineers)

where they were encouraged to share their feedback and exhibit-specific observations and learnings (Fig. 3). The group activity was preceded by a short introduction about CERN and CERN Science Gateway, during which students were also informed about the goal of the activity and how it would happen (and, of course, given the choice to withdraw if they wish). The activity concluded with an in-class discussion and Q&A session about the exhibit or more general physics topics.

In our experience, this arrangement proved to be a win–win for everyone involved. Not only did it empower students, who personally contributed to the development of a large international education project, but it also served as a nice change to the school routine complementing students' in-class science learning in an enjoyable, interactive and fun way. At the same time, as evaluation researchers, we felt lucky to get access to one of the major audiences of CERN Science Gateway—middle- and high-school students—and sufficient volume of data to analyse and thus build an evidence base for further work on the interactive exhibits.

4.3 Recommendations for Teachers

Content development for hands-on laboratories and interactive exhibitions might seem very different from formal education activities. However, we believe our experience can be useful for physics teachers and educators.

First of all, incorporating elements of formative evaluation into teaching of new, or more complex physics topics benefits students' learnings greatly (Bell and Cowie 2001). It is important to note that, in this case, evaluation should be approached somewhat differently from a more formal type of performance assessments (Allen and Peterman 2019). In our exhibit testing sessions, we always reassure students that there are no correct answers that they are supposed to know or guess. Instead, we encourage them to share their own interpretations of the topic, feel confident to admit if they do not know or understand something, and openly criticise any elements of the exhibits that do not make sense to them—or simply come across as not particularly interesting for them. There are few ways that help in achieving it: collecting students' feedback anonymously, prioritising open-ended responses over multiple-choice questions, and highlighting that students' understandings of the evaluated topic will not be graded. With a sufficient level of trust between students and an evaluator, formative and front-end evaluation is a very powerful tool for enhancing the quality of educational materials, as well as their suitability for a specific group of students.

Another option is establishing a collaboration with an evaluation team working at a science centre or science museum in close proximity to the school. As argued above, evidence-based development of science communication and science education activities is becoming more and more prominent, especially in larger, well-established educational institutions. At the same time, science communication professionals and researchers are struggling to involve audiences beyond those who have already come through the doors of a science centre (DeWitt and Archer 2017). Given there are enough resources available (such as human power, expertise, time, funding …), involving students in the development of a new exhibition or a hands-on activity has high chances of proving to be mutually enriching; our experience with developing CERN Science Gateway has definitely been so.

5 Conclusion

Contemporary physics education brings unique opportunities, not just to learn about modern physics topics but also about the Nature of Science and scientists. In this chapter, we discussed three challenges and opportunities that we encountered when developing content for the exhibitions and the education laboratories of CERN Science Gateway. Of course, only a tiny fraction of students and teachers around the world will be able to visit CERN Science Gateway and interact with its offer directly. However, we strongly believe that our learnings can inspire other informal science learning settings as well as formal education settings, for example, through the plethora of digital resources (including virtual talks, virtual reality materials, 360° images, animations, digital learning units, etc.) and classroom activities (including DIY experiments, hands-on activities, 3D-printable experiments, articles, etc.) created by CERN's outreach, education and education research teams (see, for example, visit.cern and education.cern).

References

S. Allen, K. Peterman, Evaluating informal STEM education: issues and challenges in context: evaluating informal stem education. N. Dir. Eval. **2019**(161), 17–33 (2019). https://doi.org/10.1002/ev.20354

J. Anker-Hansen, M. Andrée, In pursuit of authenticity in science education. Nord. Stud. Sci. Educ. **15**(1), 54–66 (2019). https://doi.org/10.5617/nordina.4723

S. Beerends, C. Aydin, Negotiating authenticity in technological environments. Philos. Technol. **34**(4), 1665–1685 (2021). https://doi.org/10.1007/s13347-021-00480-5

B. Bell, B. Cowie, The characteristics of formative assessment in science education. Sci. Educ. **85**(5), 536–553 (2001). https://doi.org/10.1002/sce.1022

J. Bennett, F. Lubben, S. Hogarth, Bringing science to life: a synthesis of the research evidence on the effects of context-based and STS approaches to science teaching. Sci. Educ. **91**(3), 347–370 (2007). https://doi.org/10.1002/sce.20186

J.P. Birnholtz, When do researchers collaborate? Toward a model of collaboration propensity. J. Am. Soc. Inform. Sci. Technol. **58**(14), 2226–2239 (2007). https://doi.org/10.1002/asi.20684

M. Braund, M. Reiss, Towards a more authentic science curriculum: the contribution of out-of-school learning. Int. J. Sci. Educ. **28**(12), 1373–1388 (2006). https://doi.org/10.1080/09500690500498419

CERN, *The Public Relations Service (Committee of Council Eighth Meeting)* (CERN, 1956). https://cds.cern.ch/record/28583/files/CM-P00075946-e.pdf

B. Chi, R. Dorph, L. Reisman, Evidence & impact: museum-managed STEM programs in out-of-school settings, in *National Research Council Committee on Out-of-School Time STEM* (National Research Council, Washington, DC, 2015)

I. Chompalov, J. Genuth, W. Shrum, The organization of scientific collaborations. Res. Policy **31**(5), 749–767 (2002). https://doi.org/10.1016/S0048-7333(01)00145-7

J.C. Clark, R. Tytler, D. Symington, School-community collaborations: bringing authentic science into schools. Teach. Sci. **60**(3), 28–34 (2014). https://doi.org/10.3316/informit.679312049766063

M.N. Dahlkemper, P. Klein, A. Müller, S.M. Schmeling, J. Wiener, Opportunities and challenges of using Feynman diagrams with upper secondary students. Physics **4**(4), Article 4 (2022). https://doi.org/10.3390/physics4040085

J. DeWitt, L. Archer, Participation in informal science learning experiences: the rich get richer? Int. J. Sci. Educ. Part B **7**(4), 356–373 (2017). https://doi.org/10.1080/21548455.2017.1360531

J. DeWitt, K. Bultitude, Space science: the view from European school students. Res. Sci. Educ. **50**(5), 1943–1959 (2020). https://doi.org/10.1007/s11165-018-9759-y

R. Driver, J. Leach, R. Millar, P. Scott, *Young People's Images of Science* (Open University Press, 1996)

R. Driver, A. Squires, P. Rushworth, V. Wood-Robinson, *Making Sense of Secondary Science, Research into Children's Ideas* (Routledge, 2014)

R. Duit, H. Gropengießer, U. Kattmann, M. Komorek, I. Parchmann, The model of educational reconstruction—a framework for improving teaching and learning science, in *Science Education Research and Practice in Europe* (Brill, 2012), pp. 13–37

D. Dvorzhitskaia, Formative evaluation of key CERN Science Gateway exhibitions concepts, Unpublished. Science Gateway exhibitions kick-off meeting Lot 1, CERN, 13 Oct 2020

D. Dvorzhitskaia, 'Back to the Big Bang' scenography testing, Unpublished. Internal report CERN, 2021

H. El-Kebir, M.C. Mazza, J. Liu, Perceptions of open innovation at CERN: an explorative study. CERN IdeaSquare J. Exp. Innov. 57–61 (2019). https://doi.org/10.23726/CIJ.2019.945

K. Fadigan, P.L. Hammrich, A longitudinal study of the educational and career trajectories of female participants of an urban informal science education program. J. Res. Sci. Teach. **41**(8) (2004). https://doi.org/10.1002/tea.20026

J.H. Falk, M.D. Needham, Utilizing indicator-based methods: 'measuring the impact of a science center on its community', in *Preparing Informal Science Educators*, ed. by P.G. Patrick (Springer International Publishing, 2017), pp. 261–266. https://doi.org/10.1007/978-3-319-50398-1_14

E.G. Guba, Y.S. Lincoln, *Fourth Generation Evaluation* (Sage, 1989)

C. Hampp, S. Schwan, The role of authentic objects in museums of the history of science and technology: findings from a visitor study. Int. J. Sci. Educ. **5**(2), 161–181 (2015). https://doi.org/10.1080/21548455.2013.875238

K. Hochberg, J. Kuhn, What do scientists do? Increasing awareness of social and networking aspects in everyday activities of scientists. Progr. Sci. Educ. (PriSE) **2**(1), Article 1 (2019). https://doi.org/10.25321/prise.2019.849

H.F. Hoffmann, The role of open and global communication in particle physics, in *The Globalization of Knowledge in History* (Max-Planck-Gesellschaft zur Förderung der Wissenschaften, 2012). https://doi.org/10.34663/9783945561232-32

A.K. Horvat, M. Boselli, P. Chatzidaki, M.N. Dahlkemper, R. Duggan, G. Durey, N. Herff, D. Molaro, G.W. Scheerer, S.M. Schmeling, P.G. Thill, J. Wiener, J. Woithe, S. Zoechling, The mystery box challenge: explore the nature of science. Sci. Sch. **59** (2022). https://www.scienceinschool.org/article/2022/mystery-box-challenge/

A.K. Houseal, F. Abd-El-Khalick, L. Destefano, Impact of a student-teacher-scientist partnership on students' and teachers' content knowledge, attitudes toward science, and pedagogical practices. J. Res. Sci. Teach. **51**(1), 84–115 (2014). https://doi.org/10.1002/tea.21126

P. Hutchison, Epistemological authenticity in science classrooms, Dissertation, University of Maryland, 2008. https://drum.lib.umd.edu/handle/1903/8833

E. Jensen, T. Lister, The challenges of 'measuring long-term impacts of a science center on its community': a methodological review, in *Preparing Informal Science Educators*, ed. by P.G. Patrick (Springer International Publishing, 2017), pp. 243–259. https://doi.org/10.1007/978-3-319-50398-1_13

F. Kamphorst, M.J. Vollebregt, E.R. Savelsbergh, W.R. van Joolingen, An educational reconstruction of special relativity theory for secondary education. Sci. Educ. **32**(1), 57–100 (2023). https://doi.org/10.1007/s11191-021-00283-2

O. Keller, M. Benoit, A. Müller, S. Schmeling, Smartphone and tablet-based sensing of environmental radioactivity: mobile low-cost measurements for monitoring, citizen science, and educational purposes. Sensors **19**(19), 4264 (2019). https://doi.org/10.3390/s19194264

M. Kersting, E.K. Henriksen, M.V. Bøe, C. Angell, General relativity in upper secondary school: design and evaluation of an online learning environment using the model of educational reconstruction. Phys. Rev. Phys. Educ. Res. **14**(1), 010130 (2018). https://doi.org/10.1103/PhysRevPhysEducRes.14.010130

A. Kranjc Horvat, J. Wiener, S. Schmeling, A. Borowski, Learning goals of professional development programs at science research institutions: a Delphi study with different stakeholder groups. J. Sci. Teach. Educ. **33**(1), 32–54 (2022a). https://doi.org/10.1080/1046560X.2021.1905330

A. Kranjc Horvat, J. Wiener, S.M. Schmeling, A. Borowski, What does the curriculum say? Review of the particle physics content in 27 high-school physics curricula. Physics **4**(4), 1278-1298 (2022b). https://doi.org/10.3390/physics4040082

A.E. Lawson, The nature and development of scientific reasoning: a synthetic view. Int. J. Sci. Math. Educ. **2**(3), 307–338 (2004). https://doi.org/10.1007/s10763-004-3224-2

O. Levrini, A.D. Ambrosis, S. Hemmer, A. Laherto, M. Malgieri, O. Pantano, G. Tasquier, Understanding first-year students' curiosity and interest about physics—lessons learned from the HOPE project. Eur. J. Phys. **38**(2), 1-20 (2016). https://doi.org/10.1088/1361-6404/38/2/025701

L. McGinness, S. Dührkoop, A. Jansky, O. Keller, A. Lorenz, S. Schmeling, K. Wendt, J. Woithe, 3D-printable model of a particle trap: development and use in the physics classroom. J. Open Hardw. **3**(1), 1 (2019a). https://doi.org/10.5334/joh.12

L. McGinness, S. Dührkoop, J. Woithe, A. Jansky, 3D printable quark puzzle: a model to build your own particle systems. Phys. Teach. **57**(8), 526–528 (2019b). https://doi.org/10.1119/1.5131116

M. Meyer, Researchers on display: moving the laboratory into the museum. Mus. Manag. Curatorsh. **26**(3), 261–272 (2011). https://doi.org/10.1080/09647775.2011.585800

National Research Council, *Learning Science in Informal Environments: People, Places, and Pursuits* (National Academies Press, 2009)

OECD, *PISA 2015 Results (Volume I). Excellence and Equity in Education* (2016). https://ictlogy.net/bibliography/reports/projects.php?idp=3204&lang=es

L. Saffran, S. Hu, A. Hinnant, L.D. Scherer, S.C. Nagel, Constructing and influencing perceived authenticity in science communication: experimenting with narrative. PLoS ONE **15**(1), e0226711 (2020). https://doi.org/10.1371/journal.pone.0226711

D. Schriebl, A. Müller, N. Robin, Modelling authenticity in science education. Sci. Educ. (2022). https://doi.org/10.1007/s11191-022-00355-x

S. Sjøberg, C. Schreiner, Results and perspectives from the rose project, in *Science Education Research and Practice in Europe: Retrospective and Prospective*, ed. by D. Jorde, J. Dillon (SensePublishers, 2012), pp. 203–236. https://doi.org/10.1007/978-94-6091-900-8_9

S. Spicer, The nuts and bolts of evaluating science communication activities. Semin. Cell Dev. Biol. **70**, 17–25 (2017). https://doi.org/10.1016/j.semcdb.2017.08.026

I. Stamer, H. Pönicke, F. Tirre, A. Laherto, T. Höffler, S. Schwarzer, I. Parchmann, Development & validation of scientific video vignettes to promote perception of authentic science in student laboratories. Res. Sci. Technol. Educ. **38**(2), 168–184 (2020). https://doi.org/10.1080/02635143.2019.1600491

P. Thill, J. Woithe, S. Schmeling, New educational activity about CMS air pads for education labs at CERN Science Gateway, in *Proceedings of 41st International Conference on High Energy Physics—PoS (ICHEP2022)*, vol. 414 (2022), p. 383. https://doi.org/10.22323/1.414.0383

D.F. Treagust, G. Chittleborough, T.L. Mamiala, Students' understanding of the role of scientific models in learning science. Int. J. Sci. Educ. **24**(4), 357–368 (2002). https://doi.org/10.1080/09500690110066485

P. Vannini, J.P. Williams, *Authenticity in Culture, Self, and Society* (Routledge, 2009)

S. Varga, C. Guignon, Authenticity, in *The Stanford Encyclopedia of Philosophy* (2020). https://plato.stanford.edu/archives/spr2020/entries/authenticity/

P. Veteli, K. Lassila-Perini, In pursuit of authenticity—CMS open data in education, in *Proceedings of 40th International Conference on High Energy Physics—PoS (ICHEP2020)*, vol. 390 (2021), p. 963. https://doi.org/10.22323/1.390.0963

J. Wiener, Science teachers' conceptions of atomic models. Eur. J. Math. Sci. Educ. **1**(2), 67–80 (2020). https://www.ejmse.com/science-teachers-conceptions-of-atomic-models.html

G.J. Wiener, S.M. Schmeling, M. Hopf, Can grade-6 students understand quarks? Probing acceptance of the subatomic structure of matter with 12-year-olds. Eur. J. Sci. Math. Educ. **3**(4), 313–322 (2015). https://doi.org/10.30935/scimath/9440

J. Woithe, *CLOUD CHAMBER. S'Cool LAB—Do-It-Yourself Manual. Version 7* (2016). https://scoollab.web.cern.ch/cloud-chamber

J. Woithe, Designing, measuring and modelling the impact of the hands-on particle physics learning laboratory S'Cool LAB at CERN effects of student and laboratory characteristics on high-school students' cognitive and affective outcomes, PhD Thesis, TU Kaiserslautern, 2020. https://doi.org/10.13140/RG.2.2.13201.66403

J. Woithe, M. Kersting, Bend it like dark matter! Phys. Educ. **56**, 035011 (2021). https://doi.org/10.1088/1361-6552/abe09c

J. Woithe, G.J. Wiener, F.F.V. der Veken, Let's have a coffee with the standard model of particle physics! Phys. Educ. **52**(3), 034001 (2017). https://doi.org/10.1088/1361-6552/aa5b25

J. Woithe, A. Jansky, O. Keller, T. Goncalves, A functional 3D-printable magnet model of the ATLAS toroid. Phys. Educ. **2**(3) (2020). https://doi.org/10.1142/S2661339520200024

J. Woithe, M. Boselli, P. Chatzidaki, M.N. Dahlkemper, R. Duggan, G. Durey, N. Herff, A.K. Horvat, D. Molaro, G.W. Scheerer, S. Schmeling, P.G. Thill, J. Wiener, S. Zoechling, Higgs in a box: investigating the nature of a scientific discovery. Phys. Educ. **4**(4), 2250019 (2022a). https://doi.org/10.1142/S2661339522500196

J. Woithe, A. Müller, S. Schmeling, J. Kuhn, Motivational outcomes of the science outreach lab S'Cool LAB at CERN: a multilevel analysis. J. Res. Sci. Teach. **59**(6), 930–968 (2022b). https://doi.org/10.1002/tea.21748

K. Woods-Townsend, A. Christodoulou, W. Rietdijk, J. Byrne, J.B. Griffiths, M.M. Grace, Meet the scientist: the value of short interactions between scientists and students. Int. J. Sci. Educ. Part B **6**(1), 89–113 (2016). https://doi.org/10.1080/21548455.2015.1016134

A. Zamora, From primary schools to CERN Science Gateway: front-end evaluation of primary schools' visits, Unpublished. Internal report CERN, 2020

R. Ziegler, I.R. Hedder, L. Fischer, Evaluation of science communication: current practices, challenges, and future implications. Front. Commun. **6**, 669744 (2021). https://doi.org/10.3389/fcomm.2021.669744

Artificial Intelligence: New Challenges and Opportunities in Physics Education

Rüdiger Wink and Walter M. Bonivento

Abstract Artificial intelligence is the most important recent innovation in digital technology. The availability of chatbots such as ChatGPT will create profound changes in both teaching and learning. The OpenAI technologies, once some of the errors of their infancy are going be fixed, will quickly spread worldwide, and their use will impact many aspects of our society. The potentialities and advantages will overcome the fears, and both teachers and students will have to learn to make the best out of it. The benefits of artificial intelligence for physics education are illustrated in this document with a few examples. It is important in our view that government regulations and sponsorships prevent further inequalities and long-term fixations of disparities, keeping the cost of available AI programs low.

Keywords Open AI technologies · Chatbot · Potentiality of AI in physics education · Teaching and learning

1 Introduction

Artificial intelligence, as defined by the Oxford dictionary, is: "the theory and development of computer systems able to perform tasks normally requiring human intelligence, such as visual perception, speech recognition, decision-making, and translation between languages."

Artificial intelligence allows machines to perform human-like tasks and relies on deep learning and language processing. With the ever-increasing amount of data from humans and machines to be processed, AI is seen as an aid to humans for

R. Wink (✉)
Faculty of Business Administration and Industrial Engineering, HTWK Leipzig, Leipzig, Germany
e-mail: ruediger.wink@htwk-leipzig.de

W. M. Bonivento
Istituto Nazionale di Fisica Nazionale, Sezione di Cagliari, Complesso Universitario di Monserrato, S.P. per Sestu, Km 0,700, 09042, Monserrato, Cagliari, Italy
e-mail: walter.bonivento@ca.infn.it

M. Streit-Bianchi et al. (eds.), *New Challenges and Opportunities in Physics Education*, Challenges in Physics Education, https://doi.org/10.1007/978-3-031-37387-9_27

learning, analyzing data sets, and decision-making processes. Indeed, AI can treat data at a much faster speed than humans and can find connection patterns based on finding statistical associations among training data much more quickly than humans. Limitations, however, still occur, as in contrast to human learning the identification of patterns by AI tools is not based on reflection and conscious decisions and therefore cannot be reproduced by humans. AI has left the realm of science fiction where it was either utopian or dystopian and is entering at high speed in today's real life. Artificial intelligence (AI) is indeed one of the most important innovations in digital technology which may also affect profoundly the teaching of physics at all levels of education from primary school to university.

2 Chatbot Programs

In November 2022, a new game-changer player emerged on the field. A new chatbot program was made available on the free domain named ChatGPT (based on GPT-3, generative pre-trained transformer), by the OpenAI company, founded by Sam Altman, based on AI and machine learning, and is now quickly becoming a worldwide used tool raising many animated questions (OpenAI Team 2022). Only a few weeks afterward, Google announced the integration of BARD—its own AI large language model for dialog applications—into its search engine (Pichai 2023). Microsoft as one of the biggest investors in OpenAI launched a new version of its search engine powered by OpenAI technologies, which shall be more powerful than ChatGPT (Roose 2023). The common feature of all these activities is the use of large language models (LLM), which are trained by huge amounts of data and are using transformer architectures. They are self-supervising, which means they are learning and making their predictions on the relevance of words from the data itself and are not provided with correct answers to learn in a supervised context, and they are using their previous predictions as inputs for new predictions (Kasneci et al. 2023). As they are not provided with correct answers along their learning process, these models are good at guessing correct results or generating texts but are not able to calculate and still give inaccurate answers. The famous inaccurate answer by BARD and reports on errors from ChatGPT by users illustrate these limitations (Milmo 2023). Even if the application is in its infancy (see the fast expansion of publications on AI in education since 2020 in Xu and Ouyang (2022) and the even stronger limitations of an earlier version of the chatbot in Abriata (2022)), the capability to provide correct and perhaps sophisticated solutions will quickly improve over time. It will sooner or later execute the requested tasks with high precision.

3 First Studies Using ChatGPT in Education

In the education sector, the launch of the free domain ChatGPT was recognized as a watershed moment. The most common fears refer to the termination of traditional ways to assess students' performances based on written essays, as even in the academic sector separations between original "human" papers and papers created by AI systems will not be possible (Yeadon et al. 2023; Gendron et al. 2022). Conventional teaching methods in schools based on teaching and verification will reach their limits. First studies using ChatGPT reveal the huge potential ranging from capabilities to pass exams at the university level (Terwiesch 2023) to the adjustment of language style to children's expressions as described in the following example:

Q: Explain gravity with the language of a 5-year-old child.

A: Gravity is like a big, big hug from the Earth to anything on it. So, everything on the Earth is pulled toward the center of the Earth by gravity. That is why things do not float away into space when you drop them. They just fall because the Earth is hugging them back to the ground.

US American universities already reacted to the new experiences with ChatGPT. Their reactions, however, primarily focused on the prohibition of using AI for seminar papers or other exams as well as the development and acquisition of AI technologies (see for instance the tools developed by MIT or Princeton) to verify the AI nature of a text (Huang 2023), but all preventive or prohibitive measures on the use are not going to be the solution. Indeed, the same preventive measures taken for the pocket calculator or the access to world wide web information in the classroom had no success, and the teachers were forced to accept the idea that the students have free access to all of this and to cope with their use, which in the end had a positive effect both on teaching and learning.

4 Potentials for Physics Education and Personalized Learning

What we are trying to address in this book is limited to considering if this tool as well as other tools based on AI technologies can be used efficiently and profitably by teachers in physics education soon and at all levels. Potentials to improve physics education can be recognized in at least five directions.

Firstly, teachers can be supported in structuring a lecture on a topic, suggesting experiments to explain physics topics, or finding suitable expressions on a language level for specific ages of children (Kasneci et al. 2023; Tan 2022; Horwitz 2022). In the following, an example is shown based on ChatGPT.

Q: Give me a suggestion for experiments about viscosity to do in the classroom at the undergraduate level.

A: Here are a few ideas for experiments that can be done in an undergraduate classroom to demonstrate the concept of viscosity:

Measurement of viscosity using a viscometer: Use a viscometer, such as a rotational or falling sphere viscometer, to measure the viscosity of different liquids, such as water, oil, and glycerol. Compare the results and discuss the factors that affect viscosity.

Comparison of viscosity of Newtonian and non-Newtonian fluids: Measure the viscosity of a Newtonian fluid, such as water, and a non-Newtonian fluid, such as a corn-starch solution, using a viscometer. Compare the results and discuss the differences in behavior.

Measurement of dynamic viscosity using a rheometer: Use a rheometer to measure the dynamic viscosity of a liquid over a range of shear rates or shear stresses. Compare the results for different liquids and discuss the relationship between shear rate and viscosity.

It is important to note that for these experiments, the classroom should be equipped with appropriate materials such as a viscometer, rheometer, and temperature control system, and make sure that the students have the appropriate knowledge of the theory behind the experiments.

Secondly, AI technologies using large language models can be used to support teachers in assessing writing assignments or checking for plagiarism (Kasneci et al. 2023). Teachers might be disencumbered from time-consuming and repetitive tasks and could more focus on creative interactions with their students.

Thirdly, AI technologies like ChatGPT can be used to support personalized learning processes for students (Rid 2023; on first experiences at the university level). Students can adjust their learning activities to their capabilities and preferences. The chatbot can provide tailored feedback on learning progress and suggest specific hints according to the interaction with the individual student. ChatGPT can easily show how to solve simple fundamental physics problems, certainly at the level of undergraduate, including quantum mechanics, and create software programs in different languages, from Python to C++, to solve them. After two years of remote learning and examinations during the Covid-19 pandemic, students are used to digital technologies and might not be alienated by communicating with a chatbot. As the chatbots are trained with huge amounts of data from the Internet but were not connected to the Internet, their communication was based on political correctness, and safeguard measures shall prevent their misuse. Direct interactions between chatbot and student can accelerate the learning process, and students might be motivated even to ask questions to the chatbot they would fear to ask the teacher with the whole class listening. Interactions between students and teachers can be more focused on creative elements of the lessons, while comprehension and repetition might be left to interactions with the chatbot, if the final results of these processes are corrected by the teacher. In the future, AI might even be able to help better understand the conscious learning strategies of humans and with these insights develop even more suitable tools to support students' learning processes (Molenaar et al. 2023).

Fourthly, the limitations of the current AI-powered chatbots can be turned into a constructive learning outcome by training the students to critically investigate the results provided by the chatbot. Students might find it self-assuring to reveal fallacies. Furthermore, the limitations of chatbots in using calculations can be used as case examples to show the importance of calculations and the necessity to analyze methods and preconditions to come to defendable results.

Fifthly, the inclusion of AI in physics education is not limited to chatbots (see for a systematic perspective on AI tools in education; Chen et al. 2020). Studies already emphasize the positive impact of augmented reality (AR) and virtual reality (VR) fueled by AI on physics education (Chen et al. 2020; Mogamodov 2020; Kaviyaraj and Uma 2021). Students find additional motivation by receiving more illustrative impressions based on AR and VR. New experiments can be introduced to enhance creativity in classrooms. Studies on using AR and VR in art classrooms particularly reveal the positive impact on students' concentration and deep learning periods (Rong et al. 2022).

5 The Challenges to Overcome

These potentials of AI in physics education, however, can only be realized, if at least four challenges will be successfully addressed.

Firstly, a guarantee of worldwide access to these tools for teaching independent of the economic wealth of a country, institution, or student is needed. Let us quote the UN Sustainable Development Goals for quality education "*Providing quality education for all is fundamental to creating a peaceful and prosperous world. Education gives people the knowledge and skills they need to stay healthy, get jobs, and foster tolerance.*" Quality education from now on will also include the appropriate use of AI knowledge tools. At the university level probably, this will be an easier task since teachers have more freedom of action within their lecturing programs and most of them participate as researchers in experiments where AI is already being implemented, e.g., in the data analysis. This will allow them to quickly be confident with the use of such systems. ChatGPT was launched by OpenAI on a free domain with easy access for all students and teachers. On February 1, 2023, only a few weeks after this successful publication, OpenAI already introduced ChatGPT Plus, which is only available for 20 US$ per month and which provides continuous access even in peak times, faster response times, and priority access to new features and improvements (OpenAI 2023). Therefore, it becomes obvious that the digital divide based on economic disparities could be intensified with the use of AI in physics education if sponsored accounts for poorer countries and students are not available. As research on AI depends on huge R&D investments, government regulation and sponsorship will be necessary to prevent further divides and long-term fixations of disparities.

Secondly, personalized learning programs and more creative classroom teaching might primarily provide benefits for students with a higher level of competencies, as they will be better able to analyze fallacies in chatbots' answers and be even faster and smarter with the support of AI programs. This challenge already came up with the growing availability of information via digital tools like Google Scholar, Wikipedia, and other websites and search engines during the last two decades. The accuracy of scientific information via these tools was no longer based on scientific reviews but on community discourses (in the case of Wikipedia) and algorithms focusing on citations (Google Scholar) and including non-reviewed papers like preprints, students' presentations, or other publications. For experienced and high-performing students, differentiation between reviewed and non-reviewed information is possible, and access to these databases provides huge potential for accelerating their learning processes, while weaker students face risks of being stuck to doubtful statements. The use of large language models trained by these data will multiply these challenges. Again, specific support for disadvantaged groups will be needed to prevent a divide—in this context not based on economic disparities, but on intellectual or experiential disparities between students (Kasneci et al. 2023). Teachers should focus more on these disadvantaged students if the chatbots and other AI technologies allow for more autonomous learning periods for high-performing students.

Thirdly, teachers will need support and training to use new tools and technologies in a sensitive way (Lee and Perret 2022). Compared to students acting as "digital natives," teachers have a long way to adjust their teaching methods and to be willing to experiment with new tools. Bureaucratic environments and regulations to adjust curricula further narrow the space for experiments and fast adjustments of methods based on first experiences. Accordingly, clauses will be needed to make schools a testbed for these new opportunities and to enable teachers to apply their new ideas developed in training courses. Students will not wait until the regulatory environments will be implemented.

Fourthly, ethical issues need to be solved when discussing the risks of data privacy, the necessity of transparency of data the large language models are trained with, and the risks of misuse (Kasneci et al. 2023). While traditional teaching materials and tools are completely analyzed and tested before introduction, large language models bear the risks of learning and adjustment along with their use. Again, regulatory environments are needed to cope with these risks while not closing the windows of opportunities for a more creative learning environment.

One challenge, however, might be easily overcome: the fear that youngsters fall into the same trap as the protagonist of the movie "Her," who fell in love with the voice and imagined the creation of a relationship with a supportive pleasant female from an OS operating AI machine continuously evolving, and becoming only capable of AI interactions. These are machines after all!

References

L. Abriata, Testing GPT-3 on elementary physics unveils some important problems. Towards Data Science, Jan 2022. https://towardsdatascience.com/testing-gpt-3-on-elementary-physics-unveils-some-important-problems-9d2a2e120280. Accessed 30 Jan 2023

L. Chen, P. Chen, Z. Lin, Artificial intelligence in education: a review. IEEE Access **8**, 75264–75278 (2020)

Y. Gendron, J. Andrew, C. Cooper, The perils of artificial intelligence in academic publishing. Crit. Perspect. Account. **87**, 102411 (2022)

J. Horwitz, ChatGPT, artificial intelligence, and UDL: how to harness the future to reach all students, Dec 2022. https://www.novakeducation.com/blog/chat-gpt-artificial-intelligence-and-udl-how-to-harness-the-future-to-reach-all-students. Accessed 28 Jan 2023

K. Huang, Alarmed by A.I. chatbots, universities start revamping how they teach. New York Times, Jan 2023. https://www.nytimes.com/2023/01/16/technology/chatgpt-artificial-intelligence-universities.html. Accessed 06 Feb 2023

E. Kasneci et al., ChatGPT for good? On opportunities and challenges of large language models for education. EdArXiv Preprints, Jan 2023. https://doi.org/10.35542/osf.io/5er8f. Accessed 08 Feb 2023

R. Kaviyaraj, M. Uma, A survey on future of augmented reality with AI in education, in *Conference Paper*, Mar 2021. https://doi.org/10.1109/ICAIS50930.2021.9395838. Accessed 02 Feb 2023

I. Lee, B. Perret, Preparing high school teachers to integrate AI methods in STEM classrooms, in *Conference Paper. 36th AAAI Conference on Artificial Intelligence 2022. Proceedings*, 22 Feb–1 Mar 2022, held virtually (2022), pp. 12783–12791. ISBN-10: 1-57735-876-7 and ISBN-13: 978-1-57735-876-3. https://doi.org/10.1609/aaai.v36i11.21557

D. Milmo, Why did Google's ChatGPT rival go wrong and are AI chatbots overhyped? The Guardian. https://www.theguardian.com/technology/2023/feb/09/googles-bard-demo-what-went-wrong-chatgpt-chatbots-ai. Accessed 09 Feb 2023

V.S. Mogamodov, Examining the potential of VR and AR technologies for education. J. Phys. Conf. Ser. **1691** (2020). https://doi.org/10.1088/1742-6596/1691/1/012160

I. Molenaar et al., Measuring self-regulated learning and the role of AI: five years of research using multimodal multichannel data. Comput. Hum. Behav. **139**, 107540 (2023). https://doi.org/10.1016/j.chb.2022.107540

OpenAI, Introducing ChatGPT Plus, Feb 2023. https://openai.com/blog/chatgpt-plus/. Accessed 02 Feb 2023

OpenAI Team, ChatGPT: optimizing language models for dialogue, Nov 2022. https://openai.com/blog/chatgpt/. Accessed 22 Jan 2023

S. Pichai, An important next step on our AI journey, Feb 2023. https://blog.google/technology/ai/bard-google-ai-search-updates/. Accessed 08 Feb 2023

T. Rid, Five days in class with ChatGPT, Jan 2023. https://alperovitch.sais.jhu.edu/five-days-in-class-with-chatgpt/. Accessed 25 Jan 2023

Q. Rong, Q. Lian, T. Tang, Research on the influence of AI and VR technology for students' concentration and creativity. Front. Psychol. **13** (2022). https://doi.org/10.3389/fpsyg.2022.767689

K. Roose, Bing (Yes, Bing) just made search interesting again. New York Times. https://www.nytimes.com/2023/02/08/technology/microsoft-bing-openai-artificial-intelligence.html. Accessed 08 Feb 2023

S.K. Tan, ChatGPT and its use cases for physics education. Physics Lens, Dec 2022. https://www.physicslens.com/chatgpt-and-its-use-cases-for-physics-education/. Accessed 05 Feb 2023

C. Terwiesch, *Would Chat GPT3 Get a Wharton MBA? A Prediction Based on Its Performance in the Operations Management Course* (Mack Institute for Innovation Management at the Wharton School, University of Pennsylvania, 2023)

W. Xu, F. Ouyang, The application of AI technologies in STEM education: a systematic review from 2011 to 2021. Int. J. STEM Educ. **9**, 59 (2022). https://doi.org/10.1186/s40594-022-00377-5

W. Yeadon et al., The death of the short-form physics essay in the coming AI revolution. arXiv: 2212.11661v1 (physics.ed-ph). https://doi.org/10.48550/arXiv.2212.11661. Accessed 09 Feb 2023

Epilogue from the Editors

STEM strategies and physics education research have been addressed by many governments, policymakers, and educators to equip future generations with the competences allowing them to contribute to the development of society. The availability and challenges of digital world and open science in physics and STEM both in research and in education play an important contribution to the United Nations Sustainable Development Goals (SDGs) and help children acquire and develop problem-solving skills, critical thinking, and creativity.

The goal of improving scientific literacy through research in science education, the foundation of the GIREP communities (Groupe International de Recherche sur l'Enseignement de la Physique—International Research Group on Physics Education), and other research and teacher associations has guided the discourse in all the chapters of this book. The interaction with many experts in the field of physics education has greatly enriched this book and shows the importance of interdisciplinary interactions and perspectives. We believe that the presentation of different aspects of modern physics topics knows no boundaries and is the result of a fruitful dialog between theories and experiments, and that physics research often goes beyond the world of physics, in a spirit of collaboration and openness facilitated by the WWW, invented at CERN, a real social revolution. We hope to help teachers bring this spirit into the classroom to improve the teaching of physics and to understand science as an important global enterprise.

The book contains many practical examples to help teachers overcome the difficulties of introducing basic physics concepts and explaining modern physics tasks using the tools and programs now available in the digital world.

Distance teaching and learning posed a significant challenge especially during the recent pandemic. Some of the issues raised and the presentation of programs developed in many different countries to provide physics education that explains the fundamental concepts of physics and adapts teaching to the needs of tomorrow's society are the focus of the second part of the book, "Digital Challenges for Physics Learning".

M. Streit-Bianchi et al. (eds.), *New Challenges and Opportunities in Physics Education*, Challenges in Physics Education, https://doi.org/10.1007/978-3-031-37387-9

A wide literature in physics education research evidenced that structured teaching of the main laws as answers to questions not posed does not produce awareness of concepts, and that commonsense ideas emerge as alternative interpretation with respect to the physics one. The way of working and the epistemology of physics are not assimilated to the teaching way of working, thus producing the image of physics being an abstract world of models and difficult laws that students do not know when and where could them be used. Students often leave introductory physics courses with misconceptions about Newton's laws of motion, acceleration, and even gravitational force and acceleration, despite their interest. The book offers new ways to clarify critical points discussing them on the light of the scientific way of thinking and providing useful historical overviews that can spark students' interest. The contribution and the role of history of physics for physics education are also discussed and exemplified in that context.

Digital technologies are the instruments in modern physics research for the measurements, for the predictions, and for theoretical interpretations, in addition to be used as communication facilities. These now play an important role for physics education offering students different ways to experience the physics way of working, enriching their learning goals, and providing the opportunity to gain competence in exploring phenomena and its physics interpretation. Digital technologies bridge the gap between human perception and physical reality; this is important as many phenomena, such as quantum physics or the basic building blocks of matter, cannot be fully experienced as discussed in more than one of the chapters. However, the use of digital technologies in the classroom requires appropriate competence and teacher professional development, as well as digital literacy for students, which is made evident in the many contributions and experiences reported from different country.

Active learning strategies and globally available digital information now provide opportunities to modernize teaching/learning process. The book includes many links to available platforms that enrich the proposals for school activities and help teachers and students to live the physical world. We believe it is important for teachers to enhance student learning, promote critical thinking, and satisfy their curiosity, which is happening in the basic sciences from high-energy physics to astronomy. It is also important for students to have an authentic context and to be able to talk to researchers, and many organizations and labs provide this opportunity. For this reason, we have also devoted a chapter to CERN Science Gateway, the latest example of a project that provides informal physics education in a real research environment. There are many opportunities for teachers to participate in didactic development programs and to use interactive learning and working materials to which links and information are provided in the chapters.

The role on informal learning in physics education is also highlighted in many chapters of the book. Providing students and teachers with the opportunity to engage with physics in informal contexts foster their creativity and interest in science. It is the responsibility of institutions and associations to bring science closer to people and to provide supplementary materials to improve teachers' teaching.

Finally, artificial intelligence (AI) is rapidly entering people's lives, and the availability and advancement of chatbots require the education system to learn and explore their effective use in the classroom. It will not be long before their use is mandated or regulated, as AI will inevitably have a transformative impact on society, affecting learning and education.

You keep on learning and learning and pretty soon you learn something no one has learned before. (Richard P. Feynman)

MIX
Papier aus verantwortungsvollen Quellen
Paper from responsible sources
FSC® C105338

If you have any concerns about our products,
you can contact us on
ProductSafety@springernature.com

In case Publisher is established outside the EU,
the EU authorized representative is:
Springer Nature Customer Service Center GmbH
Europaplatz 3, 69115 Heidelberg, Germany

Printed by Libri Plureos GmbH
in Hamburg, Germany